74.95 4CF

Video Display Engineering

On-Line Updates

Additional information and updates relating to video displays in general, and this book in particular, can be found at the *Standard Handbook of Video and Television Engineering* web site:

www.tvhandbook.com

The tvhandbook.com web site supports the professional video community with news, updates, and product information relating to the broadcast, post production, and business/industrial applications of digital video.

Check the site regularly for news, updated chapters, and special events related to video engineering. The technologies encompassed by *Video Display Engineering* are changing rapidly, with new developments and standards work announced each month. Changing market conditions and regulatory issues are adding to the rapid flow of news and information in this area.

Specific services found at **www.tvhandbook.com** include:

- **Video Technology News**. News reports and technical articles on the latest developments in digital television, both in the U.S. and around the world. Check in at least once a month to see what's happening in the fast-moving area of digital television.

- **Television Handbook Resource Center**. Check for the latest information on professional and broadcast video systems. The Resource Center provides updates on implementation and standardization efforts, plus links to related web sites.

- **tvhandbook.com Update Port**. Updated material for *Video Display Engineering* is posted on the site regularly. Material available includes updated sections and chapters in areas of rapidly advancing technologies.

- **tvhandbook.com Book Store**. Check to find the latest books on digital video and audio technologies. Direct links to authors and publishers are provided. You can also place secure orders from our on-line bookstore.

In addition to the resources outlined above, detailed information is available on other books in the McGraw-Hill Video/Audio Series.

www.tvhandbook.com

Video Display Engineering

Jerry C. Whitaker

McGraw-Hill
New York San Francisco Washington, D.C. Auckland Bogotá
Caracas Lisbon London Madrid Mexico City Milan
Montreal New Delhi San Juan Singapore
Sydney Tokyo Toronto

McGraw-Hill

*A Division of The **McGraw·Hill** Companies*

Copyright © 2001 by The McGraw-Hill Companies, Inc. All rights reserved. Printed in the United States of America. Except as permitted under the United States Copyright Act of 1976, no part of this publication may be reproduced or distributed in any form or by any means, or stored in a data base or retrieval system, without the prior written permission of the publisher.

1 2 3 4 5 6 7 8 9 0 DOC/DOC 0 6 5 4 3 2 1 0

P/N 137343-8
PART OF
ISBN 0-07-137342-X

The sponsoring editor for this book was Stephen S. Chapman and the production supervisor was Pamela A. Pelton. This book was typeset in Times New Roman and Helvetica by Technical Press, Morgan Hill, California.

Printed and bound by R. R. Donnelley & Sons Company.

Information contained in this work has been obtained by The McGraw-Hill Companies, Inc. ("McGraw-Hill") from sources believed to be reliable. However, neither McGraw-Hill nor its authors guarantees the accuracy or completeness of any information published herein and neither McGraw-Hill nor its authors shall be responsible for any errors, omissions, or damages arising out of use of this information. This work is published with the understanding that McGraw-Hill and its authors are supplying information but are not attempting to render engineering or other professional services. If such services are required, the assistance of an appropriate professional should be sought.

 This book is printed on recycled, acid-free paper containing a minimum of 50% recycled de-inked fiber.

For David
My favorite younger brother

Contents

Preface   xv

Chapter 1: Display System Applications   1

Introduction   1
    Display System Development   2
        Braun's CRT   2
        Zworykin: The Brains of RCA   3
        Farnsworth: The Boy Wonder   4
    Image Reproduction   5
    Transmission Standard Development   6
        Color Standard   7
    Theater Television   8
    Radar   9
    Instrumentation   9
    Importance of the Consumer Market   10
High-Definition Displays   10
    Display Resolution   12
    Production System vs. Transmission System   13
    Defining Terms   13
        Compatibility Defined   14
    HDTV Applications   14
        Business and Industrial Applications   16
        Critical Implications for the Viewer and Program Producer   16
    Image Size   18
        HDTV Image Content   19
    Motion Picture Theater Applications   19
Aural Component of Visual Realism   22
    Hearing Perception   23
    Matching Audio to Video   23
    Making the Most of Audio   25
    Ideal Sound System   25
Display Systems   26
    Format Development   27
    Viewing Environment   28
        Broad Blue Band System   28
        Yellow-Minus-Yellow System   30
        Mercury-Minus-Red System   30
    Multisensor Environment   31
    Display Technology Trends   32
        Cathode Ray Tube   33
        Flat CRT   33
        Liquid Crystal Display   34
        Plasma/Gas Discharge   34
        Electroluminescent Display   34
    Display Type Considerations   35

Advanced Display Applications	38
References	40
Bibliography	41

Chapter 2: Principles of Light, Vision, and Photometry — 45

Introduction	45
Sources of Illumination	45
The Spectrum	46
Monochrome and Color Vision	48
Visual Requirements for Video	51
Luminous Considerations in Visual Response	52
Photometric Measurements	52
Luminosity Curve	52
Luminance	54
Luminance Discrimination	55
Perception of Fine Detail	56
Sharpnes	58
Response to Intermittent Excitation	58
Photometric Quantities	60
Luminance and Luminous Intensity	60
Illuminance	61
Lambert's Cosine Law	63
Measurement of Photometric Quantities	63
Retinal Illuminance	64
Receptor Response Measurements	64
Spectral Response Measurement	65
Transmittance	66
Reflectance	68
Human Visual System	69
A Model for Image Quality	69
References	69
Bibliography	70

Chapter 3: Principles of Color Vision — 71

Introduction	71
Color Stimuli	71
Trichromatic Theory	72
Color Matching	73
Color-Matching Functions	75
Luminance Relationships	76
Vision Abnormality	77
Color Representation	77
Munsell System	78
Other Color-Order Systems	80
Color Triangle	80
Center of Gravity Law	82
Alychne	83
Spectrum Locus	84

Subjective and Objective Quantities	85
The CIE System	86
Color-Matching Functions	90
Tristimulus Values and Chromaticity Coordinates	90
Conversion Between Two Systems of Primaries	90
Luminance Contribution of Primaries	92
Standard Illuminants	92
Gamut of Reproducible Colors	95
Vector Representation	96
Refinements to the 1931 CIE Model	98
Improved Visual Uniformity	99
CIELUV	101
Colorimetry and Color Targets	104
Color Models	105
RGB Color Model	105
YIQ Model	106
CMY Model	107
HSV Model	109
References	109
Bibliography	110

Chapter 4: Application of Visual Properties — 111

Introduction	111
The Television System	111
Scanning Lines and Fields	112
Interlaced Scanning Fields	112
Synchronizing Video Signals	113
Television Industry Standards	114
Composite Video	114
Color Signal Encoding	117
Color Signal Decoding	118
Deficiencies of Conventional Video Signals	119
Comb Filtering	119
Video Colorimetry	120
Gamma	123
Display White	123
Scene White	123
Phosphor Chromaticities	124
Video System Characteristics	124
Foveal and Peripheral Vision	125
Horizontal Detail and Picture Width	127
Perception of Depth	128
Contrast and Tonal Range	128
Chrominance Properties	129
Temporal Factors in Vision	129
Continuity of Motion	130
Flicker Effects	130
Video Bandwidth	131
Bibliography	132

Chapter 5: Measuring Display Parameters — 135

- Introduction — 135
  - Visual Acuity — 135
  - Contrast Sensitivity — 137
  - Flicker — 138
- Video Signal Spectrum — 139
  - Minimum Video Frequency — 141
  - Maximum Video Frequency — 141
  - Horizontal Resolution — 143
  - Video Frequencies Arising from Scanning — 143
- Measurement of Color Displays — 145
  - Assessment of Color Reproduction — 148
  - Chromatic Adaptation and White Balance — 149
  - Overall Gamma Requirements — 149
  - Perception of Color Differences — 149
  - Display Resolution and Pixel Format — 151
    - Contrast Ratio — 151
- Applications of the Zone Pattern Signal — 152
  - Simple Zone Plate Patterns — 153
  - Producing the Zone Plate Signal — 155
    - Complex Patterns. — 156
  - The Time (Motion) Dimension — 158
- References — 159
- Bibliography — 160

Chapter 6: Cathode Ray Tube Fundamentals — 163

- Introduction — 163
  - Basic Operating System — 163
  - Classification of CRT Devices — 164
  - The CRT Envelope — 164
  - Arc Protection — 165
  - Implosion Protection — 166
- Phosphor and Screen Characteristics — 166
  - Phosphor Screen Types — 167
    - Fabrication — 168
  - Luminescent Properties — 169
    - Screen Burn and Aging — 170
    - Spectral Emission — 172
    - Chromaticity — 172
    - Persistence — 174
  - Screen Characteristics — 175
    - Contrast — 176
    - Modulation Transfer Function — 177
- Electron Gun — 178
  - Electron Motion — 178
  - Tetrode Gun — 180
  - Operating Principles — 181
- Electron Beam Focusing — 181

Electrostatic Lens	181
Practical Applications	185
Electrostatic Lens Aberrations	185
Magnetic Focusing	188
Practical Applications	190
Magnetic Lens Aberrations	190
Beam Crossover	191
Types of CRT Devices	192
Monochrome CRT	192
Monochrome Electron Gun	193
Resolution Spot Size	194
Storage CRT	196
Bistable Storage Tube	197
Computer Display Terminal Devices	198
Oscilloscope CRT Devices	198
Radar CRT Devices	199
Recording CRT Devices	199
CRT Measurement Techniques	199
Subjective CRT Measurements	200
Shrinking Raster Method	201
Line Width Method	202
TV Limiting Resolution Method	202
Application Considerations	202
Objective CRT Measurements	203
Half Power Width Method	204
Fourier Transform Methods	204
Discrete Frequency Method	205
Viewing Environment Considerations	205
Picture Monitor Alignment	206
References	208
Bibliography	209

Chapter 7: CRT Deflection Systems — 213

Introduction	213
Electrostatic Deflection	214
Principles of Operation	215
Acceleration Voltage Effects	218
Post-Deflection Accelerator	220
Electromagnetic Deflection	220
Principles of Operation	221
Flat-Face Distortion	222
Deflection Defocusing	225
Deflection Yoke	225
Yoke Selection Parameters	228
Distortion Correction Circuits	228
Flat-Face Distortion Correction	229
Dynamic Focusing	231
Pincushion Correction	234
References	235

Bibliography   236

Chapter 8: Color CRT Display Devices   237

Introduction   237
   Shadow-Mask CRT   237
   Parallel-Stripe Color CRT   239
   Beam Penetration Color CRT   241
Basics of Color CRT Design   242
   Tube Geometry   243
   Guard Band   244
   Shadow-Mask Design   245
      Round-hole Mask   245
      Slot Mask   246
   Resolution and Moiré   247
      Application Example: Delta Gun Device   247
      Application Example: In-line Gun Device   248
   Mask-Panel Temperature Compensation   249
      Tension Mask   249
   Faceplate Screening   250
   Magnetic Shielding   252
   X-radiation   253
   Screen Size   253
   Resolution Improvement Techniques   253
      High-Resolution Trinitron   255
   Flat-Face Tubes   256
Electron Gun   258
   Operating Principles   259
   Electron Gun Classifications   261
      Hybrid Lenses   263
      Trinitron   264
   Gun Arrangements   264
      Unitized Construction   265
   Guns for High-Resolution Applications   265
Deflecting Multiple Electron Beams   265
   Distortion Effects   267
   Deflection Amplifier Considerations   269
   Considerations for Flat Screen Devices   269
   Dynamic Convergence   271
   In-Line System Convergence   272
Flat CRT Devices   275
   Classic Device Designs   276
      Channel Multiplying CRT   276
      Beam Guide CRT   279
      Matrix Drive and Deflection CRT   279
      Horizontal Address Vertical Deflection CRT   280
      Practical Considerations   280
New Consumer Devices   282
References   282
Bibliography   283

Chapter 9: Projection Display Systems — 287

- Introduction — 287
  - Display Types — 287
  - Displays for HDTV Applications — 288
  - Displays for Military Applications — 288
- Geometric Optics — 289
  - Laws of Reflection and Refraction — 289
  - Refraction at a Spherical Surface in a Thin Lens — 292
  - Reflection at a Spherical Surface — 294
  - Thick (Compound) Lenses — 296
  - Lens Aberrations — 297
  - Lens Systems — 299
  - Color Beam-Splitting Systems — 299
    - Dichroic Prism — 301
    - Spectral Trim Filters — 301
  - Interference Effects — 301
  - Diffraction Effects — 303
  - Polarization Effects — 304
- Projection System Fundamentals — 305
  - Projection Requirements — 305
  - Optical Projection Systems — 309
  - Optical Distortions — 312
  - Image Devices — 313
    - Advanced CRTs for Projection Display — 314
- Projection Display Systems — 316
  - CRT Projection Systems — 316
  - Light Valve Systems — 317
    - Eidophor Reflective Optical System — 317
    - Talaria Transmissive Color System — 319
    - Laser Beam Projection Scanning System — 322
    - LCD Projection Systems — 325
    - Application Example — 326
    - Homeotropic LCLV — 328
    - Laser-Addressed LCLV — 331
    - Image Light Amplifier (ILA) — 334
    - Digital Micromirror Device — 336
    - Grating Light Valve Display — 340
    - Scanned Linear GLV Architecture — 341
    - GLV Video Processing Architecture — 342
  - Discrete Element Display Systems — 344
    - Luminescent Panels — 345
    - Flood Beam CRT — 346
    - Matrix Addressing — 346
  - Projectors for Cinema Applications — 349
  - Operational Considerations — 349
    - Screen Considerations — 351
    - Mounting Considerations — 352
- References: — 353
- Bibliography — 355

Chapter 10: Flat Panel Displays — 359

- Introduction — 359
- Liquid Crystal Displays — 359
  - Principles of Operation — 359
    - Display Addressing — 363
  - AMLCD Performance Parameters — 366
    - AMLCD Applications — 367
  - Plasma-Address Liquid Crystal Display — 367
    - Application Considerations — 368
    - HD Display Device — 369
- Plasma Displays — 370
  - Gas Discharge Characteristics — 371
  - Memory-Type ac Plasma Display — 373
  - Hybrid ac-dc Plasma Display — 375
  - Color Plasma Displays — 376
  - Performance Issues for Video Applications — 378
  - Advanced XGA Device — 379
- Field Emission Displays — 379
  - Operational Elements — 381
- Fixed Pixel Pattern — 382
- References — 382

Listing of Figures and Tables — 385

Listing of References — 397

Subject Index — 415

About the Author — 423

On the CD-ROM — 425

Preface

The world's first digital electronic computer was built using 18,000 vacuum tubes. It occupied an entire room, required 140 kW of ac power, weighed 50 tons, and cost about $1 million. Today, an entire computer can be built within a single piece of silicon about the size of a child's fingernail. And you can buy one at the local parts house for less than $10.

Within our lifetime, the progress of technology has produced dramatic changes in our lives and respective industries. Impressive as the current generation of computer-based equipment is, we have seen only the beginning. New technologies promise to radically alter the electronics business as we know it. Display systems are a key element in this revolution. In the past, displays were added to electronics systems almost as an afterthought. Today, systems are increasingly being designed around, and literally built around, display devices.

The need for improved video and data displays increases each year across a broad range of applications. Customers demand displays that are smaller, larger, brighter, cheaper, more efficient, and higher performance than current devices. Higher resolution and better color rendition are two of the most often requested attributes of a new display system.

The electronic display industry is dynamic, as technical advancements are driven by an ever-increasing customer demand. Two areas of intense interest include high resolution computer graphics and high-definition television. In fact, the two are likely to become tightly intertwined.

Consumers worldwide have demonstrated an insatiable appetite for new electronic tools. The personal computer has redefined the office environment, and HDTV promises to redefine home entertainment. Furthermore, the needs of industry and national defense for innovation in display system design have grown enormously. Technical advances are absorbed as quickly as they roll off the production lines.

In the future, display products will increasingly be designed for more than a single market. For example, video hardware intended for use at a television station or post production facility should also be applicable in some form to the teleconferencing market. By the same token, high-resolution computer monitors should have spin-off technology that can be used in consumer HDTV sets. The wider customer base that cross-market products offer provides the promise of greater return on investment for equipment manufacturers, and lower prices for individual users.

This increasing pace of development represents a significant challenge to standardizing organizations around the world. Nearly every element of the electronics industry has standardization horror-stories in which the introduction of products with incompatible interfaces forged ahead of standardization efforts. The end result is often needless expense for the end-user, and the potential for slower implementation of a new technology. No one wants to purchase a piece of equipment that may not be supported in the future by the manufacturer or the industry. This dilemma is becoming more of a problem as the rate of technical progress accelerates.

In simpler times, simpler solutions would suffice. Legend has it that George Eastman (who founded the Eastman Kodak Company) first met Thomas Edison during a visit to Edison's New Jersey laboratory in 1907. Eastman asked Edison how wide he wanted the film for his new cameras to be. Edison held his thumb and forefinger about 1 3/8-in (35 mm) apart and said, "about so wide." With that, a standard was developed that has endured for nearly a century.

This successful standardization of the most enduring imaging system yet devised represents the ultimate challenge for all persons involved in display engineering. While technically not an electronic imaging system, film has served as the basis of comparison for nearly all electronic display systems. The performance of video displays are invariably described in relation to 35 mm film.

With the changes taking place in display engineering, up-to-date information on technologies and trends is critically important. This book examines all facets of video display:

- Chapter 1 provides a perspective on display technology, with special emphasis on the historical developments that brought us to the point we have reached today. Furthermore, the impact of HDTV on the display industry is discussed.

- Chapters 2 through 5 examine various aspects of the human visual system, and analyze how display devices can be designed to capitalize on the properties of the eye. Considerable detail is devoted to the properties of vision because of its importance in understanding the design of display systems.

- Chapters 6 through 8 examine the operating principles of the cathode ray tube. Because the CRT is the most common display device in use today, electron optics, electron beam deflection, and phosphor screen types are discussed in detail.

- Chapter 9 discusses projection systems, including CRT- and light-valve-based, and the rapidly moving area of large screen solid state imaging.

- Chapter 10 examines flat-panel display technologies, with special emphasis on high-resolution, high-brightness devices.

Information display has become an indispensable tool in modern life. Desktop computers, pocket-sized television sets, stadium displays, big-screen HDTV, flight simulator systems, high resolution graphics workstations, and countless other applications rely on advanced display technologies.

This book is intended to provide the reader with a clear understanding of the video display options available to a wide range of end-users. We have entered an era of multiple options for a given display application. Before you can make an informed decision, you need to understand the options. This book is dedicated to that effort.

Jerry C. Whitaker
October 2000

For more information on this book and related titles, visit the author's web site:

www.technicalpress.com

Video Display Engineering

Chapter 1

Display System Applications

1.1 Introduction

Information display is a key element in the advancement of electronics technology. The continued rapid growth of computer applications and embedded computer controllers in a wide variety of products have pushed forward the requirements for display components of all types. The basic display technologies in common use include the following:

- Cathode ray tube (CRT)
- CRT-based large screen projection
- Light valve-based large screen projection
- Light emitting diode (LED)
- Plasma display panel (PDP)
- Electroluminescence (EL)
- Liquid crystal display (LCD)

These technologies find use in a broad spectrum of applications, including:

- Consumer television
- Computer display
- Medical imaging
- Industrial process control
- Test and measurement
- Automotive instrumentation
- Aerospace instrumentation
- Military devices and systems
- Camera imaging

- Printing
- Textiles
- Office equipment
- Telecommunications systems
- Advertising and point-of-purchase

Additional uses are identified at a rapid rate.

1.1.1 Display System Development

The cathode ray tube is the grandfather of all display devices. Invention of the CRT is generally attributed to Karl Ferdinand Braun (Germany), who demonstrated the device as early as 1896. Braun's crude instrument included a cathode, deflection elements, and a phosphor screen. Braun originally developed the device as a method of indicating the output of an ac generator.

With the development of television, the CRT became a commercial product. Most historians credit the design of the first all-electronic television system to Philo T. Farnsworth, who demonstrated a closed-circuit system in the U.S. in 1927, and John Baird in the United Kingdom. At about the same time, Allen B. DuMont was working to perfect the picture tube. Serious development aimed at making television a commercial enterprise began in 1929, when Vladimir Zworykin joined the Radio Corporation of America. With the support of then RCA chief David Sarnoff, Zworykin and his colleagues brought the technology of television to consumers.

The mass communications media of television is one of the most significant technical accomplishments of the twentieth century. The ability of persons not only across the country, but around the world to see and communicate with each other, and experience each other's cultures and ideas, is a monumental development. It is also the development that-more than anything else-has pushed the technology of image display. The technical sophistication that consumers enjoy today, however, required many decades to mature.

Braun's CRT

Work by Karl Braun on the cathode ray tube was spurred by Wilhelm Konrad Roentgen's efforts to identify and understand X-rays. In addition, the city of Strassbourg, where Braun lived, had a new system for generating electricity. The system produced alternating current, whose periodic nature presented measurement problems because the current reversed polarity with every cycle. Braun had a long-standing interest in oscillatory phenomena, having written his dissertation on the oscillations of strings and elastic rods. Braun therefore set out to construct a device that would enable him to make precise measurements of alternating current; not simply the low-frequency current commercially produced, but the higher-frequency 5,000-10,000 Hz current that was needed for X-ray research.

Braun developed a tube similar to Roentgen's, but concentrated his efforts on the beam of electrons emitted from the negative electrode at the narrow end of the device. This negative electrode (cathode) consisted of a plane disk perpendicular to the axis of the tube. The electrons emitted from the cathode passed down the tube, which gradually widened out into a cone, finally striking the face of the tube at the other end. The electrons passed by an anode, situated on the side of the tube so as to not obstruct their path. However, before the electrons could strike the face of the tube, two things happened in Braun's device. First, he placed an insulated metal diaphragm slightly more than halfway from the cathode to the screen. A 2 mm hole in the foil diaphragm allowed only a small beam of electrons to pass through the obstruction, resulting in a fine beam of electrons, which made a single spot on the screen. The screen itself was another of Braun's contributions. Inside the wide end of the tube he placed a translucent mica plate, coated with a layer of mineral substance (phosphor), such as barium platinocyanide or zinc silicate. When the beam of electrons hit the mica plate they caused the phosphor to glow, producing a spot on the screen.

Outside the tube itself, Braun placed a coil around the device, located close to the screen. This coil, connected to the current being studied, would deflect the beam. In this manner the spot produced by the beam would alter its position on the screen. The motion of the spot was too rapid for the naked eye, so the spot appeared on the phosphor screen as a vertical line. To add the horizontal dimension Braun used a rotating mirror placed in front of the tube, thus "scanning" the spot and transforming the line into a curve.

With this crude instrument the first cathode ray tube was born, a critical element for what was to become television. Boris Rosing of Russia first proposed using the new device for television in 1907.

Zworykin: The Brains of RCA

A Russian immigrant, Vladimir Zworykin came to the United States after World War I and went to work for Westinghouse in Pittsburgh. During his stay at the company—1920 until 1929—Zworykin performed some of his early electronic imaging experiments. Zworykin had emigrated from Russia to America in order to develop his dream: television. His conception of the first practical TV camera tube, the *iconoscope* (1923), and his subsequent development of the *kinescope* picture tube formed the basis for further advances in the field. Zworykin is credited by most historians as the *father of television*.

Zworykin's iconoscope (from Greek for "image" and "to see") consisted of a thin aluminum-oxide film supported by a thin aluminum film and coated with a photosensitive layer of potassium hydride. With this crude camera tube and a CRT as the picture reproducer, he had the essential elements for television.

Continuing his pioneering work, Zworykin developed an improved iconoscope six years later that employed a relatively thick, one-sided target area. In the meantime, he had continued his work on improving the quality of the CRT, and presented a paper on his efforts to the Eastern Great Lakes district convention of the Institute of Radio Engineers (IRE) on November 18, 1929. This presentation attracted the attention of another Russian immigrant, David Sarnoff, then vice president and general manager of RCA. Sarnoff persuaded Zworykin to join RCA Victor in Camden, New Jersey, where he was made director of RCA's electronics research laboratory. The company provided the management and financial backing that enabled Zworykin (and the RCA scientists working with him) to develop television into a practical system.

By 1931, with the iconoscope and CRT well-developed, electronic television was ready to be launched and Sarnoff and RCA were ready for the new industry.

Farnsworth: The Boy Wonder

Legend has it that Philo Farnsworth conceived of a system of all-electronic television when he was a 15-year-old high school sophomore in Rigby, Idaho, a small town about 200 miles north of Salt Lake City. Farnsworth met a financial expert by the name of George Everson in Salt Lake City when he was 19 years old and persuaded him to try and secure venture capital for an all-electronic television system.

Everson was successful in securing financial investors for this unproven young man with unorthodox ideas after he and Farnsworth were able to convince the backers that they were the only experimenters on the trail of a total electronic television system. The investors were concerned because many people were interested in capturing the control over patents of a vast new field for profit. If no one was working on this method, then Farnsworth had a clear field. If, on the other hand, other companies were working in secret, without publishing their results, then Farnsworth would have little chance of receiving the patent awards and the royalty income that would surely result.

Farnsworth established his laboratory first in Los Angeles, and later in San Francisco at the foot of Telegraph Hill. Farnsworth was the proverbial lone basement experimenter. It was at his Green Street (San Francisco) laboratory that Farnsworth gave the first public demonstration in 1927 of the television system he had dreamed of for six years. He was not yet 21 years of age.

Farnsworth was quick to develop the basic concepts of an electronic television system, giving him an edge on most other inventors in the race for patents. His patents included the principle of blacker-than-black synchronizing signals, linear sweep, and the ratio of forward sweep to retrace time. Zworykin won a patent for the principle of field interlace.

Farnsworth's original "broadcast" included the transmission of graphic images, clips of a Dempsey/Tunney fight, and scenes of Mary Pickford combing her hair (from her role in the *Taming of the Shrew*). In his early systems, Farnsworth could transmit pictures with 100- to 150-line definition at a repetition rate of 30 lines per second. This pioneering demonstration set in motion the progression of technology that would lead to commercial broadcast television a decade later.

Farnsworth held many patents for television, and through the mid-1930s remained RCA's fiercest competitor in developing new technology. Indeed, Farnsworth's thoughts seemed to be directed toward cornering patents for the field of television and protecting his ideas. In the late 1930s, fierce patent conflicts between RCA and Farnsworth flourished. They were settled in September 1939 when RCA capitulated and agreed to pay continuing royalties to Farnsworth for the use of his patents. This action ended a long, bitter period of litigation. By that time, Farnsworth held an impressive list of key patents for electronic television.

Unsuccessful attempts were made to use pickup devices without storage capability for studio applications. The most ambitious was the Allen B. DuMont Laboratories' experiments in the 1940s with an electronic *flying-spot* camera. The set in the studio was illuminated with a projected raster frame of scanning lines from a cathode-ray tube. The light from the scene was gathered by a single photocell to produce a video signal.

The artistic and staging limitations of the dimly-lit studio are all too obvious. Nevertheless, while useless for live pickup, it demonstrated the flying-spot principle, a technology that is widely used today for a variety of imaging applications.

General Electric also played an early role in the development of television. In 1926, Ernst Alexanderson, a young engineer at the company, developed a mechanical scanning disc for video transmission. He gave a public demonstration of the system two years later. Coupled with the GE experimental TV station, WGY (Schenectady, NY), Alexanderson's system made history on September 11, 1928, by broadcasting the first dramatic program on television. It was a 40-minute play titled, *The Queen's Messenger*. The program consisted of two characters performing before three simple cameras.

There was a spirited race to see who could begin bringing high-quality television programs to the public first. In fact, the 525 line 60 Hz standards promoted in 1940 and 1941 were known as "high-definition television," as compared with some of the experimental systems of the 1930s. The original reason for the 30 frame per second rate was the simplified receiver design that it afforded. With the field scan rate the same as the power system frequency, ac line interference effects were minimized in the reproduced picture.

Both Zworykin and Farnsworth were members of the committee that came up with proposed standards for a national (U.S.) system. The standard was to be in force before any receiving sets could be sold to the public.

The two men knew that to avoid flicker, it would be necessary to have a minimum of 40 complete pictures per second; this was known from the motion picture industry. Although film is exposed at 24 frames per second, the projection shutter is opened twice for each frame, giving a net effect of 48 frames per second. If 40 complete pictures per second were transmitted, even with 441 lines of horizontal segmentation (which was *high-definition* TV prior to WW II), the required bandwidth of the transmitted signal would have been greater than the hardware of the day could handle. The *interlace* scheme was developed to overcome the technical limitations faced by 1940s technology.

1.1.2 Image Reproduction

From the start of commercial television in the 1940s, until the emergence of color as the dominant programming medium in the mid-1960s, nearly all receivers were the direct-view monochrome type. A few large-screen projection receivers were produced, primarily for viewing in public places by relatively small audiences. Initially the screen sizes were 10- to 12-in diagonal.

In the old days, viewers were advised to sit at least one foot away from the screen for every inch of screen size as measured diagonally. Thus, for a 12-in screen TV set, consumers were supposed to sit 12 ft away. In those early days, the electron beam scan of the CRT phosphors revealed with crisp clearness the individual scanning lines in the raster. In fact, the focus of the electron beam was sometimes purposely set soft so the scan lines were not as easily seen.

Following commercialization of monochrome television, it was realized-and expected-that color television was possible, and would likely soon follow. From experience in photography and early work in color matching, it was known that three *primary colors*-red, green, and blue-mixed in correct proportions, could essentially match the spectrum of colors. Color signals were first produced by optically combining the images from three color tubes, one for each of the red, green, and blue primary transmitted colors. This early *trinescope*, as it was called by RCA, demonstrated the feasibility of color television. The approach was, however, too cumbersome and costly to be a practical solution for viewing in the home.

The problem was solved by the invention of the shadow-mask picture tube in 1953. The first successful tube used a triad assembly of electron guns to produce beams that scanned a screen composed of groups of red, green, and blue phosphor dots. The dots were small enough not to be perceived as individual light sources at normal viewing distances. Directly behind the screen a metal mask, perforated with small holes approximately the size of each dot triad, was aligned so that each hole was behind an RGB dot cluster.

The three beams were aligned by *purity* magnetic fields so that the mask shadowed the green and blue dots from the beam driven by the red signal. Similarly, the mask shadowed the red and blue dots from the green beam, and the red and green dots from the blue beam.

The concept of the shadow-mask color tube was first proposed by Werner Flechsig of Germany in 1938. In a patent application, Flechsig proposed not only a three-gun tube with a common deflection yoke, but also a single-switch beam tube, and a tube wherein mask shadowing is replaced by mask focusing to increase electron-beam transmission.

1.1.3 Transmission Standard Development

In 1936, the Radio Manufacturers Association (RMA), the forerunner of the Electronics Industries Association (EIA), set up a committee to recommend standards for a commercial TV broadcasting service in the U.S. In December 1937, the committee advised the Federal Communications Commission (FCC) to adopt the RCA 343-line/30-frame/s system that had been undergoing intensive development since 1931. The RCA system was the only one tested under both laboratory and field conditions. A majority of the RMA membership objected to the RCA system because of the belief that rapidly advancing technology would soon render this marginal system obsolete, and, perhaps more importantly, would place them at a competitive disadvantage (RCA was prepared to immediately start manufacturing TV equipment and sets). Commercial development of television was put on hold.

At an FCC hearing in January 1940, a majority of the RMA was willing to embrace the RCA system, now improved to 441 lines. However, a strong dissenting majority (Zenith, Philco, and DuMont) was still able to block any action. The result was that the *National Television Systems Committee* (NTSC) functioned essentially as a forum to investigate various options. DuMont proposed a 625-line/15-frame/4-field interlaced system. Philco advocated a 605-line/24-frame/s system. Zenith took the stance that it was still premature to adopt a national standard. Not until June 1941 did the FCC accept the consensus of a 525-line/30-frame/s (60 Hz) black and white system, which still exists today with minor modifications.

Television was formally launched in July 1941 when the FCC authorized the first two commercial TV stations to be constructed in the United States. However, the growth of early television was ended by the licensing freeze that accompanied World War II. By the end of 1945 there were just nine commercial TV stations authorized, with six of them on the air. The first post-war full-service commercial license was issued to WNBW, the NBC-owned station in Washington, D.C.

Color Standard

During the early development of commercial television systems—even as early as the 1920s—it was assumed that color would be demanded by the public. Primitive field sequential systems were demonstrated in 1929. In the early 1940s, Peter Goldmark of the Columbia Broadcasting System (CBS) showed a field sequential (color filter wheel) system and promoted it vigorously during the post-war years. Despite the fact that it was incompatible with existing receivers, had limited picture size possibilities, and was mechanically noisy, the CBS system was adopted by the FCC as the national color television standard in October 1950.

The engineering community felt betrayed (CBS excepted). Monochrome TV was little more than 3 years old with a base of 10 to 15 million receivers; broadcasters and the public were faced with the specter of having much of their new, expensive equipment become obsolete. The general wisdom was that color must be an adjunct to the 525/30 monochrome system so that existing terminal equipment and receivers could accept color transmissions.

The FCC decision was based on engineering tests presented in early 1950. Contenders were the RCA dot sequential, the CTI (Color Television Incorporated) line sequential, and the CBS field sequential systems. The all-electronic compatible approach was in its infancy and there were no suitable display devices. Thus, for a decision made in 1950 based on the available test data, the commission's move to embrace the color wheel system was reasonable. CBS, however, had no support from other broadcasters or manufacturers; indeed, the company had to purchase Hytron-Air King to produce color TV sets (which would also receive black and white NTSC). Two hundred sets were manufactured for public sale.

The proponents of compatible, all-electronic color systems were, meanwhile, making significant advances. RCA had demonstrated a tri-color delta-delta kinescope. Hazeltine demonstrated the *constant luminance* principle, as well as the *shunted monochrome* concept. General Electric introduced the frequency interlaced color system. Philco showed a color signal composed of wideband luminance and two color difference signals encoded by a quadrature-modulated subcarrier. These, and other manufacturers, met in June 1951 to reorganize the National Television Systems Committee for the purpose of pooling their resources in the development of a compatible system. By November, a system employing the basic concepts of today's NTSC color system was demonstrated.

A major problem, however, remained: the color kinescope. It was expensive and could only be built to yield a 9 × 12-in picture. Without the promise of an affordable large screen display, the future of color television would be uncertain. Then came the development of a method of directly applying the phosphor dots on the faceplate together with a curved shadow mask mounted directly on the faceplate, a breakthrough developed by the CBS-Hytron company.

The achievement of sending three pictures in the same bandwidth as a monochrome NTSC picture was monumental for the time. The direct-view color tube was equally impressive. The first commercial tube, the 15GP22, had semiround sides with a 12-in display area. It used an internal pack consisting of a flat screen plate, a stretched-tension aperture mask, and a steel frame separating the two elements. Hytron called it the *Colortron*.

The commission adopted the color standard on December 17, 1953. It is interesting to note that the *phase alternation line* (PAL) principle was tried, but practical hardware to implement the scheme would not be available until 10 years later. The PAL principle, in a

625-line/50 Hz version, was later adopted for use in the United Kingdom and much of Europe.

1.1.4 Theater Television

On July 30, 1930, a large television screen was installed at an RKO theater in Schenectady, New York. After a successful test a reporter for a leading entertainment magazine wrote: "With this successful experiment, the technical arrangements are virtually complete for projecting (television) on normal-sized motion picture screens... Television will be a regular feature in large theaters before the new year." The writer had, unfortunately, considerably underestimated the difficulties that lay ahead.

RKO had hopes of supplementing the motion pictures they regularly featured in theaters with televised vaudeville acts. The "large-sized" television screen was just 5 ft high. Nevertheless, theater television had been launched.

From these humble beginnings, the major motion picture studios began to develop plans for dealing with this new medium. Only one week before the RKO demonstration, David Sarnoff wrote on the science pages of the New York Times that television would soon serve as "a theater in every household." He foresaw great cultural benefits, with educational programs for children and an electronic art gallery for the family. The motion picture industry did not care much for Sarnoff's "theater in every household" concept.

From the start, the motion picture industry viewed television with suspicion. It is interesting, then, that in 1928 RCA—under Sarnoff's direction—became heavily involved in movies through the *Photophone* sound recording system, and the purchase of two theater chains to form RKO.

Nevertheless, the Academy of Motion Picture Arts and Sciences was sufficiently concerned about television to appoint a research council in 1938 to study the issue. The committee suggested that motion picture companies install television systems in their theaters. Members of the panel reasoned that the major movie corporations could "take over" television through their ownership of first-run theaters offering theater television.

During 1938, Paramount took the lead by purchasing a significant interest in DuMont Laboratories. Paramount put the first television station in Chicago on the air in 1940, and established another in Los Angeles in 1943. With DuMont's stations in New York and Washington, Paramount had an interest in four of the nine first television stations in the U.S.

At the end of World War II, the other major motion picture corporations sought to get into the television business. There was even talk of alliances with NBC and CBS. None of these plans ever came to fruition because of the movie industry's mounting record of antitrust convictions. The FCC ruled that it would not grant a television license to a corporation convicted of any monopolistic practices. In 1948, the U.S. Supreme Court affirmed the FCC action, effectively stopping the motion picture industry's efforts to "take over" television.

Disappointed but undaunted, the movie corporations returned to the idea of theater television proposed a decade earlier. Led again by Paramount, television rooms were added to existing selected theaters, which would continue to run feature films. The television portion of the operation was intended to provide newsreels and sporting events, which could be covered better and faster with television than with film. Theater television offered the opportunity to deliver news and sports on the screen with a speed formerly associated only with radio.

There was no shortage of promising theater television projection systems in the late 1940s. Many were exotic, and many never made it out of the laboratory. The most intriguing idea came from Paramount. Called the Paramount *intermediate film system*, the concept was to distribute audio and video signals to theaters via electronic means, and then—upon reception—transcribe the video images and sound onto motion-picture film. The benefits were that both standard television receiving systems and standard motion picture projectors could be used. The conversion process from video and audio to film took 66 seconds.

The equipment to accomplish this task, however, was elaborate and expensive. The film had to be recorded, developed, and projected in one continuous process. While the system worked, the quality of the reproduced image was considered only "fair." Other systems were tried, but the quality of the reproduced image never compared favorably with 35 mm film.

Still, numerous theater television systems were installed across the country with some degree of commercial success. By May 1952, more than 300 different events had been presented. Networks were formed, using telephone company facilities, to offer special attractions to the local theaters. Perhaps best remembered was the Joe Walcott versus Rocky Marciano heavyweight championship fight in September 1952. Fifty movie houses in 30 cities were wired together for the event.

As home television receivers became affordable, the interest in theater television began to diminish. It was no longer necessary to travel to the nearest movie house to view a news broadcast or sporting event. As television grew and matured, entertainment programming appeared as well, marking the end of theater television

1.1.5 Radar

As early as 1922, Albert Taylor of the U. S. Naval Research Laboratory conducted experiments in *radar* (radio detection and ranging). Subsequent tests confirmed the military value of this technology, and development continued until 1940 when mass production of radar equipment was underway.

Parallel work was conducted in Britain from the early 1930s until war broke out with Germany, when full-scale production began. One of the earliest successful instruments was invented by a Scots physicist, Sir Robert Watson-Watt, who demonstrated a working device in 1934. Pulsed electromagnetic waves propagated by a directional antenna reflected off objects that intercepted them. The received reflected waves were then processed and displayed on a CRT.

The early efforts of Watson-Watt were severely limited by the available technology, but the British persisted, expanding upon their initial lead. After war began, Winston Churchill offered Franklin Roosevelt an unprecedented measure of cooperation, allowing American scientists access to British findings and methods of production.

1.1.6 Instrumentation

The development of the first commercially-successful, high-quality oscilloscope by Tektronix in 1947 opened yet another application area for the CRT. Well before the end of World War II, Howard Vollum (who would later go on to found Tektronix) had conceived the idea of a new oscilloscope. His work with radar during the war had shown that commercially available instruments of the day were simply too elementary and too limited to be of value for the coming generation of electronic circuits. Prewar scopes could measure only repetitive or continuous events. During the war, numerous special devices were built in the laboratory to perform specialized jobs, but no attempt was made to design a general-purpose instrument. The model 501 changed all that. The instrument, huge by today's standards, measured nearly 3 ft high by 3 ft deep by 18-in wide. It was a monster by any measure of comparison. While offering limited features, the instrument showed great promise.

A repackaged and redesigned version of the 501, the 511, was introduced in May of 1947 and became an overnight commercial success. The 511 weighted about 50 pounds, making it "portable," after a fashion. Still, it was smaller, lighter, and provided more features than any of its competitors.

The instrument was the first oscilloscope to have a calibrated amplifier and a calibrated time base. The vertical amplifier offered nine calibrated ranges, from 0.25 V/cm through 200 V/cm. The time base was calibrated in five ranges from 0.01 s/cm to 0.1 µs/cm. In addition to vertical and horizontal calibration, the 511 was able to repeat a measurement and measure exactly the same part of the displayed waveform by means of a trigger circuit.

1.1.7 Importance of the Consumer Market

Display technology in general, and the CRT in particular, developed around television. Although applications for the CRT are wide and varied—and were many decades ago—much of the developmental work has been directed toward the television receiver business-where most of the money is. Commercialization and subsequent widespread use of a new display technology does not just happen; it is driven by need, and it is intricately tied with the cost of the end product. Development of any new technology is expensive, and the best way to offset research costs is through high volume production.

The current-day equivalent of this process can be seen in the development of high-resolution displays for computer applications. Significant resources are being directed toward this portion of the display business. Laptop and portable computer applications are two of the market forces driving the development of solid state and other flat-panel displays. Here again, the applied use of display technology is driven by the needs (and wishes) of the end user.

The next stage of display system development is focused on the emergence of widescreen (16:9 aspect ratio) *high-definition television* (HDTV) systems. While HDTV appears to be basically a technology for video consumers, it has also found a number of applications in business and industry, medical imaging, and the military.

1.2 High-Definition Displays

The term *high definition television* applies more to a class of technology than a single system. Because of the many uses of HDTV outside of commercial terrestrial television, the name is somewhat of a misnomer. Still, it is in general use and will be used in this book to describe the technology of high-definition video. HDTV can be logically divided into two basic applications:

- Closed-loop systems

- Broadcast systems

Each system has its own applications and its own markets, although both share the same basic technology.

Japanese professional video equipment manufacturers, under the direction of NHK (the Japanese national broadcasting company), launched a major HDTV development program long before either European or American organizations gave HDTV serious consideration. Early Japanese efforts to establish common international standards for HDTV were largely responsible for stimulating development projects both in North America and in Europe.

The applications for HDTV are numerous and varied. Some of the closed-circuit uses having a significant need for video imaging of higher definition than conventional 525- or 625-line systems are:

- Printing and publishing
- Computer graphics
- Medical diagnosis and treatment
- Biochemistry
- Image and document storage and retrieval
- Teleconferencing
- Process control
- Sales promotion/point-of-sale
- Flight simulation and other types of training

Thus, the future applications of high-definition television imaging extend far beyond entertainment programming. Indeed, HDTV already has. A wide variety of industrial and educational fields are finding HDTV a necessary technology to meet the continuing need for improved video imaging.

In the development of HDTV display systems, two major categories have been studied:

- Direct view
- Projection

A number of display approaches are practical, including the following:

- Cathode ray tube
- Thin-film technology liquid crystal display (TFT-LCD)
- Electroluminescence (EL)
- Gas discharge (plasma display panels)
- Light emitting diode (LED)
- Oil film projection

Table 1.1 Relative Merits of HDTV Display Technologies (*After* [1].)

Technology	Advantages	Disadvantages
Direct view CRT	Highest overall quality for medium size (less than 34-in)	Depth of set; practical size limit 40-in
Projection CRT	High overall quality for large size display; cost effective method for large size display	Depth of set; viewing angle limited
Liquid crystal display	Flat panel; high quality	Small size; high cost
Electroluminescence	Flat panel	Cost considerations; size considerations
Gas discharge (plasma panel)	Flat panel	Cost considerations
LED	Flat panel	High cost
Oil film projection (light valve)	Very large size (greater than 90-in); high quality	Luminous efficiency; display brightness; cost considerations

The principal advantages and disadvantages of each technology are summarized in Table 1.1.

1.2.1 Display Resolution

The resolution of the displayed picture is the most basic attribute of any HDTV system. The image of the most common HDTV video production system, as defined in SMPTE[1] standard 260M (1125 lines and 60 frames/s, interlaced), has approximately twice as much luminance definition vertically as the 525-line NTSC system or the 625-line PAL and SECAM systems. The total number of luminance picture elements (*pixels*) in the image is, therefore, four times as great. The wider aspect ratio of HDTV adds even more visual information.

Increased vertical definition is achieved by employing more than 1000 lines in the scanning patterns. Increased luminance detail in the image is achieved by utilizing a video bandwidth approximately five times that of conventional (NTSC, PAL, or SECAM) systems.

The HDTV image is 25 percent wider than the conventional video image (for a given screen height). The ratio of image width to height in HDTV is 16:9, or 1.777. The conventional NTSC image has a 4:3 aspect ratio.

The HDTV image may be viewed more closely than is customary in conventional video systems. Full visual resolution of the detail of conventional television is available when the image is viewed at a distance equal to about six or seven times the height of the display. The HDTV image can be viewed from a distance of about three times picture height for the full detail of the scene to be resolved.

---

1 Society of Motion Picture and Television Engineers

1.2.2 Production System vs. Transmission System

Bandwidth is the basic point that separates production HDTV systems from transmission-oriented systems for broadcasting. A closed-circuit system does not suffer the same restraints imposed upon a video image that must be transported by radio frequency or other means from an origination center to consumers. It is this distinction that has led to the development of widely varied systems for production and transmission applications. Conventional terrestrial broadcasting is restricted to a video baseband that is 4.2 MHz wide (a total channel bandwidth of 6 MHz). The required bandwidth for full resolution HDTV, however, is on the order of 30 MHz. Video compression algorithms are available that can reduce the required bandwidth without noticeable deterioration. The most common of these algorithms, MPEG-2, is utilized in the ATSC[2] digital televison system and the European DVB digital terrestrial broadcasting system.

Video compression involves compromises. In one case, a tradeoff is made between higher definition and precise rendition of moving objects. It is possible, for example, to defer the transmission of image detail, spreading the signal over a longer time period and thus reducing the required bandwidth. If motion is present in the scene over this longer interval, however, the deferred detail does not occupy its proper place. Smearing, ragged edges, and other types of distortion can occur.

1.2.3 Defining Terms

The terrestrial transmission of digital signals in general, and HDTV in particular, have been the subject of considerable debate worldwide for some years. The following terms are commonly used to describe HDTV systems intended for production and transmission functions:

- *Aspect ratio*. The ratio of picture width to picture height.
- *Advanced systems*. In the broad sense, all systems other than conventional television systems (sometimes referred to as *advanced-definition television*, or ADTV).
- *Contrast*. The range of brightness in the displayed image.
- *Conventional systems*. The NTSC, PAL, and SECAM television systems as standardized prior to the development of advanced systems.
- *Distribution systems*. Terrestrial broadcast, cable, satellite, video cassette, and video disc methods of bringing video programs to end-users.
- *Horizontal resolution*. The number of elements separately present in the picture width.
- *Persistence of vision*. The retention of an image by the human eye after the exciting light is removed.

---

[2] Advanced Television Systems Committee

- *Production systems*. HDTV systems intended for use in the production of video programs, but not necessarily in their distribution by terrestrial broadcasters.
- *Simulcast systems*. The transmission of conventional NTSC, PAL, or SECAM on existing channels and HDTV (or digital standard-definition) transmission of the same programs on one or more additional channels.
- *Vertical resolution*. The number of picture elements separately present in the picture height.

Compatibility Defined

Whenever a new system is proposed, the question of its impact on the utility and value of existing equipment and services is raised. The question of compatibility, therefore, must be addressed. A new system of video image capture, transmission, and/or display, is defined as *compatible* with an old one when the devices of the old system retain their utility when used with the new system. Few electronic systems are totally compatible; some impairment in the performance of old equipment usually occurs because the new system is designed to meet a different set of objectives. A compatible system is fundamentally concerned with preserving the performance of existing equipment.

There is general agreement that an old system-and the equipment developed to support it-should be protected while the new service establishes its technical validity and shows its economic strength in the marketplace. The alternative, a sudden change from the old to the new system, would have economic consequences—and potentially political consequences—that no one cares to face.

An instructive example can be found in the history of television services in the British Isles. Public broadcasting began there in 1936 using a 405-line monochrome system. When the 625-line PAL color system was introduced in Britain in 1967, it was incompatible with the 405-line service. To preserve the existing service, a new means of distribution had to be identified for the color transmissions. The required spectrum was found in the then largely unused ultra-high-frequency (UHF) channels. The color and monochrome systems continued as separate services operating in parallel.

The 405-line service had a life span of 50 years. When it was finally disconnected in 1986, so few 405-line receivers remained in use-and the cost of serving them was so high-that Parliament considered providing the 405-line audience with 625-line receivers free of charge. This solution, however, was deemed unacceptable politically.

1.2.4 HDTV Applications

In the late 1960s, the Japanese Broadcasting Corporation (NHK) undertook an in-depth, long-term examination of what might constitute a totally new "viewing experience" using the electronic medium of television.[3] A research team (led by Fujio of the NHK Technical Research Labs) began a search for the technical parameters that would define a TV system

---

3 Portions of this section were contributed by Lawrence Thorpe, Sony Corporation of America.

of radically improved capability. This TV system was envisaged to attain a quality level sufficient to address the needs of a future sophisticated "information society". It would, thus, embrace:

- Industrial applications
- Motion picture production
- Broadcasting (future applications and services)
- High quality information capture, storage, and retrieval
- Educational, medical, and cultural applications
- Community viewing, including electronic theaters

The term *new viewing experience*-from the standpoint of the human viewer-is worthy of elaboration. As targeted by the NHK research group, this new experience would comprise: (1) a visual image of substantially higher quality, and (2) high quality multi-channel sound. The separate qualities to realize a higher performance video image would, in turn, be an ensemble of the following physical and electrical characteristics:

- A larger (and wider) picture
- Increased horizontal and vertical resolution
- Considerably enhanced color rendition and resolution
- Elimination of current video system artifacts

A larger screen is important for this new viewing experience to achieve the sensation of reality. To quantify this factor, measurements were made of the viewers' tendency to unconsciously tilt their bodies as the displayed picture inclined [2]. It was found that the sensation of reality begins to become apparent when the size of the picture in the field of view (the *viewing angle*) is 20°, and becomes conspicuous at 30°. Furthermore, the preferred viewing distance for a still picture of high-definition was found to be from 2-3 times the screen height; the preferred distance for moving pictures was 4 times the picture height because of the additional factor of dizziness.

Previous research (NHK) had shown that a contrast ratio of 30:1 was sufficient under typical viewing conditions for good picture quality for conventional video. Film, on the other hand, provides a contrast ratio in excess of 100:1. For HDTV, a contrast ratio of 50:1 was established as satisfactory.

Picture resolution in excess of 1000 TV lines was identified as an early objective of HDTV system designers. This number was not simply chosen at random. The smallest unit that can be distinguished by a person with normal visual acuity of 1.0 is one minute of arc in the visual field. The number of scanning lines required at a viewing distance of 3 times the picture height and a vertical viewing angle of 20° is approximately 1100.

The field rate and use of interlace scanning for HDTV have been the subject of some debate in technical circles. A field rate of 60 Hz was found to be sufficient to enable smooth reproduction of 24°/s movement, which is nearly equal to the maximum speed at which the human visual system can follow a subject. The use of interlace relates to overall system bandwidth and hardware practicality. During the early stages of HDTV development, the use of progressive scan presented serious technical problems, not the least of which was the significantly increased bandwidth that it demanded. As hardware has improved and digital image compression has become available, progressive scanning has become practical.

Business and Industrial Applications

While the primary goal of early work on HDTV equipment development centered around those system elements essential to support program production, a considerably broader view of HDTV anticipated important advances in its wider use. The application of video imaging has already branched out in many directions from the original exclusive over-the-air broadcast system intended to bring entertainment programming to the home. Throughout the past three decades, television has been increasingly applied to a vast array of teaching, training, scientific, corporate, and industrial applications. Within this same era there has also emerged an extensive worldwide infrastructure of independent production and post-production facilities (more than 1000 in the U.S. alone) to support these needs.

As the Hollywood film industry became increasingly involved in supplying prime time programming for television (on an international basis) via 35 mm film origination, video technology was harnessed to support off-line editing of these film originals, and to provide creative special effects. Meanwhile, as the world of computer-generated imaging grew at an explosive pace, it too began penetrating countless industrial, scientific, and corporate applications, including film production. The video industry has, in essence, splintered into disparate (although at certain levels, overlapping) specialized industries, as illustrated in Figure 1.1. Any of these video application sectors is gigantic in itself. It was into this environment that HDTV was born.

HDTV brings an entirely new dimension to imaging. Where 525/625 television was inadequate for many applications in the clarity of the image it portrays, the HDTV image contains almost six times more total electronic information. Thus, it dramatically alters the scope of the potential uses envisaged by industry.

Quite apart from the issues of terrestrial (or cable/direct-broadcast satellite) distribution of HDTV entertainment programs, there exists today the very large issue of electronic imaging as a whole. Video technology is being applied to a vast diversity of applications. For some of these applications, 525 NTSC has been adequate, for some barely adequate, and, for others, woefully inadequate.

Critical Implications for the Viewer and Program Producer

In its search for a "new viewing experience," NHK conducted an extensive psychophysical research program in the early 1970s. A large number of attributes were studied. Non-technical people were exposed to a wide variety of electronic images, whose many parameters were then varied over a wide range. A definition of those imaging parameters was being sought, the aggregate of which would satisfy the average viewer that the image portrayal produced an emotional stimulation similar to a large screen cinema experience.

Central to this effort was the pivotal fact that the image portrayed would be large-considerably larger than current NTSC television receivers. Some of the key definitions being

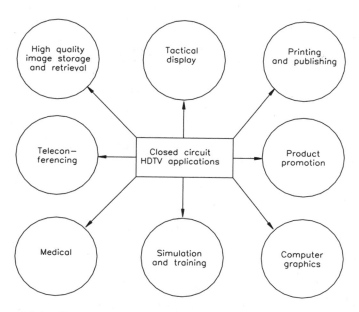

Figure 1.1 Applications for high-definition imaging in business and industrial facilities.

sought by NHK were precisely how large, how wide, how much resolution, and what the optimum viewing distance of this new video image would be.

A substantial body of research gathered over the years has established that the average U.S. consumer views the television receiver from a distance of approximately seven picture heights. This translates to, for example, a 27-in NTSC screen viewed from a distance of about 10 ft. At this viewing distance most of the NTSC artifacts are invisible, with perhaps the exception of *cross color*. Certainly the scanning lines are invisible. The luminance resolution is satisfactory on camera closeups. A facial closeup on a modern, high performance, 525-line NTSC receiver, viewed from a distance of 10 ft, is quite a realistic and pleasing portrayal. But the system quickly fails on many counts when dealing with more complex scene content.

Wide-angle shots (such as jersey numbers on football players) represent one simple and familiar example. Television camera shooting, however, has long adapted to this inherent restriction of 525-line NTSC, as witnessed by the continuous zooming-in for closeups on most sporting events. The camera operator accommodates for the technical shortcomings of the present video system and delivers an image that meets the capabilities of NTSC, PAL, and SECAM quite reasonably. There is however, a penalty, as illustrated in Figure 1.2. The average home viewer is presented with a narrow angle of view, on the order of 10°. The video image has been rendered "clean" of many inherent disturbances by the 10 ft viewing distance, and made adequate in resolution by the action of the camera operator. In the process, however, the scene has become a small "window". The now "acceptable" video image pales in comparison with the sometimes awesome visual stimulation of the cinema. The pri-

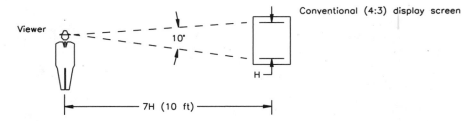

Figure 1.2 Viewing angle vs. screen distance for a conventional video image.

mary limitation of current video systems is, therefore, one of image size. A direct consequence is further limitation of image content; the angle of view is constantly constricted by the need to provide adequate resolution. There is significant, necessary, and unseen intervention by the TV program director in the establishment of the image content that can be passed on to the viewer with acceptable resolution.

Compared to the 525-line NTSC signal (or the marginally better PAL and SECAM systems), the North American HDTV studio standard—SMPTE 260M—and its subsequent variations offers a vast increase in total information contained within the visual image. If all this information is portrayed on an appropriate HDTV studio monitor (available in 19, 28, 38, and 42-in diagonal sizes), the dramatic technical superiority of SMPTE 260M over current technology can easily be seen. The additional visual information, coupled with the component[4] video-based design of HDTV systems, portrays a video image almost totally free (subjectively) of visible artifacts, even when viewed at a close distance.

HDTV, on a direct-view CRT monitor, displays a technically superb picture. The information density is high; the picture has a startling clarity. However, when viewed from a distance of approximately seven picture heights, it is virtually indistinguishable from a good NTSC portrayal. The wider aspect ratio is the most dramatic change in the viewing experience at normal viewing distances.

1.2.5 Image Size

If HDTV is to find a home with the consumer, it will be in the living room. If consumers are to retain the average viewing distance of 10 ft, then the minimum image size required for an HDTV screen (according to NHK research) for the average definition of a totally new viewing experience is about a 75-in diagonal. This represents an image area considerably in excess of present "large" 27-in NTSC (and PAL/SECAM) television receivers. In fact, as indicated in Figure 1.3, the viewing geometry translates into a viewing angle close to 30°, and a distance of only three picture heights between the viewer and the HDTV screen.

---

4 Discrete GRB signals as opposed to a composite color signal.

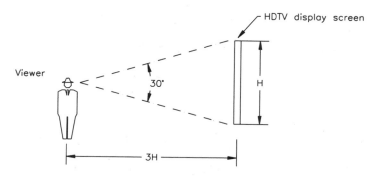

Figure 1.3 Viewing angle vs. screen distance for an HDTV image.

HDTV Image Content

There is more to the enhanced viewing experience than merely elevating picture size. The larger, artifact-free imaging capability of HDTV allows a new image portrayal that capitalizes on the attributes of the larger screen. As previously mentioned, if the camera operator appropriately fills the 525 (or 625) scanning system, the resulting image (from a resolution viewpoint) is actually quite satisfactory. However, if, the same scene is shot with an HDTV camera, and the angle of view of the lens is adjusted to portray the same resolution (in the picture center) as the 525 camera when capturing a closeup of an individual element for the 525 screen, a vital difference between the two pictures emerges.

The larger HDTV image contains considerably more information, as illustrated in Figure 1.4. This HDTV picture shows a far wider field of view. The HDTV image is, thus, radically different than that of the NTSC portrayal. The individual elements are portrayed with the same resolution on the retina—at the same viewing distance—but a totally different viewing experience is provided. The essence of HDTV imaging is this greater sensation of reality. The real dramatic impact of HDTV for the consumer will be realized only when two key ingredients are offered:

- Presentation of an image size of approximately 75-in diagonal.
- Presentation of image content that capitalizes on new camera freedom in formatting larger, wider, and more true to life angles of view.

1.2.6 Motion Picture Theater Applications

As previously discussed, the concept of theater television is nothing new. The technology of high definition video, however, has brought the concept closer to reality.

All of the components for the electronic production and distribution of images that equal the performance of 35 mm film currently exist. It is only the display that limits the development of an all-electronic cinema. Although film under ideal conditions is capable of higher resolution than HDTV systems demonstrated as of this writing, the projected image in a cin-

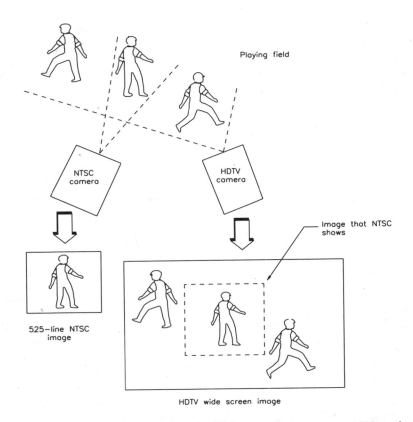

Figure 1.4 Illustration of the differences in screen capture capabilities of convention and HDTV images.

ema generally has a limiting resolution of less than 700 TV lines on a test chart because of the poor dimensional stability of the film and the limits of the optics typically used [2]. Buckling under the high-light flux density, plus jump and weave, limit the performance of a film projector. Most projectors use a double shutter (48 Hz) for a frame rate of 24 frames/s. Fast moving objects either appear to jerk or to split into a double image if the eye tracks the object. For this reason, cinematographers limit the velocity of pan when shooting on film. Further, the brightness of the image in a cinema is somewhat less than desirable for good color perception. However, flicker perception depends on brightness. If the brightness were increased, the 48 cycle flicker would be objectionable. Other film artifacts include grain, dust and scratches.

While many advances have been made in electronic image capture, processing, and storage, the display of those images on a large screen remains limited. For an electronic projector to match the performance of a 35 mm film projector in the center of a typical theater, the display would require the general specifications shown in Table 1.2. Table 1.3 lists the re-

Table 1.2 Display Requirements for Cinema Projection of HDTV Images (*After* [2].)

Parameter	Small Theater Display	Large Theater Display
Vertical active lines	1000 lines minimum	1000 lines minimum
Test chart resolution	800 TV lines minimum	800 TV lines minimum
Light output	2000 lumens minimum	10,000 lumens minimum
Cost	Less than $40,000	Less than $150,000
Response time	Less than 10 ms	Less than 10 ms
Power consumption	Less than 2000 W	Less than 10,000 W
Contrast ratio	Greater than 50:1	Greater than 50:1
Small area uniformity	±0.25 percent	±0.25 percent

Table 1.3 Ultimate Requirements for Cinema Projection of HDTV Images (*After* [2].)

Parameter	Small Theater Display	Large Theater Display
Vertical active lines	2000 lines minimum	2000 lines minimum
Test chart resolution	1500 TV lines minimum	1500 TV lines minimum
Light output	2000 lumens minimum	20,000 lumens minimum
Cost	Less than $20,000	Less than $100,000
Response time	Less than 10 ms	Less than 10 ms
Power consumption	Less than 1000 W	Less than 10,000 W
Contrast ratio	Greater than 100:1	Greater than 100:1
Small area uniformity	± 0.25 percent	± 0.25 percent

quirements for projection to clearly exceed the capabilities of 35 mm film. The resolution and small area uniformity specified in the tables are based on contrast sensitivity measurements of human vision. The contrast ratio is based on the maximum perceptible contrast ratio at low spatial frequencies.

If the flicker frequency of the displayed image can be increased to 60 Hz from 48 Hz, as would be the case in a move from film to HDTV, screen brightness can be increased to 35-50 ft·L without perceptible flicker. The higher screen brightness would increase the perceived performance of the visual display system for resolution, color, and motion rendition.

Because of the requirements of power consumption and cost, only technologies that can provide over one lumen per watt are suitable candidates for electronic cinema projectors. Table 1.4 lists the efficiencies of several technologies that could be used for the electronic cinema. Light valve and solid-state laser technologies are particularly promising.

Despite the inroads that video projectors have made, and will certainly make in the near-term future, cinema applications must take into consideration the concurrent progress being made to improve film. It is easy to argue that if film performance is fixed, any electronic video standard on par with the SMPTE 260M HDTV system would provide adequate

Table 1.4 Electrical to Optical Efficiency of Various Electronic Projection Display Technologies (*After* [2].)

Display System	Efficiency (lumens/W)	Power Required (W)	
		Small Theater Display (2,000 lumens)	Large Theater Display (10,000 lumens)
Gas laser scanner	0.03	66,000	330,000
CRT projector	0.5	4,000	20,000
Light valve[1]	1.0	2,000	10,000
High efficiency light valve[1]	4.0	500	2,500
Solid state laser	10	200	1,000

[1] Light source metal halide or xenon high intensity arc

resolution in a cinema situation. Indeed, the subjective picture quality of HDTV projection—leaving aside the question of image brightness—is impressive, especially when compared with a scratched and/or dirty cinema print.

Film quality is not fixed, however. Minor improvements continue to be made each year, and major improvements are being investigated. One improved 35 mm format (Kinotron) removes the optical sound track from the print and increases the frame height. The sound track is carried by an auxiliary compact disc machine that is locked to the film by timecode. Another approach involves shooting in 70 mm and copying down to 35 mm for release. This preserves most of the original resolution because positive stock resolves three times more lines than negative stock. The resolution limit is set by weave and flutter effects of the projector itself, where improvements are also likely to come.

It is tempting for persons involved in electronic imaging to discount the role of film in capture and display. Film, however, is a sophisticated system that serves as the standard by which video capture and display devices are judged. It cannot be discounted or ignored.

1.3 Aural Component of Visual Realism

The realism of a video image greatly depends on the realism of the accompanying sounds. Particularly in the close viewing of HDTV images, if the audio system is monophonic, the sounds appear to be confined to the center of the screen. The visual and aural senses then convey conflicting information. From the beginning of HDTV system design, it has been clear that stereophonic sound must be used. The generally-accepted quality standard for high fidelity audio has been set by the digital compact disk. This medium covers audio frequencies from below 30 Hz to above 20 kHz, with a dynamic range of 90 dB or greater.

Sound is an important element in the viewing environment. For the greatest realism for the viewer, the picture and the sound must be complementary, both technically and editorially. The sound system should match the picture in terms of positional information and offer the producer the opportunity to use the spatial field creatively. The sound field can be used effectively to enlarge the picture. A *surround sound* system can further enhance the viewing experience.

1.3.1 Hearing Perception

There is a large body of scientific knowledge regarding how humans localize sound. Most of the research has been conducted with earphone listening to monophonic signals to study *lateralization*. *Localization* in stereophonic listening with loudspeakers is less well understood; however, the influence of two factors is dominant:

- Interaural amplitude differences
- Interaural time delay

Of these two properties, time delay is the more influential factor. Over intervals related to the time it takes for a sound wave to travel around the head from one ear to the other, interaural time cues determine where a listener will perceive the location of sounds. Interaural amplitude differences have a lesser influence. An amplitude effect is simulated in stereo music systems by the action of the stereo balance control, which adjusts the relative gain of the left and right channels. It is also possible to implement stereo balance controls based on time delays, but the required circuitry is more complex.

A listener positioned along the line of symmetry between two loudspeakers will hear the center audio as a *phantom* or *virtual image* at the center of the stereo stage. Under such conditions, dialog, for example, will be spatially coincident with the on-screen image. Unfortunately, this coincidence is lost if the listener is not positioned properly with respect to the loudspeakers. Figure 1.5 illustrates the sensitivity of listener positioning to aural image shift. As illustrated, if the loudspeakers are placed six feet apart and the listener is positioned ten feet back from the speakers, an image shift will occur if the listener changes position (relative to the centerline of the speakers) by just 16-in. The data shown in the figure is approximate, and will yield different results for different types and sizes of speakers. Also, the effects of room reverberation are not factored into the data. Still, the sensitivity of listener positioning can be clearly seen. Listener positioning is most critical when the loudspeakers are widely spaced, and less critical when they are closely spaced. To limit loudspeaker spacing, however, runs counter to the purpose of wide-screen displays. The best solution involves the addition of a third audio channel dedicated exclusively to the transmission of center-channel signals for reproduction by a center loudspeaker positioned at the video display, and placement of left and right speakers apart from the display to emphasize the wide-screen effect.

1.3.2 Matching Audio to Video

It has been demonstrated that even with no picture to provide visual cues, the ear/brain combination is sensitive to the direction of sound, particularly in an arc in front of and immediately to the rear of the listener. Even at the sides, listeners are able to locate-with reasonable accuracy-direction cues. With a large screen display, visual cues make the accuracy of sound positioning even more important.

If the number of frontal loudspeakers and associated channels is increased, the acceptable viewing/listening area can be enlarged. Three channel frontal sound using three loudspeakers provides good stereo listening for three or four viewers, while a four channel pre-

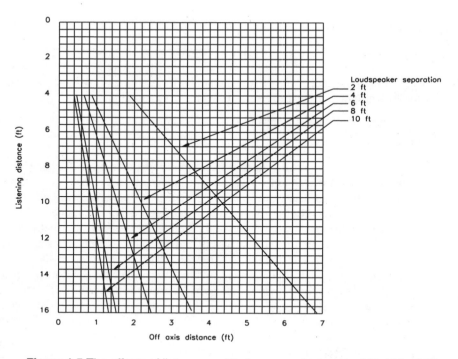

Figure 1.5 The effects of listener positioning on center image shift. (*After* [3].)

sentation increases the area still further. The addition of one or more rear channels permits surround sound effects.

Surround sound presentations, when done correctly, can provide a significant addition to the viewing experience. For example, consider the presentation of a concert or similar performance in a public hall. Members of the audience, in addition to hearing the direct performance sound from the stage, also receive reflected sound, usually delayed slightly and perhaps diffused, from the building surfaces. These acoustic elements give a hall its tonal quality. If the spatial quality of the reflected sound can be made available to the home viewer, the experience will be greatly enhanced. The viewer will see the stage performance in high-definition, and hear both the direct and indirect sound, all of which will add to the feeling of being present at the performance.

Several methods have been successfully used to convey the surround sound channel(s) in conventional NTSC broadcasts. For HDTV broadcasts, the ATSC standard provides for 5.1 channels of audio information using Dolby AC-3 coding. The system includes provisions for the transmission of left, right, center, surround left, surround right, and a low-frequency effects (LFE) channel. All channels convey the full audio bandwidth, with the exception of the LFE channel, which is band-limited to 150 Hz and below, thus the designation 5.1 channels.

In sports coverage, much use can be made of positional information. In a tennis match, for example, the umpire's voice would be located in the center sound field, in line with his observed position, and crowd and ambient sounds would emanate from left or right.

The application of these guidelines is not limited to entertainment and news/sports programming for consumers. Many commercial, industrial, and military applications can benefit from using the audio portion of a presentation to enhance the visual element. For example, the learning experience that can be gained in an aircraft flight simulator is greatly enhanced if the aural cues match the visual cues.

Conventional practice treats the visual and aural components of a presentation as separate entities. This approach however, fails to take full advantage of the potential of each element. As the drive for realism in a presentation—call it *virtual reality* if you like—increases, the need also increases for improved audio production and presentation.

1.3.3 Making the Most of Audio

In any video production there is a great deal of sensitivity to the visual power punch: special effects, acting, direction, and the other elements that build the image. But audio tends to become separated from the visual element. Achieving a good audio product is difficult because of its subjective content. In the visual area there are subtleties video specialists understand and use to their advantage that an audio specialist might not be aware of. By the same token, there are psychoacoustic subtleties relating to how humans hear and experience the world around them that audio specialists can manipulate to their advantage.

Reverb, for example, is poorly understood; it is more than just echo. This tool can be used creatively to trigger certain psychoacoustic responses in an audience. The brain will perceive a voice with some degree of reverb louder. Echo has been used effectively for years to change positions and dimensions in audio mixes.

The use of such psychoacoustic tools is a delicate, specialized area, and audio is a subjective discipline that is short on absolute answers. One of the reasons it is difficult to achieve good quality sound is because it is hard to define what that is. It is easier to quantify video than it is audio. Most people would, given the same video image, come away with the same perception of it. With audio, however, harmony is not so easy to come by. Musical instruments, for example, are harmonically rich and distinctive devices. A violin is not a pure tone; it is a complex balance of textures and harmonics. In audio there is an incredible palette available, and it is acceptable to be different. Most video images have any number of absolute references by which images can be judged. This reference, by and large, does not exist in audio.

When an audience is experiencing a program—be it a television program or an aircraft simulator training session—there is a balance of aural and visual cues. If the production is done right, the audience will be drawn into the program and put themselves into the events occurring on the screen. This *suspension of disbelief* is the key to effectively reaching the audience.

1.3.4 Ideal Sound System

Based on the experience of the film industry, HDTV sound will have the greatest impact when it incorporates (at minimum) a five-channel system with a center channel and sur-

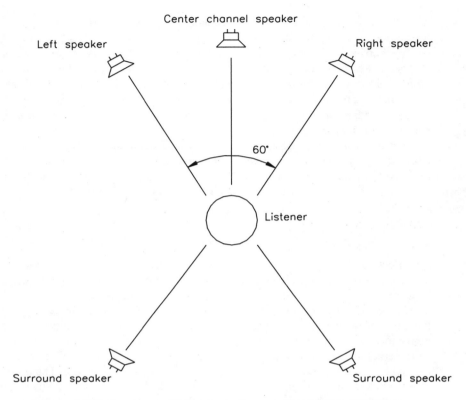

Figure 1.6 Optimum speaker system placement for HDTV. (*After* [4].)

round sound. Figure 1.6 illustrates the optimum speaker placement for enhancement of the viewing experience.

1.4 Display Systems

Color video displays may be classified under the following general categories:

- Direct view cathode ray tube.
- Large screen display, optically projected from a CRT.
- Large screen display, projected from a modulated light beam.
- Large area display of individually-driven light-emitting CRTs or incandescent picture elements.
- Flat panel matrix of transmissive or reflective picture elements.
- Flat panel matrix of light-emitting picture elements.

The CRT remains the dominant type of display for both consumer and professional applications. Light-valve systems using a modulated light source have found widespread application for presentations to large audiences in theater environments, particularly where high screen brightness is required. Matrix-driven flat-panel displays are used in increasing numbers for small-screen personal computers and for portable projector units. For the future, these types of displays are expected to be refined into large-screen wall-mounted panels with a resolution capability suitable for viewing high-definition images.

Flat panel devices—in fact—have now (finally) emerged from the lab and into retail stores. The common joke about large flat-panel displays—the "hang it on the wall like a picture" types of devices—was that the technology was about five years away, and has been five years away for the past thirty years. While a bit of an over-simplification, there is more than just a grain of truth in the statement. At this writing, however, real progress had been made in developing large-screen flat panel devices that were practical, if not entirely affordable.

1.4.1 Format Development

Established procedures in the program production community provide for the 4:3 aspect ratio of video productions and motion-picture films shot specifically for video distribution. This format convention has-by and large-been adopted by the computer industry for desktop computer systems.

In the staging of motion-picture films intended for theatrical distribution, generally no provision is made for the limitations of conventional video displays. Instead, the full screen, in wide aspect ratios-such as CinemaScope-is used by directors for maximum dramatic and sensory impact. Consequently, cropping of essential information may be encountered more often than not on the video screen. This problem is particularly acute in wide-screen features where cropping the sides of the film frame is necessary in producing a print for video transmission. This objective is met in one of the following ways:

- *Letter-box* transmission with blank areas above and below the wide screen frame. Audiences in North America and Japan have not accepted this presentation format, primarily because of the reduced size of the picture images and the aesthetic distraction of the blank screen areas.

- Print the full frame height and crop equal portions of the left and right sides to provide a 4:3 aspect ratio. This process frequently is less than ideal because, depending upon the scene, important visual elements may be eliminated.

- Program the horizontal placement of a 4:3 aperture to follow essential picture information. Called *pan and scan*, this process is used when producing a print or making a film-to-tape transfer for video viewing. Editorial judgment is required to determine the scanning cues for horizontal positioning and, if panning is used, the rate of horizontal movement. This is an expensive and laborious procedure which, at best, compromises the artistic judgments made by the director and the cinematographer in staging and shooting, and by the film editor in post production.

These considerations are also important to the computer industry, which currently has a keen interest in multimedia technology.

One of the reasons for moving to a 16:9 format is to take advantage of consumer acceptance of the 16:9 aspect ratio commonly found in motion picture films. Actually, however, motion pictures are produced in several formats, including:

- 4:3 (1.33)
- 2.35, used for 35 mm anamorphic CinemaScope film
- 2.2 in a 70 mm format

However, the 16:9 aspect ratio is still commonly supported by the motion picture industry.

1.4.2 Viewing Environment

The environment in which a display device is viewed is an important criterion for critical viewing situations.[5] Applications where color purity and adherence to set standards are important require a standardized (or at least consistent) viewing environment. For example, textile colors viewed on a display with a white surround appear differently than the same colors viewed with a black surround. By the same token, different types of ambient lighting make identical colors appear different on a display.

Some applications require a viewing environment that offers optimum display-device viewing conditions and a minimum of operator fatigue. Radar centers provide a convenient illustration.

Traditionally, operations are performed in semi-darkened rooms, that are sufficiently illuminated for viewing the CRT screen without compromising the necessary contrast. Considerable care is taken to avoid reflections from the face of the display from reaching the operator's eyes. Lighting systems are used both in air defense and air traffic control operational centers that attempt to reduce glare, minimize operator fatigue, and maintain the *daylight* or normal color of objects. Polarized filters (90° out of phase) are sometimes employed, one over the display face and another over the room light. This arrangement prevents glare, provided care is taken to situate the light source properly with respect to each operator. The interaction of these two variables is illustrated in Figure 1.7. General overhead illumination cannot be used because such light tends to become depolarized upon striking the walls, ceiling, and other objects in the room, thus reducing the effectiveness of the display filter.

Broad Blue Band System

The *broad blue band* system is a common method of lighting air defense or air traffic control operations centers. The system permits the shorter visible wavelengths to be used for general room illumination and the longer ones for the display, without undue interference. Under this scheme, the following steps are taken:

---

5 This section is based on: C. J. Richards, *Electronic Displays and Data Systems: Constructional Practice*, McGraw-Hill, London, pp. 97–101, 1973.

Display System Applications   29

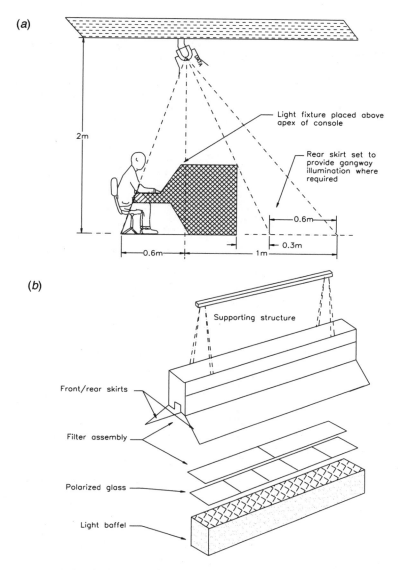

Figure 1.7 Lighting methods for optimum viewing conditions: (a) position of lighting components, (b) detail of light assembly. (*After* [5].)

- Fluorescent lighting is covered with a blue filter (400-500 millimicrons). The display screen is covered with a yellow-orange filter, which absorbs wavelengths below 540 millimicrons.
- Operators wear goggles with the same yellow-orange filters

This arrangement permits a general increase in room illumination because the blue light is prevented from reaching the display, and specular reflections from the display cover are eliminated by the operator's goggles. Special paint is required, however, on controls and indicator markings, so that the operator wearing goggles can see them effectively.

The advantages of this system are good target/background contrast, high filter transmission and elimination of specular reflections from the display face by the goggles regardless of room light position. In addition, filtered blue light is effective in exciting fluorescing colors and permits personnel other than the operator to see well for most tasks. Disadvantages of this approach include:

- Color coding is limited by the range of the filters, although fluorescent paints can be used to improve this situation.

- Monochromatic illumination can affect some people adversely, especially when combined with a moving platform (ship or aircraft), and other adverse environmental conditions, such as poor ventilation.

- Not all phosphors can be used with this system because the display filter will not pass blue or green light.

Yellow-Minus-Yellow System

In the *yellow-minus-yellow* system a sodium light is used with a didymium filter over the display. This arrangement is required because narrow *sodium-D* lines fall at the narrow cut-off point of the didymium filter, as illustrated in Figure 1.8. This coincidence prevents sodium light from reaching the display face, yet allows sufficient transmission. Didymium goggles can be worn by the operators to prevent reflections from the display from reaching their eyes. The advantages of this system include:

- All personnel can see to move about in the room

- There is good target-to-background contrast

- Specular reflections can be eliminated for all who wear goggles

The disadvantages include loss of color and the fact that goggles are required. In addition, the range of potential phosphors is limited.

Mercury-Minus-Red System

In the *mercury-minus-red* system, a mercury vapor lamp is used with two filters, one (a *Noviol filter*) at the room light to remove ultraviolet energy and the other (*red plexiglass 160*) at the display to prevent the shorter wavelengths from reaching it. Red goggles can be worn by the operators to prevent reflections from reaching their eyes. The advantages of this system, like that of the sodium-vapor arrangement, include:

- All personnel can see to move about the room

- There is good target-to-background contrast

- Specular reflections can be eliminated for all who wear goggles

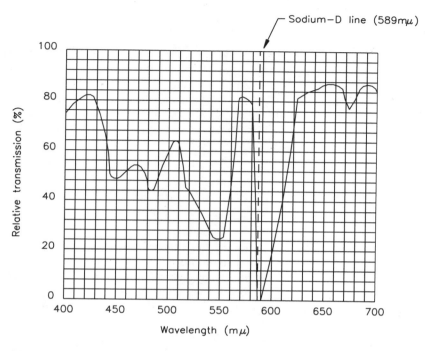

Figure 1.8 Filter cutoff characteristics for yellow-minus-yellow lighting system. (*After* [5].)

In addition, limited color perception is possible with this system. The disadvantages are the limited range of color perception and the fact that goggles are required. In addition, as with the broad blue band and yellow-minus-yellow systems, only selective phosphors can be used.

Table 1.5 shows various combinations of filters and light sources and gives an indication of the display light transmission loss resulting from the display filter used. The data applies to non-goggle-wearing operators; a further loss is associated with the wearing of goggles.

1.4.3 Multisensor Environment

Many different types of imaging sensors exist, each sensitive to a different portion of the electromagnetic spectrum [6]. Passive sensors, which collect energy transmitted by or reflected from a source, include television (visible light), night-vision devices (intensified visible and near-infrared light), passive millimeter wave sensors, and thermal imaging (infrared sensors). Active sensors, when objects are irradiated and the energy reflected from those objects is collected, include the various types of radar.

Each of these systems was developed because of their ability to increase the probability of identification, or detection of objects under difficult environmental conditions. To best present this information to an operator, image processing algorithms have been developed which *fuse* the information into a single coherent image containing information from more

Table 1.5 Comparison of Common Radar-Room Lighting Systems (*After* [5].)

System	Light Source			CRT Display		Operator	
	Type	Filter	Emission (Å)	Filter	Absorption (Å)	Filter	Brightness loss (%)
Yellow-minus-yellow	Sodium	None	5900	Didymium	5900	Didymium	60
Mercury-minus-red	Mercury	Noviol	Below 5900	Red	Below 6000	Red	90/77[1]
Broad blue band	Fluorescent	Blue	Below 5400	Orange	Below 5400	Orange	38/15[1]
Cross-polarization	Incandescent or fluorescent	Polaroid		Polaroid		None	65

[1] The upper value is for P-7 and the lower value is for P-19; other types yield different values.

than one sensor. Such systems are referred to as *multisensor* displays. The process of combining the display data is known as *sensor fusion*. New sensor integration hardware and software can be evaluated both relatively, by determining which combinations are better than others, and absolutely, by comparing operator performance to theoretical expectations.

Typically, as the environmental conditions change in which an individual sensor operates, so does the information content of that image. The information content of the image can further be *scaled* by the operator's ability to perform a target identification or discrimination task. It is reasonable to expect task performance with a sensor fusion display formed from two low information content (and, hence, poor performance) images to still be relatively poor. Similarly, two high information content (high performance) sensor images should yield good performance when combined into a sensor fusion display. Assuming that there was some independent information in the two individual sensor images, it is logical to expect performance with the sensor fusion display to be better than either of the two individual sensors alone. This property results in a three-dimensional performance space.

The greater the number of sensor inputs, the greater the overall performance of the fusion display. However, as the number of inputs grow, the task of designing a display that provides coherent information to the operator is significantly increased. This area of display technology represents perhaps the greatest single technical challenge for the imaging industry. While a wealth of sensor data is available in a tactical situation, the ability of the operator to integrate the data on a split-second basis remains limited. As the computer industry has already discovered, gathering data is relative easy—determining how to display the data so the operator can understand it is the real challenge.

1.4.4 Display Technology Trends

Advanced display system design is an area of great technological interest across a broad range of industries. As a result, considerable engineering expertise is being directed toward improved displays of all types, from consumer television to specialized aeronautical applications. Table 1.6 summarizes the primary display parameters and modes.

Key evaluation metrics for any display include the following [7]:

Table 1.6 Content Descriptors for Common Computer and Video Systems (*After* [7].)

Common Descriptor	Pixel Arrangement	Aspect Ratio (H:V)
VGA	640 × 480	4:3
SVGA	800 × 600	4:3
XGA	1024 × 768	4:3
SXGA	1280 × 1024	5:4
UXGA	1600 × 1280	4:3
NTSC	484 × 450	4:3
U.S. HDTV	1280 × 720 (progressive scan)	16:9
	1920 × 1080 (interlace scan)	16:9

- Overall luminous efficiency
- Viewability (brightness and contrast)
- Uniformity of reproduction, both large- and small-area
- Gray scale
- Color capability, gamut, and accuracy
- Life expectancy and reliability
- Cost of the display device and supporting circuitry

Important technology trends for principal display technologies are outlined in the following sections.

Cathode Ray Tube

In a cathode ray tube, a deflected electron beam is used to excite a cathodoluminescent phosphor. In this very mature technology, continued emphasis is being placed on achieving higher resolution, lower cost, sunlight viewability, and longer life, for both direct-view and projection devices. Improved computer modeling will lead to smaller and more intense electron beams. Additional trends include continued emphasis on achieving flatter faceplates and wider deflection angles.

Flat CRT

The flat CRT is similar in principal to the conventional CRT only that it is configured in a flat (or flatter, relative to the conventional CRT) design. This type of device may or may not use a deflection system. A number of different electron sources are used. Deflected beam versions of classic flat CRT designs have allowed low-cost, portable, small-size displays. Large-area multiplexed versions offering full color and gray scale reproduction have also been produced. Improvements in LCD display technologies, however, have eroded what market existed for flat CRT designs in small-sized screens.

Liquid Crystal Display

In a liquid crystal display (LCD), an electric field is applied across a material having both liquid and crystalline properties. This field is used to modulate light by controlling the amplitude, wave vector, or phase vector of the device. There are three primary LCD types or *modes* [7]:

- *Absorption mode*, which uses amplitude control to produce an image
- *Scattering mode*, which uses wave vector control to produce an image
- *Polarization mode*, which uses phase vector control to produce an image

LCDs are likely to dominate in low-cost vector graphic applications, particularly if low power consumption and overall physical size are important. For large area information display, the future depends on continued progress of active matrix concepts. The large number of companies pursuing LCDs give this technology a significant advantage over other non-CRT display systems.

Plasma/Gas Discharge

In the plasma/gas discharge display technology, an electric field is applied across a gas atmosphere, which creates an avalanche effect. Photons are emitted when the excited atoms return to the ground state.

This technology can be divided logically into two basic configurations: ac and dc-based. Current technology trends for each approach include the following [7]:

- *AC plasma display panel* (AC PDP). Developmental work is concentrating on large, high information content applications, particularly for harsh environments. As color and gray scale performance improves, new application areas will develop. This technology shows considerable promise for HDTV applications. Circuitry and panel costs remain high at this writing, but improvements are likely with volume production.

- *DC plasma display panel* (DC PDP). Efforts are primarily related to applications requiring large size, good color rendition and gray scale representation, such as conventional and advanced television. DC plasma displays face stiff competition in moderate size *alphanumeric and graphics* (A/N&G) applications from other display technologies. While DC PDP holds promise for flat panel HDTV display applications, panel complexity—and therefore cost—are greater than for the AC PDP.

Electroluminescent Display

In an electroluminescent (EL) display device, an electric field applied across a polycrystalline phosphor stimulates the material and light energy is subsequently emitted. This technology can be divided into two basic classes [7]:

- *High field type*. This class of electroluminescent device includes *ac powder*, *dc powder*, *ac thin film*, *dc thin film*, and combinations of these schemes. There are both memory and nonmemory variants.

- *Low field type*. This class of device (LEDs) include organic and inorganic.

Table 1.7 Comparison of Technology Problems Facing High Information Content Display Systems (*After* [7].)

Problem	AC PDP	DC PDP	Flat CRT[1]	Matrix Flat CRT	ACTFEL	DCPEL	Intrinsic LCD	Active Matrix LCD
Luminous efficiency	◇	◇						
Matrix address uniformity			◇	◇		◇	◇	
Indoor viewability								
Outdoor viewability	◆	◆	◆	◇	◇	◆		
Gray scale					◇		◇	
Multicolor capability					◇	◇		
Large screen size[2]			◆	◆	◇	◇	◇	◆
Driver cost	◇	◇		◇	◇	◇	◇	◇
Panel cost	◇	◆	◇	◇	◇	◇	◇	◇

◇ Current problem area.
◆ Probable long term problem area.
[1] Deflected beam device.
[2] Screen size greater than 30-in diagonal.

It is intuitive that with the large number of basic EL technologies, there will be a large number of applied devices. At this writing, *ac thin film* (ACTF) is the most advanced. The future of this technology is highly dependent on the ability of developers to achieve full color, acceptable gray scale, larger display size, and lower cost. It is likely that drivers and decoding logic will be integrated on the display panel.

1.4.5 Display Type Considerations

For each technology, a trade-off must be made between panel complexity and electronics complexity. Typically, technologies with the simplest addressing techniques have the most complex structures, and vice-versa. Technologies requiring high voltage drive (EL, PDP, and most CRTs) use relatively expensive drivers. However, many of these same technologies require fewer drivers for a given panel size. Currently, electronics cost is often viewed as a more significant problem to overcome than panel cost. The cost of drivers alone can be significantly more than the cost of an entire equivalent performance CRT monitor. Table 1.7 provides a comparison of technical challenges for various display devices.

While each display device usually was developed for a specific market, as the technology improves, the range of potential markets expands. The range of applications-in fact-overlap

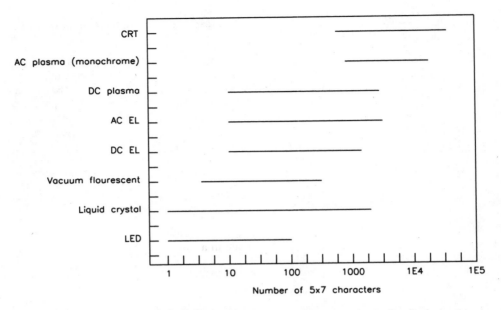

Figure 1.9 Information display capability of various common technologies.

to a great extent. The number of characters capable of being display is a convenient method of classification. Figure 1.9 compares the relative merits of the leading display technologies in this regard.

Despite significant progress in solid-state display systems, the conventional CRT remains the most common display device. The primary advantages of the CRT over competing technologies include the following:

- Low cost for high information content
- Full color available (greater than 256 colors)
- High resolution, high pixel count displays readily available
- Direct view displays of up to 40-in diagonal practical
- Devices available in high volume
- It is a mature, well-understood technology

The CRT, however, is not without its drawbacks, which include:

- High voltages required for operation
- Relatively high power consumption
- Excessive weight for large-screen tubes
- Limited brightness under high-ambient light conditions

- Conventional tubes have a long neck, making the overall display somewhat bulky

Flat-panel devices are not expected to dislodge conventional CRTs any time in the near-term future for video applications, including HDTV. The reasons for this continued dominance of conventional devices include:

- More than 95 percent of all TV sets sold in the world have screens no larger than 34-in diagonal. (It is fair to point out that this percentage will likely change with the appearance of HDTV sets.)
- Initial flat-panel HDTV screens are typically projection systems using CRT sources. A major barrier to the LCD flat-panel projection display is the typically short average lifetime for the light source.
- Long experience in producing CRT television sets make them far less costly and easier to manufacture than plasma and LCD technologies.

Progress, however, continues to be made in alternative display systems for video in general, and HDTV in particular.

Projection systems using light valve devices are capable of the resolution and brightness required for HDTV. High purchase and maintenance expenses, however, have priced such systems out of the reach of most segments of the consumer market, for the present time anyway.

Projection LCD systems offer high-resolution and medium brightness. The compact optical system and constantly improving LCD cells are expected to permit such systems to reach acceptable consumer pricing levels.

It is possible-and perhaps likely-that CRT manufacturers will merge research and development resources for consumer and personal computer applications. Historically, television CRT devices have been treated differently because they did not require the high-resolution of computer workstations. HDTV will change all that, however, and the combined consumer video and desktop computer markets will provide significant opportunities for manufacturers operating in both areas. The end result should be improved products and lower prices for end-users.

The size of the television market is staggering: more than 100,000,000 sets worldwide (black-and-white plus color) per year. Color devices represent the vast majority of devices produced (on the order of 75 percent). The market for monochrome devices has remained strong because of new applications for CRT display, typically computer control of processes and systems.

As in the consumer market, dominance of the CRT for computer applications is likely to continue. Various flat panel display technologies, however, are beginning to erode the traditional stronghold of the CRT: desktop computer systems. The advantages of smaller footprint, light weight construction, increased brightness, and reduced power consumption have made such displays attractive to users, despite a considerable price premium relative to a comparably sized CRT.

In entertainment, communications, and computer systems, the display usually represents the single most expensive component. It is often the product differentiator as well. Offering a variety of attributes, flat panel displays (FPDs) are becoming the platform of choice for new information systems. Flat panel display systems are of interest to design engineers be-

38   Video Display Engineering

Figure 1.10 Flat panel LCD display for desktop computer applications. (*Courtesy of ViewSonic.*)

cause of their favorable operational characteristics relative to CRT devices. These advantages include:

- Portability
- Low occupied volume
- Low weight
- Modest power requirements

Within the last few years, a great deal of progress has been made on improving the performance of FPDs. A 15-in LCD unit intended for desktop computer applications is shown in Figure 1.10.

1.4.6 Advanced Display Applications

As noted previously in this chapter, the applications for information display technologies are wide and varied. Aerospace is one area where different display technologies must be merged, and where significant development is being conducted on an on-going basis. More information must be presented in the cockpit than ever before, and in greater varieties as well. The "glass cockpit" has already become a reality, with electronic displays—mostly CRTs—replacing electromechanical devices. With the need for more detailed information, the size and resolution of the display devices must increase. Full-color

Figure 1.11 Overview of the MD-11 "glass cockpit." (*Courtesy of Boeing.*)

palates and multilevel gray scales are a definite requirement. Reliability, naturally, is of the utmost importance.

The cockpit will grown into an interactive space that may include a heads-up display and/or helmet visor display. Alternatively, a "big picture" approach has been suggested in which the entire cockpit display area is one large reconfigurable element that integrates system readouts and controls. Reconfiguration is important where flight mode determines the cockpit display needs.

The current generation of commercial airliners provides an excellent example of the progress made in display technology for aviation. The MD-11 (Boeing) is typical of the new technology glass cockpit. The instrument panel contains six large (8-in × 8-in) high-resolution CRTs, as shown in Figure 1.11. Multiple display formats can be selected, depending upon the requirements of the situation. The normal configuration is as follows:

- Pilot primary flight display
- Pilot navigation display
- Primary engine and aircraft status display
- Aircraft systems display
- Copilot navigation display

40   Video Display Engineering

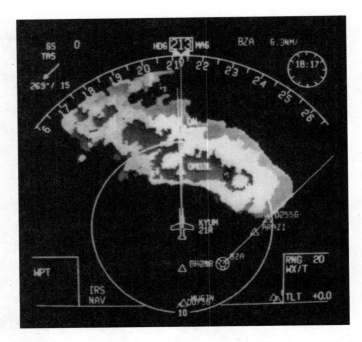

Figure 1.12 Navigation display from the MD-11 aircraft. (*Courtesy of Boeing.*)

- Copilot primary flight display

The navigation and systems displays may be switched as required to provide different status reports. The navigation unit integrates data from the flight plan, weather radar, and traffic collision avoidance systems into a single map image. (See Figure 1.12.) The six full-color displays are driven by two active controllers. A spare controller is available in the event of a failure in one of the primary units.

In the next generation of cockpit systems, more advanced control techniques may include the use of touch screens, track balls, and related pointing devices. Clearly, this is an area of continuing development.

1.5 References

1. Tong, Hua-Sou: "HDTV Display—A CRT Approach," in *Display Technologies*, Shu-Hsia Chen and Shin-Tson Wu (eds.), Proc. SPIE, SPIE, Bellingham, Wash., pp. 2–4, 1992.
2. Glenn, William E.: "Display Requirements for the High-Definition Electronic Cinema," *SID 91 Digest*, Society for Information Display, San Jose, Calif., pp. 144–145, 1991.
3. Torick, Emil L.: "HDTV: High Definition Video—Low Definition Audio?," *1991 HDTV World Conference Proceedings*, National Association of Broadcasters, Washington, D.C., April 1991.

4. Keller, Thomas B.: "Proposal for Advanced HDTV Audio," *1991 HDTV World Conference Proceedings*, National Association of Broadcasters, Washington, D.C., April 1991.
5. Richards, C. J.: *Electronic Display and Data Systems: Constructional Practice*, McGraw-Hill, London, pg. 98, 1973.
6. Foyle, David C.: "Proposed Evaluation Framework for Assessing Operation Performance with Multisensor Displays," *Human Vision, Visual Processing, and Digital Display III*, Bernice E. Rogowitz ed., Proc. SPIE 1666, SPIE, Bellingham, Wash., pp. 514–525, 1992.
7. Goede, Walter F: "Electronic Information Display Perspective," *SID Seminar Lecture Notes*, Society for Information Display, San Jose, Calif., vol. 1, pp. M-1/3–M1/49, May 17, 1999.

1.6 Bibliography

Baldwin, M., Jr.: "The Subjective Sharpness of Simulated Television Images," *Proceedings of the IRE*, vol. 28, July 1940.

Battison, John:, "Making History," *Broadcast Engineering*, Intertec Publishing, Overland Park, Kan., June 1986.

Belton, J.: "The Development of the CinemaScope by Twentieth Century Fox," *SMPTE Journal*, SMPTE, White Plains, N.Y., vol. 97, September 1988.

Benson, K. B., and D. G. Fink: *HDTV: Advanced Television for the 1990s*, McGraw-Hill, New York, N.Y., 1990.

Benson, K. B., and J. C. Whitaker (eds.): *Television Engineering Handbook*, revised ed., McGraw-Hill, New York, N.Y., 1991.

Benson, K. B., and J. C. Whitaker,:*Television and Audio Handbook For Engineers and Technicians*, McGraw-Hill, New York, N.Y., 1989.

Bock, Wolfgang: "Some European Capabilities in Satellite Cinema Exhibition," in *Large-Screen Projection Displays II*, William P. Bleha, Jr., (ed.), Proc. SPIE 1255, SPIE, Bellingham, Wash., pp. 13–20, 1990.

Casteloano, Joseph A.: *Handbook of Display Technology*, Academic, New York, N.Y., 1992.

"Dr. Vladimir K. Zworykin: 1889-1982," *Electronic Servicing and Technology*, Intertec Publishing, Overland Park, Kan., October 1982.

Fink, D. G.: *Color Television Standards*, McGraw-Hill, New York, N.Y., 1986.

Fink, D. G., et al.: "The Future of High-Definition Television," *SMPTE Journal*, SMPTE, White Plains, N.Y., vol. 89, February/March 1980.

Fleschig, W.: German Patent No. 736,575, July 2, 1938.

Fujio, T., J. Ishida, T. Komoto and T. Nishizawa: "High-Definition Television Systems-Signal Standards and Transmission," *SMPTE Journal*, SMPTE, White Plains, N.Y., vol. 89, August 1980.

Fyler, N. F., et al.: "The CBS Colortron," *Proc. IRE*, vol 42, pp. 326–334, January 1954.

Geer, C. W.: U.S. Patent No. 2,480,848, July 11, 1944.

Gomery, Douglas: "Theater Television: A History," *SMPTE Journal*, SMPTE, White Plains, N.Y., February 1989.

Hamasaki, Kimio: "How to Handle Sound with Large Screen," *Proceedings of the ITS, International Television Symposium*, Montreux, Switzerland, 1991.

Kaplan, Sam H.: "The History of Color Picture Tubes and Some Future Projections," *SMPTE Journal*, SMPTE, White Plains, N.Y., pp. 396–400, May 1990.

Lagadec, Roger: "Audio for Television: Digital Sound in Production and Transmission," *Proceedings of the ITS, International Television Symposium*, Montreux, Switzerland, 1991.

Law, H. B.: "A Three-Gun Shadow Mask Color Kinescope, *Proc. IRE*, vol. 39, pp. 1186–1194, October 1951.

Lee, Marshall M.: *Winning with People: The First 40 Years of Tektronix*, Tektronix, Beaverton, Ore., 1986.

Lincoln, Donald: "TV in the Bay Area as Viewed from KPIX," *Broadcast Engineering*, Intertec Publishing, Overland Park, Kan., May 1979.

McCroskey, Donald: "Setting Standards for the Future," *Broadcast Engineering*, Intertec Publishing, Overland Park, Kan., May 1989.

Miller, Howard: "Options in Advanced Television Broadcasting in North America," *Proceedings of the ITS, International Television Symposium*, Montreux, Switzerland, 1991.

Mitsuhashi, Tetsuo: "HDTV and Large Screen Display, in *Large-Screen Projection Displays II*, William P. Bleha, Jr., ed., Proc. SPIE 1255, SPIE, Bellingham, Wash., pp. 2—12, 1990.

Morizono, M.: "Technological Trends in High Resolution Displays Including HDTV," *SID International Symposium Digest*, Society for Information Display, San Jose, Calif., paper 3.1, May 1990.

Pitts, K., and N. Hurst: "How Much Do People Prefer Widescreen (16 × 9) to Standard NTSC (4 × 3)?," *IEEE Transactions on Consumer Electronics*, IEEE, New York, N.Y., August 1989.

Ratliff, Earl: "A Survey of Display Technologies for Military Aircraft Cockpit Applications, in *High-Resolution Displays and Projection Systems*, Elliott Schlam and Marko Slusarczuk (eds.), Proc. SPIE, SPIE, Bellingham, Wash., vol 1664, pp.66–89, 1992.

Schow, Edison: "A Review of Television Systems and the Systems for Recording Television," *Sound and Video Contractor*, Intertec Publishing, Overland Park, Kan., May 1989.

Silverman, Alan: "The Future of the Electronic Cinema, New Math, Old Economics, and Restaurant Cuisine, *SID 91 Digest*, Society for Information Display, San Jose, Calif., 1991.

Slamin, Brendan: "Sound for High-Definition Television," *Proceedings of the ITS, International Television Symposium*, Montreux, Switzerland, 1991.

Slobodin, David E.: "ARPA High-Definition Systems Program," in *Advanced Flat Panel Display Technologies*, Peter S. Freidman (ed.), Proc. SPIE 2174, SPIE, Bellingham, Wash. pp. 2–3, 1994.

"Suitable Sound Systems to Accompany High Definition and Enhanced Television Systems," Report 1072, Recommendations and Reports to the CCIR, 1986, Broadcast Service-Sound, International Telecommunications Union, Geneva, 1986.

Summers, Christopher J.: "The Phosphor Technology Center of Excellence: Research, Education, Industrial Interactions," in *Advanced Flat Panel Display Technologies*, Peter S. Freidman (ed.), Proc. SPIE 2174, SPIE, Bellingham, Wash., pp. 9–15, 1994.

"Television Pioneering," *Broadcast Engineering*, Intertec Publishing, Overland Park Kan., May 1979.

"The Role of Film in TV Programming," *Broadcast Engineering*, Intertec Publishing, Overland Park, Kan., May 1979.

Toet, A.: "Hierarchical Image Fusion," *Machine Vision and Applications*, SPIE, Bellingham, Wash., vol. 3, pp. 111, 1990.

Ulbrich, E., and Walters C.: "Controls and Displays for Douglas Aircraft for the 1990s," in *High-Resolution Displays and Projection Systems*, Elliott Schlam and Marko Slusarczuk (eds.), Proc. SPIE, SPIE, Bellingham, Wash., vol. 1664, pp. 96–106, 1992.

van Raalte, John A.: "CRT Technologies for HDTV Applications," *1991 HDTV World Conference Proceedings*, National Association of Broadcasters, Washington, D.C., April 1991.

Wolfe, Richard M.: "HDTV for the Cinema: Friend or Foe?," *SID 91 Digest*, Society for Information Display, San Jose, Calif., 1991.

Chapter 2

Principles of Light, Vision, and Photometry

2.1 Introduction

Vision results from stimulation of the eye by light and consequent interaction through connecting nerves with the brain.[1] In physical terms, light constitutes a small section in the range of electromagnetic radiation, extending in wavelength from about 400 to 700 nanometers (nm) or billionths (10^{-9}) of a meter. (See Figure 2.1.)

Under ideal conditions, the human visual system can detect:

- Wavelength differences of 1 milllimicron (10 Å, 1 Angstrom unit = 10^{-8} cm)

- Intensity differences as little as 1 percent

- Forms subtending an angle at the eye of 1 arc-minute, and often smaller objects

Although the range of human vision is small compared with the total energy spectrum, human discrimination—the ability to detect differences in intensity or quality—is excellent.

2.2 Sources of Illumination

Light reaching an observer usually has been reflected from some object. The original source of such energy typically is radiation from molecules or atoms resulting from internal (atomic) changes. The exact type of emission is determined by:

- The ways in which the atoms or molecules are supplied with energy to replace what they radiate

- The physical state of the substance, whether solid, liquid, or gaseous

The most common source of radiant energy is the thermal excitation of atoms in the solid or gaseous state.

---

1 Portions of this chapter were adapted from: Jerry C. Whitaker and K. B. Benson (eds.), *Standard Handbook of Video and Television Engineering*, 3rd ed., McGraw-Hill, New York, N.Y., 1999. Used with permission.

46 Video Display Engineering

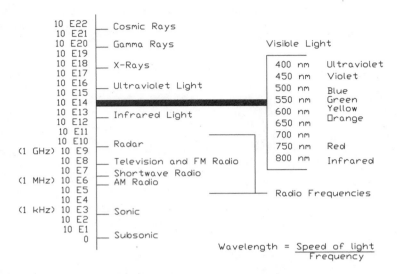

Figure 2.1 The electromagnetic spectrum.

2.2.1 The Spectrum

When a beam of light traveling in air falls upon a glass surface at an angle, it is *refracted* or bent. The amount of refraction depends upon the wavelength, its variation with wavelength being known as *dispersion*. Similarly, when the beam, traveling in glass, emerges into air, it is refracted (with dispersion). A glass prism provides a refracting system of this type. Because different wavelengths are refracted by different amounts, an incident white beam is split up into several beams corresponding to the many wavelengths contained in the composite white beam. This is how the spectrum is obtained.

If a spectrum is allowed to fall upon a narrow slit arranged parallel to the edge of the prism, a narrow band of wavelengths passes through the slit. Obviously, the narrower the slit, the narrower the band of wavelengths or the "sharper" the spectral line. Also, more dispersion in the prism will cause a wider spectrum to be produced, and a narrower spectral line will be obtained for a given slit width.

It should be noted that purples are not included in the list of spectral colors. The purples belong to a special class of colors; they can be produced by mixing the light from two spectral lines, one in the red end of the spectrum, the other in the blue end. Purple (magenta is a more scientific name) is therefore referred to as a *nonspectral color*.

A plot of the power distribution of a source of light is indicative of the watts radiated at each wavelength per nanometer of wavelength. It is usual to refer to such a graph as an *energy distribution curve*.

Individual narrow bands of wavelengths of light are seen as strongly colored elements. Increasingly broader bandwidths retain the appearance of color, but with decreasing purity, as if white light had been added to them. A very broad band extending throughout the visible spectrum is perceived as white light. Many white light sources are of this type, such as the fa-

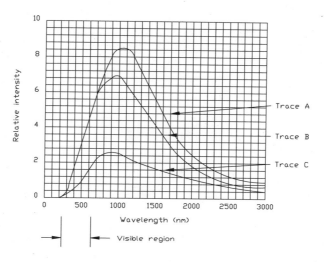

Figure 2.2 The radiating characteristics of tungsten: (trace *A*) radiant flux from 1 cm² of a blackbody at 3000 K, (trace *B*) radiant flux from 1 cm² of tungsten at 3000 K, (trace *C*) radiant flux from 2.27 cm² of tungsten at 3000 K (equal to curve *A* in the visible region). (*After* [1].)

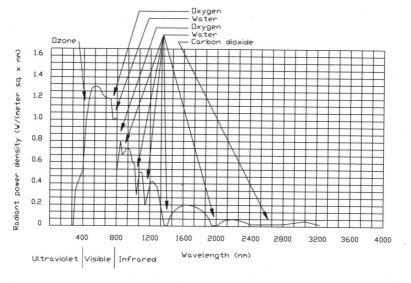

Figure 2.3 Spectral distribution of solar radiant power density at sea level, showing the ozone, oxygen, and carbon dioxide absorption bands. (*After* [1].)

miliar tungsten-filament electric light bulb (see Figure 2.2). Daylight also has a broad band of radiation, as illustrated in Figure 2.3. The energy distributions shown in Figures 2.2 and 2.3 are quite different and, if the corresponding sets of radiation were seen side by side,

48 Video Display Engineering

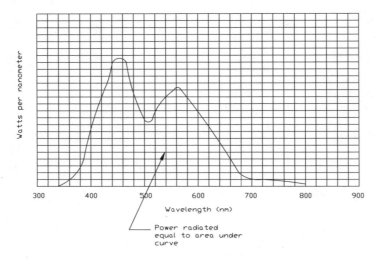

Figure 2.4 Power distribution of a monochrome video picture tube light source. (*After* [2].)

would be different in appearance. Either one, particularly if seen alone, would represent a very acceptable white. A sensation of white light can also be induced by light sources that do not have a uniform energy distribution. Among these is fluorescent lighting, which exhibits sharp peaks of energy through the visible spectrum. Similarly, the light from a monochrome (black-and-white) video cathode ray tube is not uniform within the visible spectrum, generally exhibiting peaks in the yellow and blue regions of the spectrum; yet it appears as an acceptable white (see Figure 2.4).

2.2.2 Monochrome and Color Vision

The color sensation associated with a light stimulus can be described in terms of three characteristics:

- Hue
- Saturation
- Brightness

The spectrum contains most of the principal hues: red, orange, yellow, green, blue, and violet. Additional hues are obtained from mixtures of red and blue light. These constitute the purple colors. Saturation pertains to the strength of the hue. Spectrum colors are highly saturated. White and grays have no hue and, therefore, have zero saturation. Pastel colors have low or intermediate saturation. Brightness pertains to the intensity of the stimulation. If a stimulus has high intensity, regardless of its hue, it is said to be bright.

The psychophysical analogs of hue, saturation, and brightness are:

- Dominant wavelength

Table 2.1 Psychophysical and Psychological Characteristics of Color

Psychophysical Properties	Psychological Properties
Dominant wavelength	Hue
Excitation purity	Saturation
Luminance	Brightness
Luminous transmittance	Lightness
Luminous reflectance	Lightness

- Excitation purity
- Luminance

This principle is illustrated in Table 2.1.

By using definitions and standard response functions, which have received international acceptance through the International Commission on Illumination, the dominant wave length, purity, and luminance of any stimulus of known spectral energy distribution may be determined by simple computations. Although roughly analogous to their psychophysical counterparts, the psychological attributes of hue, saturation, and brightness pertain to observer responses to light stimuli and are not subject to calculation. These sensation characteristics—as applied to any given stimulus—depend in part on other visual stimuli in the field of view and upon the immediately preceding stimulations.

Color sensations arise directly from the action of light on the eye. They are normally associated, however, with objects in the field of view from which the light comes. The objects themselves are therefore said to have color. *Object colors* may be described in terms of their hues and saturations, such as with light stimuli. The intensity aspect is usually referred to in terms of lightness, rather than brightness. The psychophysical analogs of lightness are *luminous reflectance* for reflecting objects and *luminous transmittance* for transmitting objects.

At low levels of illumination, objects may differ from one another in their lightness appearances, but give rise to no sensation of hue or saturation. All objects appear as different shades of gray. Vision at low levels of illumination is called *scotopic vision*. This differs from *photopic vision*, which takes place at higher levels of illumination. Table 2.2 compares the luminosity values for photopic and scotopic vision.

Only the rods of the retina are involved in scotopic vision; cones play no part. Because the fovea centralis is free of rods, scotopic vision takes place outside the fovea. Visual acuity of scotopic vision is low compared with photopic vision.

At high levels of illumination, where cone vision predominates, all vision is color vision. Reproducing systems such as black-and-white photography and monochrome video cannot reproduce all three types of characteristics of colored objects. All images belong to the series of grays, differing only in relative brightness.

The relative brightness of the reproduced image of any object depends primarily upon the luminance of the object as seen by the photographic or video camera. Depending upon the camera pickup element or the film, the dominant wavelength and purity of the light may also

Table 2.2 Relative Luminosity Values for Photopic and Scotopic Vision

Wavelength, nm	Photopic Vision	Scotopic Vision
390	0.00012	0.0022
400	0.0004	0.0093
410	0.0012	0.0348
420	0.0040	0.0966
430	0.0116	0.1998
440	0.023	0.3281
450	0.038	0.4550
460	0.060	0.5670
470	0.091	0.6760
480	0.139	0.7930
490	0.208	0.9040
500	0.323	0.9820
510	0.503	0.9970
520	0.710	0.9350
530	0.862	0.8110
540	0.954	0.6500
550	0.995	0.4810
560	0.995	0.3288
570	0.952	0.2076
580	0.870	0.1212
590	0.757	0.0655
600	0.631	0.0332
610	0.503	0.0159
620	0.381	0.0074
630	0.265	0.0033
640	0.175	0.0015
650	0.107	0.0007
660	0.061	0.0003
670	0.032	0.0001
680	0.017	0.0001
690	0.0082	
700	0.0041	
710	0.0021	
720	0.00105	
730	0.00052	
740	0.00025	
750	0.00012	
760	0.00006	

be of consequence. Most films and video pickup elements currently in use exhibit sensitivity throughout the visible spectrum. Consequently, marked distortions in luminance as a function of dominant wavelength and purity are not encountered. However, their spectral sensitivities seldom conform exactly to that of the human observer. Some brightness distortions, therefore, do exist.

2.2.3 Visual Requirements for Video

The objective in any type of visual reproduction system is to present to the viewer a combination of visual stimuli that can be readily interpreted as representing, or having, a close association with a real viewing situation. It is by no means necessary that the light stimuli from the original scene be duplicated. There are certain characteristics in the reproduced image, however, that are necessary and others that are highly desirable. Only a general discussion of such characteristics will be given here.

In monochrome video, images of objects are distinguished from one another and from their backgrounds as a result of luminance differences. In order that details in the picture be visible and that objects have clear, sharp edges, it is necessary for the video system to be capable of rapid transitions from areas of one luminance level to another. While this degree of resolution need not match what is possible in the eye itself, too low an effective resolution results in pictures with a fuzzy appearance and lacking fineness of detail.

Luminance range and the transfer characteristic associated with luminance reproduction are also of importance in monochrome television. Objects seen as white usually have minimum reflectances of approximately 80 percent. Black objects have reflectances of approximately 4 percent. This gives a luminance ratio of 20/1 in the range from white to black. To obtain the total luminance range in a scene, the reflectance range must be multiplied by the illumination range. In outdoor scenes, the illumination ratio between full sunlight and shadow may be as high as 100/1. The full luminance ranges involved with objects in such scenes cannot be reproduced in normal video reproduction equipment. Video systems must be capable of handling illumination ratios of at least 2, however, and ratios as high as 4 or 5 are desirable. This implies a luminance range on the output of the receiver of at least 40, with possible upper limits as high as 80 or 100.

Monochrome video transmits only luminance information, and the relative luminances of the images should correspond at least roughly to the relative luminances of the original objects. Red objects, for example, should not be reproduced markedly darker than objects of other hues but of the same luminance. Exact luminance reproduction, however, is by no means a necessity. Considerable distortion as a function of hue is acceptable in many applications. Luminance reproduction is probably of primary consequence only if the detail in some hues becomes lost.

Images in monochrome video are transmitted one point, or small area, at a time. The complete picture image is repeatedly scanned at frequent intervals. If the frequency of scan is not sufficiently high, the picture appears to flicker. At frequencies above a *critical frequency* no flicker is apparent. The critical frequency changes as a function of luminance, being higher for higher luminance. The basic requirement for monochrome television is that the *field frequency* (the rate at which images are presented) be above the critical frequency for the highest image luminances.

Images of objects in color television are distinguished from one another by luminance differences or by differences in hue or saturation. Exact reproduction in the image of the original scene differences is not necessary or even attainable. Nevertheless, some reasonable correspondence must prevail because the luminance gradation requirements for color are essentially the same as those for monochrome video.

2.3 Luminous Considerations in Visual Response

Vision is considered in terms of physical, psychophysical, and psychological quantities. The primary stimulus for vision is radiant energy. The study of this radiant energy in its various manifestations, including the effects on it of reflecting, refracting, and absorbing materials, is a study in physics. The response part of the visual process embodies the sensations and perceptions of seeing. Sensing and perceiving are mental operations and therefore belong to the field of psychology. Evaluation of radiant-energy stimuli in terms of the observer responses they evoke is within the realm of psychophysics. Because observer response sensations can be described only in terms of other sensations, psychophysical specifications of stimuli are made according to sensation equalities or differences.

2.3.1 Photometric Measurements

Evaluation of a radiant-energy stimulus in terms of its brightness-producing capacity is a photometric measurement. An instrument for making such measurements is called a *photometer*. In visual photometers, which must be used in obtaining basic photometric measurements, the two stimuli to be compared are normally directed into small adjacent parts of a viewing field. The stimulus to be evaluated is presented in the *test field*; the stimulus against which it is compared is presented in the *comparison field*. For most high-precision measurements, the total size of the combined test and comparison fields is kept small, subtending about 2° at the eye. The area outside these fields is called the *surround*. Although the surround does not enter directly into the measurements, it has adaptation effects on the retina. Thus, it affects the appearances of the test and comparison fields and also influences the precision of measurement.

2.3.2 Luminosity Curve

A *luminosity curve* is a plot indicative of the relative brightnesses of spectrum colors of different wavelength or frequency. To a normal observer, the brightest part of a spectrum consisting of equal amounts of radiant flux per unit wavelength interval is at about 555 nm. Luminosity curves are, therefore, commonly normalized to have a value of *unity* at 555 nm. If, at some other wavelength, twice as much radiant flux as at 555 nm is required to obtain brightness equality with radiant flux at 555 nm, the luminosity at this wavelength is 0.5. The luminosity at any wavelength λ is, therefore, defined as the ratio P_{555}/P_{λ}, where P_{λ} denotes the amount of radiant flux at the wavelength λ, which is equal in brightness to a radiant flux of P_{555}.

The luminosity function that has been accepted as standard for photopic vision is given in Figure 2.5. Tabulated values at 10 nm intervals are given in Table 2.2. This function was agreed upon by the International Commission on Illumination (CIE) in 1924. It is based upon considerable experimental work that was conducted over a number of years. Chief reliance in arriving at this function was based on the step-by-step equality-of-brightness method. Flicker photometry provided additional data.

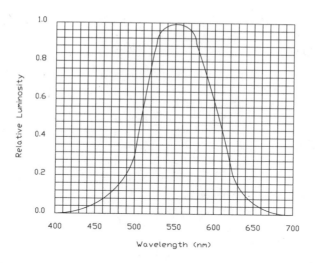

Figure 2.5 The photopic luminosity function. (*After* [2].)

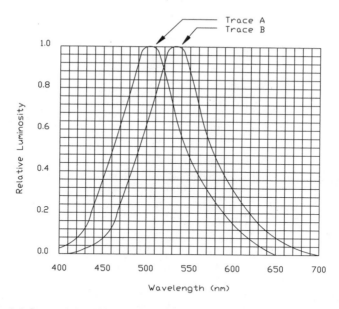

Figure 2.6 Scotopic luminosity function (trace *A*) as compared with photopic luminosity function (trace *B*). (*After* [2].)

In the scotopic range of intensities, the luminosity function is somewhat different from that of the photopic range. The two curves are compared in Figure 2.6. Values are listed in

Table 2.2. While the two curves are similar in shape, there is a shift for the scotopic curve of about 40 nm to the shorter wavelengths.

Measurements of luminosity in the scotopic range are usually made by the *threshold-of-vision* method. A single stimulus in a dark surround is used. The stimulus is presented to the observer at a number of different intensities, ranging from well below the threshold to intensities sufficiently high to be visible. Determinations are made as to the amount of energy at each chosen wavelength that is reported visible by the observer a certain percentage of the time, such as 50 percent. The reciprocal of this amount of energy determines the relative luminosity at the given wavelength. The wavelength plot is normalized to have a maximum value of 1.00 to give the scotopic luminosity function.

In the intensity region between scotopic and photopic vision, called the *Purkinje* or *mesopic region*, the measured luminosity function takes on sets of values intermediate between those obtained for scotopic and photopic vision. Relative luminosities of colors within the mesopic region will therefore vary, depending upon the particular intensity level at which the viewing takes place. Reds tend to become darker in approaching scotopic levels; greens and blues tend to become relatively lighter.

2.3.3 Luminance

Brightness is a term used to describe one of the characteristics of appearance of a source of radiant flux or of an object from which radiant flux is being reflected or transmitted. Brightness specifications of two or more sources of radiant flux should be indicative of their actual relative appearances. These appearances will greatly depend upon the viewing conditions, including the state of adaptation of the observer's eye.

Luminance, as previously indicated, is a psychophysical analog of brightness. It is subject to physical determination, independent of particular viewing and adaptation conditions. Because it is an analog of brightness, however, it is defined to relate as closely as possible to brightness.

The best established measure of the relative brightnesses of different spectral stimuli is the luminosity function. In evaluating the luminance of a source of radiant flux consisting of many wavelengths of light, the amounts of radiant flux at the different wavelengths are weighted by the luminosity function. This converts radiant flux to luminous flux. As used in photometry, the term *luminance* applies only to extended sources of light, not to point sources. For a given amount (and quality) of radiant flux reaching the eye, brightness will vary inversely with the effective area of the source.

Luminance is described in terms of luminous flux per unit projected area of the source. The greater the concentration of flux in the angle of view of a source, the brighter it appears. Therefore, luminance is expressed in terms of amounts of flux per unit solid angle or *steradian*.

In considering the relative luminances of various objects of a scene to be captured and reproduced by a video system, it is convenient to normalize the luminance values so that the "white" in the region of principal illumination has a relative luminance value of 1.00. The relative luminance of any other object then becomes the ratio of its luminance to that of the white. This white is an object of highly diffusing surface with high and uniform reflectance throughout the visible spectrum. For purposes of computation it may be idealized to have 100 percent reflectance and perfect diffusion.

Principles of Light, Vision, and Photometry 55

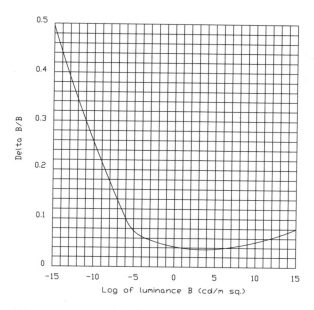

Figure 2.7 Weber's fraction $\Delta B/B$ as a function of luminance B for a dark-field surround. (*After* [3].)

2.3.4 Luminance Discrimination

If an area of luminance B is viewed side by side with an equal area of luminance $B + \Delta B$, a value of ΔB may be established for which the brightnesses of the two areas are just noticeably different. The ratio of $\Delta B/B$ is known as *Weber's fraction*. The statement that this ratio is a constant, independent of B, is known as *Weber's law*.

Strictly speaking, the value of Weber's fraction is not independent of B. Furthermore, its value depends considerably on the viewer's state of adaptation. Values as determined for a dark-field surround are shown in Figure 2.7. It is seen that, at very low intensities, the value of $\Delta B/B$ is relatively large; that is, relatively large values of ΔB, as compared with B, are necessary for discrimination. A relatively constant value of roughly 0.02 is maintained through a brightness range of about 1 to 300 cd/m^2. The slight rise in the value of $\Delta B/B$ at high intensities as given in the graph may indicate lack of complete adaptation to the stimuli being compared.

The plot of $\Delta B/B$ as a function of B will change significantly if the comparisons between the two fields are made with something other than a dark surround. The greatest changes are for luminances below the adapting field. The loss of power of discrimination proceeds rapidly for luminances less by a factor of 10 than that of the adapting field. On the high-luminance side, adaptation is largely controlled by the comparison fields and is relatively independent of the adapting field.

Because of the luminance discrimination relationship expressed by Weber's law, it is convenient to express relative luminances of areas from either photographic or video images in logarithmic units. Because $\Delta(\log B)$ is approximately equal to $\Delta B/B$, equal small changes in $(\log B)$ correspond reasonably well with equal numbers of brightness discrimination steps.

2.3.5 Perception of Fine Detail

Detail is seen in an image because of brightness differences between small adjacent areas in a monochrome display or because of brightness, hue, or saturation differences in a color display. Visibility of detail in a picture is important because it determines the extent to which small or distant objects of a scene are visible, and because of its relationship to the "sharpness" appearance of the edges of objects.

"Picture definition" is probably the most acceptable term for describing the general characteristic of "crispness," "sharpness," or image-detail visibility in a picture. Picture definition depends upon characteristics of the eye, such as visual acuity, and upon a variety of characteristics of the picture-image medium, including its resolving power, luminance range, contrast, and image-edge gradients.

Visual acuity may be measured in terms of the visual angle subtended by the smallest detail in an object that is visible. The *Landolt ring* is one type of test object frequently employed. The ring, which has a segment cut from it, is shown in any one of four orientations, with the opening at the top or bottom, or on the right or left side. The observer identifies the location of this opening. The visual angle subtended by the opening that can be properly located 50 percent of the time is a measure of visual acuity.

Test-object illuminance, contrast between the test object and its background, time of viewing, and other factors greatly affect visual-acuity measurements. Up to a visual distance of about 20 ft (6 m) acuity is partially a function of distance, because of changes in the shape of the eye lens when focusing. Beyond 20 ft, it remains relatively constant. Visual acuity is highest for foveal vision, dropping off rapidly for retinal areas outside the fovea.

A black line on a light background is visible if it has a visual angle no greater than 0.5 s. This is not, however, a true measure of visual acuity. For visual-acuity tests of the type described, normal vision, corresponding to a Snellen 20/20 rating, represents an angular discrimination of about 1 min. Separations between adjacent cones in the fovea and resolving-power limitations of the eye lens give theoretical visual-acuity values of about this same magnitude.

The extent to which a picture medium, such as a photographic or a video system, can reproduce fine detail is expressed in terms of *resolving power* or *resolution*. Resolution is a measure of the distance between two fine lines in the reproduced image that are visually distinct. The image is examined under the best possible conditions of viewing, including magnification.

Two types of test charts are commonly employed in determining resolving power, either a wedge of radial lines or groups of parallel lines at different pitches for each group. For either type of chart, the spaces between pairs of lines usually are made equal to the line widths. Figure 2.8 shows a test signal electronically generated by a video measuring test set.

Resolution in photography is usually expressed as the maximum number of lines (counting only the black ones or only the white ones) per millimeter that can be distinguished from one another. In addition to the photographic material itself, measured values of resolving power depend upon a number of factors. The most important ones typically are:

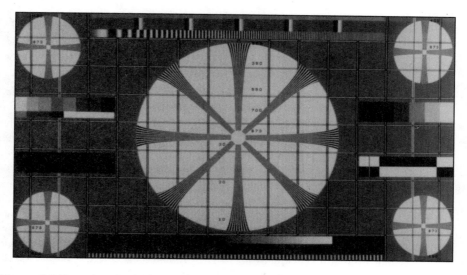

Figure 2.8 Test chart for high-definition television applications produced by a signal waveform generator. The electronically-produced pattern is used to check resolution, geometry, bandwidth, and color reproduction. (*Courtesy of Tektronix.*)

- Density differences between the black and the white lines of the test chart photographed
- Sharpness of focus of the test-chart image during exposure
- Contrast to which the photographic image is developed
- Composition of the developer

Sharpness of focus depends upon the general quality of the focusing lens, image and object distances from the lens, and the part of the projected field where the image lies. In determining the resolving power of a photographic negative or positive material, the test chart employed generally has a high-density difference, such as 3.0, between the black-and-white lines. A high-quality lens is used, the projected field is limited, and focusing is critically adjusted. Under these conditions, ordinary black-and-white photographic materials generally have resolving powers in the range of 30 to 200 line-pairs per millimeter. Special photographic materials are available with resolving powers greater than 1000 line-pairs per millimeter.

Resolution in a video system is expressed in terms of the maximum number of lines (counting both black and white) that are discernible when viewing a test chart. The value of horizontal (vertical lines) or vertical (horizontal lines) resolution is the number of lines equal to the dimension of the raster. Vertical resolution in a well-adjusted system equals the number of scanning lines, roughly 500 in conventional television. In normal broadcasting and reception practice, however, typical values of vertical resolution range from 350 to 400 lines. The theoretical limiting value for horizontal resolution (R_H) in a 525 line, 30 Hz frame rate system is given by:

$$R_H = \frac{2(0.75)(\Delta f)}{30(525)} = 0.954 \times 10^{-4} \Delta f$$

Where:
Δf = the available bandwidth frequency in Hz

The constants 30 and 525 represent the frame and line frequencies, respectively, in the conventional U.S. NTSC television system. A factor of 2 is introduced because in one complete cycle both a black and a white line are obtainable. Factor 0.75 is necessary because of the receiver aspect ratio; the picture height is three-fourths of the picture width. There is an additional reduction of about 15 percent (not included in the equation) in the theoretical value because of horizontal blanking time during which retrace takes place. A transmission bandwidth of 4.25 MHz-typically that of the conventional terrestrial television system-thus makes possible a maximum resolution of about 345 lines.

2.3.6 Sharpnes

The appearance evaluation of a picture image in terms of the edge characteristics of objects is called *sharpness*. The more clearly defined the line that separates dark areas from lighter ones, the greater the sharpness of the picture. Sharpness is, naturally, related to the transient curve in the image across an edge. The average gradient and the total density difference appear to be the most important characteristics. No physical measure has been devised, however, that predicts the sharpness (appearance) of an image in all cases.

Picture resolution and sharpness are to some extent interrelated, but they are by no means perfectly correlated. Pictures ranked according to resolution measures may be rated somewhat differently on the basis of sharpness. Both resolution and sharpness are related to the more general characteristic of picture definition. For pictures in which, under particular viewing conditions, effective resolution is limited by the visual acuity of the eye rather than by picture resolution, sharpness is probably a good indication of picture definition. If visual acuity is not the limiting factor, however, picture definition depends to an appreciable extent on both resolution and sharpness.

2.3.7 Response to Intermittent Excitation

The brightness sensation resulting from a single, short flash of light is a function of the duration of the flash and its intensity. For low-intensity flashes near the threshold of vision, stimuli of shorter duration than about 1/5 s are not seen at their full intensity. Their apparent intensities are nearly proportional to the action times of the stimuli.

With increasing intensity of the stimulus, the time necessary for the resulting sensation to reach its maximum becomes shorter. A stimulus of 5 mL reaches its maximum apparent intensity in about 1/10 s; a stimulus of 1000 mL reaches its maximum in less than 1/20 s. Also, for higher intensities, there is a brightness overshooting effect. For stimulus times longer

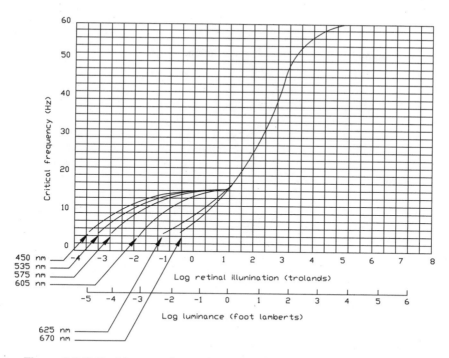

Figure 2.9 Critical frequencies as they relate to retinal illumination and luminance (1 ft·L ≅ cd/m^2; 1 troland = retinal illuminance per square millimeter pupil area from the surface with luminance of 1 cd/m^2). (*After* [4].)

than what is necessary for the maximum effect, the apparent brightness of the flash is decreased. A 1000 mL flash of 1/20 s will appear to be almost twice as bright as a flash of the same intensity that continues for 1/5 s. These effects are essentially the same for colors of equal luminances, independent of their chromatic characteristics.

Intermittent excitations at low frequencies are seen as successive individual light flashes. With increased frequency, the flashes appear to merge into one another, giving a coarse, pulsating *flicker effect*. Further increases in frequency result in finer and finer pulsations until, at a sufficiently high frequency, the flicker effect disappears.

The lowest frequency at which flicker is not seen is called the *critical fusion frequency* or simply the *critical frequency*. Over a wide range of stimuli luminances, the critical fusion frequency is linearly related to the logarithm of luminance. This relationship is called the *Ferry-Porter law*. Critical frequencies for several different wavelengths of light are plotted as functions of retinal illumination (*trolands*) in Figure 2.9. The second abscissa scale is plotted in terms of luminance, assuming a pupillary diameter of about 3 mm. At low luminances, critical frequencies differ for different wavelengths, being lowest for stimuli near the red end of the spectrum and highest for stimuli near the blue end. Above a retinal illumination of about 10 trolands (0.4 ft·L) the critical frequency is independent of wavelength. This is in the critical frequency range above approximately 18 Hz.

The critical fusion frequency increases approximately logarithmically with the increase in retinal area illuminated. It is higher for retinal areas outside the fovea than for those inside, although fatigue to flicker effects is rapid outside the fovea.

Intermittent stimulations sometimes result from rapid alternations between two color stimuli, rather than between one color stimulus and complete darkness. The critical frequency for such stimulations depends upon the relative luminance and chromatic characteristics of the alternating stimuli. The critical frequency is lower for chromatic differences than for luminance differences. Flicker photometers are based upon this principle. The critical frequency also decreases as the difference in intensity between the two stimuli becomes smaller. Critical frequency depends to some extent upon the relative time amounts of the component stimuli, and the manner of change from one to another. Contrary to what might be expected, smooth transitions such as a sine-wave characteristic do not necessarily result in the lowest critical frequencies. Lower critical frequencies are sometimes obtained when the transitions are rather abrupt in one direction and slow in the opposite.

When intermittent stimuli are seen at frequencies above the critical frequency, the visual effect is a single stimulus that is the mean, integrated with respect to time, of the actual stimuli. This additive relationship for intermittent stimuli is known as the *Talbot-Plateau* law.

2.4 Photometric Quantities

The study of visual response is facilitated by division of the subject into its physical, psychophysical, and psychological aspects. The subject of photometry is concerned with the psychophysical aspects: the evaluation of radiant energy in terms of equality or differences for the human observer. Specifically, photometry deals with the luminous aspects of radiant energy, or-in other words-its capacity to evoke the sensation of brightness.

2.4.1 Luminance and Luminous Intensity

By international agreement, the standard source for photometric measurements is a *blackbody* heated to the temperature at which platinum solidifies, 2042 K, and the luminance of the source is 60 candelas per square centimeter of projected area of the source.

Luminance is defined as:

$$B = K_m \int \frac{V(\lambda)P(\lambda)}{\varpi \alpha \cos\theta} d\lambda$$

Where:
K_m = maximum luminous efficiency of radiation (683 lumens per watt)
V = relative efficiency, or luminosity function
$P/(\varpi \alpha \cos\theta)$ = radiant flux (P) per steradian (ϖ) per projected area of source (α) cos θ

Upon first examination, this appears to be an unnecessarily contrived definition. Its usefulness, however, lies in the fact that it relates directly to the sensation of brightness, although there is no strict correspondence.

Table 2.3 Conversion Factors for Luminance and Retinal Illuminance Units (After[2].)

Multiply Quantity Expressed in Units of X by Conversion Factor to Obtain Quantity in Units of Y

X \ Y	Candelas per square centimeter	Candelas per square meter	Candelas per square inch	Candelas per square foot	Lamberts	Millilamberts	Footlamberts	Trolands†‡
Candelas per square centimeter	1	1×10^4	6.452	9.290×10^2	3.142	3.142×10^3	2.919×10^3	7.854×10^3
Candelas per square meter (nit)§	1×10^{-4}	1	6.452×10^{-4}	9.290×10^{-2}	3.142×10^{-4}	3.142×10^{-1}	2.919×10^{-1}	7.854×10^{-1}
Candelas per square inch	1.550×10^{-1}	1.550×10^3	1	1.440×10^2	4.869×10^{-1}	4.869×10^2	4.524×10^2	1.217×10^3
Candelas per square foot	1.076×10^{-3}	1.076×10	6.944×10^{-3}	1	3.382×10^{-3}	3.382	3.142	8.454
Lamberts	3.183×10^{-1}	3.183×10^3	2.054	2.957×10^2	1	1×10^3	9.290×10^2	2.5×10^3
Millilamberts	3.183×10^{-4}	3.183	2.054×10^{-3}	2.957×10^{-1}	1×10^{-3}	1	9.290×10^{-1}	2.500
Footlamberts	3.426×10^{-4}	3.426	2.210×10^{-3}	3.183×10^{-1}	1.076×10^{-3}	1.076	1	2.691
Trolands‡	1.273×10^{-4}	1.273	8.213×10^{-4}	1.183×10^{-1}	4.000×10^{-4}	4.000×10^{-1}	3.716×10^{-1}	1

†In converting luminance to trolands it is necessary to multiply the conversion factor by the square of the pupil diameter in millimeters.
‡In converting trolands to luminance it is necessary to divide the

§As recommended at Session XII in 1951 of the International Commission on Illumination, one nit equals one candela per square meter.

Other luminous quantities are similarly related to their physical counterparts, for example, luminous flux F is defined by:

$$F = K_m \int V(\lambda) P(\lambda) d\lambda$$

When P is given in watts, F is given in lumens.

When the source is far enough away that it may be considered a *point source*, then the luminous intensity I in a given direction is:

$$I = \frac{F}{\varpi}$$

Where:
F = luminous flux in lumens
ϖ = the solid angle of the cone (in steradians) through which the energy is flowing

Conversion factors for various luminance units are listed in Table 2.3. Luminance values for a variety of objects are given in Table 2.4.

2.4.2 Illuminance

In the discussion up to this point, the photometric quantities have been descriptive of the luminous energy emitted by the source. When luminous flux reaches a surface, the surface is illuminated, and the illuminance E is given by $E = F/S$, where S is the area over which the luminous flux F is distributed. When F is expressed in lumens and S in square meters, the illuminance unit is *lumens per square meter*, or *lux*.

Table 2.4 Typical Luminance Values (After [2].)

illumination	Illuminance, ft-L
Sun at zenith	4.82×10^8
Perfectly reflecting, diffusing surface in sunlight	9.29×10^3
Moon, clear sky	2.23×10^3
Overcast sky	$9\text{-}20 \times 10^2$
Clear sky	$6\text{-}17.5 \times 10^2$
Motion-picture screen	10

Table 2.5 Conversion Factors for Illuminance Units (After [2].)

Parameter	Lux	Phot	Footcandle
Lux (meter-candle); lumens per square meter	1.00	1×10^{-4}	9.290×10^{-2}
Phot; lumens per square centimeter	1×10^4	1.00	9.290×10^2
Footcandle; lumens per square foot	1.06×10	1.076×10^{-3}	1.00
Multiply the quantity expressed in units of X by the conversion factor to obtain the quantity in units of Y.			

An element of area S of a sphere of radius r subtends an angle ω at the center of the sphere where $\omega = S/r^2$. For a source at the center of the sphere and r sufficiently large, the source, in effect, becomes a point source at the apex of a cone with S (considered small compared with r^2) as its base. The luminous intensity I for this source is given by $I = F/(S/r^2)$. It follows that

$$I = Fr^2/S \text{ and } I = Er^2$$

Therefore, the illuminance E on a spherical surface element S from a point source is $E = I/r^2$. The illuminance thus varies inversely as the square of the distance. This relationship is known as the *inverse-square law*.

As previously indicated, the unit for illuminance E may be taken as lumen per square meter or lux. This value is also expressed in terms of the *metercandle*, which denotes the illuminance produced on a surface 1 meter distant by a source having an intensity of 1 candela. Similarly, the *footcandle* is the illuminance produced by a source of 1 candela on a surface 1 ft distant and is equivalent to 1 lumen per square foot. Conversion factors for various illuminance units are given in Table 2.5.

The expression given for illuminance, $E = I/r^2$, involves the solid angle S/r^2, which therefore requires that area S is normal to the direction of propagation of the energy. If the area S is situated so that its normal makes the angle θ with the direction of propagation, then the solid angle is given by $(S\cos\theta)/r^2$, as shown in Figure 2.10. The illuminance E is given by $E = I\cos\theta/r^2$.

Principles of Light, Vision, and Photometry

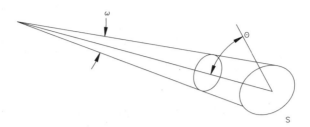

Figure 2.10 Solid angle ω subtended by surface S with its normal at angle θ from the line of propagation. (*After* [2].)

2.4.3 Lambert's Cosine Law

Luminance was previously defined by its relationship to radiant flux because it is the fundamental unit for all photometric quantities. Luminance may also be defined as:

$$B_\theta = \frac{I_\theta}{\alpha \cos\theta}$$

Where:
I_θ = the luminous intensity from a small element α of the area S at an angle of view θ, measured with respect to the normal of this element

For luminous intensities expressed in candelas (or candles), luminance may be expressed in units of candelas (or candles) per square centimeter.

A special case of interest arises if the intensity I_θ varies as the cosine of the angle of view, that is, $I_\theta = I \cos\theta$. This is known as *Lambert's cosine law*. In this instance $B = I/\alpha$ so that the luminance is independent of the angle of view θ. Although no surfaces are known which meet this requirement of "complete diffusion" exactly, many materials conform reasonably well. Pressed barium sulfate is frequently used as a comparison standard for diffusely reflecting surfaces. Various milk-white glasses, known as opal glasses, are used to provide diffuse transmitting media.

The luminous flux emitted per unit area F/α is called the *luminous emittance*. For a perfect diffuser whose luminance is 1 candela per square centimeter, the luminous emittance is π lumens per square centimeter. Or, if an ideal diffuser emits 1 lumen per square centimeter, its luminance is $1/\pi$ candelas per square centimeter. The unit of luminance equal to $1/\pi$ candelas per square centimeter is called the *lambert*. When the luminance is expressed in terms of $1/\pi$ candelas per square foot, the unit is called a *footlambert*.

The physical unit corresponding to luminous emittance is *radiant emittance*, measured in watts per square centimeter. *Radiance*, expressed in watts per steradian per square centimeter, corresponds to luminance.

2.4.4 Measurement of Photometric Quantities

Of the photometric quantities luminous flux, intensity, luminance, and illuminance, the last is, perhaps, the most readily measured in practical situations. However, where the light source in question can be placed on a laboratory photometer bench, the intensity can be determined by calculation from the inverse-square law by comparing it with a known standard.

Total luminous flux may be determined from luminous intensity measurements made at angular intervals of a few degrees over the entire area of distribution. It can also be found by inserting the source within an integrating sphere and comparing the flux received at a small area of the sphere wall with that obtained from a known source in the sphere.

In many practical situations the illuminance produced at a surface is of greatest interest. Visual and photoelectric photometers have been designed for such measurements. For most situations, a photoelectric instrument is more convenient to use because it is portable and easily read. Because the spectral sensitivity of the cell differs from that of the eye, the instrument must not be used for sources differing in color from what the instrument was calibrated for. Filters are also available to make the cell sensitivity conform more closely to that of the eye.

Visual measurements of illumination are generally more suitable where the light is colored. The general procedure is to convert the flux incident on the surface of interest to a luminance value that can be compared with the luminance of a surface within the photometer.

2.4.5 Retinal Illuminance

A psychophysical correlate of brightness is the measure of luminous flux incident on the retina (*retinal illuminance*). One unit designed to indicate retinal stimulation is the *troland*, formerly called a *photon* (not to be confused with the elementary quantum of radiant energy). The troland is defined (under restricted viewing conditions) as the visual stimulation produced by a luminance of 1 candela per square meter filling an entrance pupil of the eye whose area is 1 mm². If luminance B is measured in millilamberts and pupil diameter δ in millimeters, then the retinal illuminance i is given approximately by:

$$i = 2.5\iota^2 B$$

For more accurate evaluation, the actual pupil area must be corrected to the effective pupil area, to take into account that brightness-producing efficiencies of light rays decrease as the rays enter the eye at increasing distances from the central region of the pupil (the *Stiles-Crawford effect*). Variations in transmittance of the ocular media among individuals also prevent the complete specification of the visual stimulus on the retina.

2.5 Receptor Response Measurements

The eye, photographic film, and video cameras are receptors that respond to radiant energy. The video camera tube, for example, exhibits a photoelectric response. Photons of energy absorbed by the photosensitive surface cause ejection of electrons from this surface. The resulting change in electrical potential in the surface gives rise to electrical signals either directly or through the scanning process.

The initial response of a photographic film is photoelectric. Photons absorbed by the silver halide grains cause the ejection of electrons with a consequent reduction of positive silver ions to silver atoms. Specks of atomic silver are thus formed on the silver halide grains.

Conversion of this "latent image" into a visible one is accomplished by chemical development. Grains with the silver specks are reduced by the developer to silver; those without the silver specks remain as silver halide. Chemical reactions occurring simultaneously or subsequently to this primary development determine whether the final image will be negative or positive, and whether it will be in color or black-and-white.

The direct response of the eye is either photoelectric, photochemical, or both. Absorption of light by the eye receptors causes neural impulses to the brain with a resulting sensation of seeing.

Each of these receptors, the eye, photographic film, and video camera, responds differentially to different wavelengths of light. Determination of these receptor responses for photographic film and the video camera provides a basis for correlating the reproduced image with that incident upon the receptor. Interpretation in terms of visual effects is made through a similar analysis of the eye response.

2.5.1 Spectral Response Measurement

In the photoelectric effect of releasing electrons from metals or other materials, light behaves as if it travels in discrete packets, or *quanta*. The energy of a single quantum, or photon, equals $h\upsilon$, where h is Planck's constant and υ is the frequency of the radiation. To release an electron, the photon must transfer sufficient energy to the electron to enable it to escape the potential-energy barrier of the material surface. For any material there is a minimum frequency, called the *threshold frequency*, of radiant energy that provides sufficient energy for an electron not already in an excited state to leave the material. Because of thermal excitation, some electrons may be ejected at frequencies below the threshold frequency. The number of these is usually quite small in comparison with those ejected at frequencies above the threshold frequency. It is because of the relationship between frequency and energy that ultraviolet light usually has a greater photoelectric effect than visible light, and that visible light has a greater effect than infrared light.

For any given wavelength distribution of incident radiation, the number of electrons emitted from a photocathode is proportional to the intensity of the incident radiation. Photoelectric emission is, therefore, linear with irradiation. In practical applications where there are space-charge effects, secondary emissions, or other complicating factors, this linear relationship does not always apply to the current actually collected.

Spectral response measurements are made by exposing the photosensitive surface to narrow-wavelength bands of light. The ratio of the emission current to the incident radiant power is a measure of the sensitivity for this wavelength region. A plot of this ratio as a function of wavelength gives the spectral response curve. If the photo-emissive device is a linear one, the intensity of the incident radiation in each spectral region may be taken at any convenient value without affecting the resulting curve.

If the electric output of the photoelectric device is not linear with the intensity of illumination, the intensities of the spectral irradiations must be more carefully controlled. The electrical output for each spectral region should be the same. A plot of the reciprocal of the incident irradiance as a function of wavelength then gives the spectral response distribution. Common response distributions for camera pickup devices are shown in Figure 2.11.

The eye is a precise measuring device for judging the equality and nonequality of two stimuli, if they are viewed side by side. It cannot be depended upon to give accurate results in

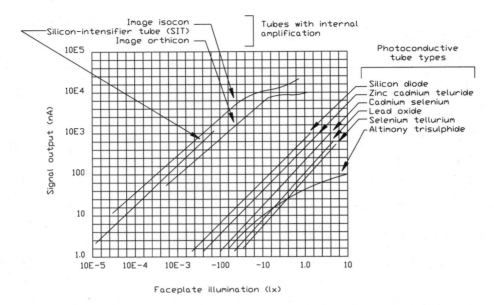

Figure 2.11 Light-transfer characteristics of typical video camera tubes.

ascertaining the amount of difference between two stimuli that are not alike. Therefore, in determining the spectral response characteristics of the eye, it is essential that measurements be made at equal response levels. For color response measurements, amounts of three primary stimuli are found which, in combination, identically match the fourth stimulus being evaluated. The relative amounts of the three primary stimuli necessary for the match are indicative of the response elicited by the test stimulus.

For spectral-luminance response measurements, the evaluations must be made at a common response level of equal brightness. The loss of precision associated with such measurements where chromatic differences exist is minimized by means of step-by-step comparisons or by means of flicker photometry. The brightness response characteristics, or *luminosity function*, of the spectrum colors are the reciprocals of the amounts of energy of these colors, all of which have the same brightness.

2.5.2 Transmittance

Light incident upon an object is either reflected, transmitted, or absorbed. The transmittance of an object may be measured as illustrated in Figure 2.12. Light from the source S passes through the object O and is collected at the receiver R. The spectral transmittance $t\lambda$ of the object is defined as:

$$t_\lambda = \frac{P_\lambda}{P_{O\lambda}}$$

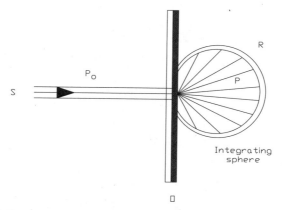

Figure 2.12 The measurement of diffuse transmittance. (*After* [2].

Where:
P_λ = the radiance at wavelength λ reaching the receiver through the object
$P_{O\lambda}$ = the radiance reaching the receiver with no object in the beam path

The spectral density D_λ of the object at wavelength λ is defined as:

$$D_\lambda = -\log t_\lambda = \log \frac{P_{O\lambda}}{P_\lambda}$$

An object with a transmittance of 1.00, or 100 percent, has a density of zero. One with a transmittance of 0.1, or 10 percent, has a density of 1.0.

Objects that transmit light also generally scatter the light to some extent. Consequently, the transmittance measurement depends in part upon the geometrical conditions of measurements. In Figure 2.12, the light incident upon the film is shown as a narrow collimated, commonly called *specular*, beam. The receiver, in the form of an integrating sphere, is placed in contact with the object so that all the transmitted energy is collected. The transmittance measured in this fashion is called *diffuse transmittance*. The corresponding density value is referred to as *diffuse density*. The same results are obtained if the incident light is made completely diffuse and only the specular component evaluated.

If the incident beam is specular and only the specular component of the transmitted light is evaluated, the measurement is called *specular transmittance*. The corresponding density value is *specular density*. A smaller portion of the transmitted energy is collected in a specular measurement than in a diffuse measurement. The specular transmittance of an object is always less than the diffuse transmittance, unless the object does not scatter light, in which case the two transmittances are equal. Specular densities are equal to, or larger than, diffuse densities. The ratio of the specular density to the diffuse density is a measure of the scatter of the object. It is defined as the *Callier Q* coefficient.

68 Video Display Engineering

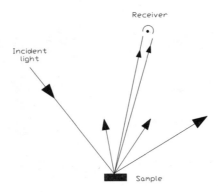

Figure 2.13 The measurement of reflectance. (*After* [2].)

Transmittance measurements made with both the incident and collected beam diffuse are known as *doubly diffuse transmittances*.

2.5.3 Reflectance

Reflectance of an object may be measured as illustrated in Figure 2.13. Following reflection from the object, a portion of the light reaches the receiver. Spectral reflectance r_λ, is defined as:

$$r_\lambda = \frac{P_\lambda}{P_{O\lambda}}$$

Where:
P_λ = the radiance at wavelength λ reaching the receiver from the object
$P_{O\lambda}$ = the radiance reaching the receiver when the sample object is replaced by a standard comparison object

Because of its high reflectance and diffusing properties, a surface of barium sulfate is frequently used as a standard. White paints that have satisfactory reflectance characteristics also are available.

The geometrical arrangement of the light source, sample, and receiver greatly influences reflectance measurements. The incident beam may be either specular or diffuse. If specular, it may be incident upon the object surface perpendicularly or at any angle up to nearly 90° from normal. Essentially all, or only a part, of the reflected light may be collected by the receiver. The effects of various combinations of these choices on the reflectance measurement will depend considerably upon the surface characteristics of the object being measured.

2.6 Human Visual System

The *human visual system* (HVS) is powerful and exceeds the performance of artificial visual systems in almost all areas of comparison. Researchers have, therefore, studied the HVS extensively to ascertain the most efficient and effective methods of communicating information to the eye. An important component of this work has been the development of models of how humans see.

2.6.1 A Model for Image Quality

The classic approach to image-quality assessment involves presenting a group of test subjects with visual test material for evaluation and rating. The test material may include side-by-side display comparisons, or a variety of perception-threshold presentations. One common visual comparison technique is called the *pair-comparison* method. A number of observers are asked to view a specified number of images at two or more distances. At each distance, the subjects are asked to rank the order of the images in terms of overall quality, clearness, and personal preference.

An image acquisition, storage, transmission, and display system need not present more visual information to the viewer than the viewer can process. For this reason, image quality assessment is an important element in the development of any new video system. For example, in the development of a video compressing algorithm, the designer needs to know at what compression point impairments are visible to the "average" viewer.

Evaluation by human subjects, while an important part of this process, is also expensive and time-consuming. Numerous efforts have been made to reduce the human visual system and its interaction with a display device to one or more mathematical models [5–9]. The benefits of this research to design engineers is obvious: more timely feedback on new product offerings. In the development of advanced displays, it is important to evaluate the visual performance of the system well in advance of the expensive fabrication process. Display design software, therefore, is valuable. Because image quality is the primary goal in many display designs, tools that permit the engineer to analyze and emulate the displayed image assist in the design process. Such tools provide early feedback on new display techniques and permit a wider range of prospective products to be evaluated.

After the system or algorithm has successfully passed the minimum criteria established by the model, it can be subjected to human evaluation. The model simulation requires the selection of many interrelated parameters. A series of experiments is typically conducted to improve the model in order to more closely approximate human visual perception.

2.7 References

1. *IES Lighting Handbook*, Illuminating Engineering Society of North America, New York, N.Y., 1981.
2. Fink, D. G.: *Television Engineering*, 2nd ed., McGraw-Hill, New York, N.Y., 1952.
3. Hecht, S.:"The Visual Discrimination of Intensity and the Weber-Fechner Law," *J. Gen Physiol.*, vol. 7, pg. 241, 1924.

4. Hecht, S., S. Shiaer, and E. L. Smith: "Intermittent Light Stimulation and the Duplicity Theory of Vision," Cold Spring Harbor Symposia on Quantitative Biology, vol. 3, pg. 241, 1935.
5. Grogan, Timothy A.: "Image Evaluation with a Contour-Based Perceptual Model," *Human Vision, Visual Processing, and Digital Display III*, Bernice E. Rogowitz ed., Proc. SPIE 1666, SPIE, Bellingham, Wash., pp. 188–197, 1992.
6. Barten, Peter G. J.: "Physical Model for the Contrast Sensitivity of the Human Eye," *Human Vision, Visual Processing, and Digital Display III*, Bernice E. Rogowitz ed., Proc. SPIE 1666, SPIE, Bellingham, Wash., pp. 57–72, 1992.
7. Daly, Scott: "The Visible Differences Predictor: An Algorithm for the Assessment of Image Fidelity, *Human Vision, Visual Processing, and Digital Display III*, Bernice E. Rogowitz ed., Proc. SPIE 1666, SPIE, Bellingham, Wash., pp. 2–15, 1992.
8. Reese, Greg: "Enhancing Images with Intensity-Dependent Spread Functions," *Human Vision, Visual Processing, and Digital Display III*, Bernice E. Rogowitz ed., Proc. SPIE 1666, SPIE, Bellingham, Wash., pp. 253–261, 1992.
9. Martin, Russel A., Albert J. Ahumanda, Jr., and James O. Larimer: "Color Matrix Display Simulation Based Upon Luminance and Chromatic Contrast Sensitivity of Early Vision," *Human Vision, Visual Processing, and Digital Display III*, Bernice E. Rogowitz ed., Proc. SPIE 1666, SPIE, Bellingham, Wash., pp. 336–342, 1992.

2.8 Bibliography

Benson, K. B., and J. C. Whitaker: *Television Engineering Handbook*, revised ed., McGraw-Hill, New York, N.Y., 1991.

Boynton, R. M.: *Human Color Vision*, Holt, New York, N.Y., 1979.

Committee on Colorimetry, Optical Society of America: *The Science of Color*, New York, N.Y., 1953.

Davson, H.: *Physiology of the Eye*, 4th ed., Academic, New York, N.Y., 1980.

Evans, R. M., W. T. Hanson, Jr., and W. L. Brewer: *Principles of Color Photography*, Wiley, New York, N.Y., 1953.

Kingslake, R. (ed.): *Applied Optics and Optical Engineering*, vol. 1, Academic, New York, N.Y., 1965.

Polysak, S. L.: *The Retina*, University of Chicago Press, Chicago, Ill., 1941.

Richards, C. J.: *Electronic Displays and Data Systems: Constructional Practice*, McGraw-Hill, New York, N.Y., 1973.

Schade, O. H.: "Electro-optical Characteristics of Television Systems," *RCA Review*, vol. 9, pp. 5–37, 245–286, 490–530, 653–686, 1948.

Wright, W. D.: *Researches on Normal and Defective Colour Vision*, Mosby, St. Louis, Mo., 1947.

Wright, W. D.: *The Measurement of Colour*, 4th ed., Adam Hilger, London, England, 1969.

Chapter 3
Principles of Color Vision

3.1 Introduction

Visible light is a form of electromagnetic radiation whose wavelengths fall into the relatively narrow band of frequencies to which the human visual system (HVS) responds: the range from approximately 380 nm to 780 nm.[1] These wavelengths of light are readily measurable. The perception of color, however, is a complicated subject. Color is a phenomenon of physics, physiology, and psychology. The perception of color depends on factors such as the surrounding colors, the light source illuminating the object, individual variations in the HVS, and previous experiences with an object or its color.

Colorimetry is the branch of color science that seeks to measure and quantify color in this broader sense. The foundation of much of modern colorimetry is the CIE system developed by the Commission Internationale de l'Eclairage (International Commission on Illumination). The CIE colorimetric system consists of a series of essential standards, measurement procedures, and computational methods necessary to make colorimetry a useful tool for science and industry.

3.2 Color Stimuli

The sensation of color is evoked by a physical stimulus consisting of electromagnetic radiation in the visible spectrum. The stimulus associated with a given object is defined by its *spectral concentration of radiance $L_e(\lambda)$*:

$$L_e(\lambda) = \frac{1}{\pi} E_e(\lambda) R(\lambda)$$

---

[1] Portions of this chapter were adapted from: Jerry C. Whitaker and K. B. Benson (eds.), *Standard Handbook of Video and Television Engineering*, third ed., McGraw-Hill, New York, N.Y., 1999. Used with permission.

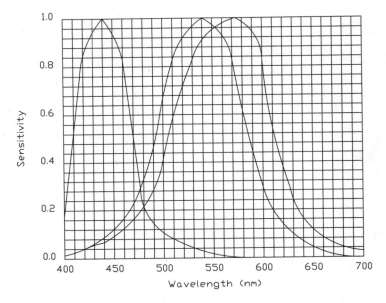

Figure 3.1 Spectral sensitivities of the three types of cones in the human retina. The curves have been normalized so that each is unity at its peak. (*After* [1, 2])

Where:
$E_e(\lambda)$ = the *spectral irradiance*
$R(\lambda)$ = the *spectral reflectance factor*

The spectral reflectance $p(\lambda)$ is sometimes used instead of $R(\lambda)$. Because $p(\lambda)$ is a measure of the total flux reflected by the object, whereas $R(\lambda)$ is a measure of the flux reflected in a specified direction, the use of $p(\lambda)$ implies that the object reflects uniformly in all directions. For most objects this is approximately true, but for some, such as mirrors, it is not.

3.2.1 Trichromatic Theory

Color vision is a complicated process. Full details of the mechanisms are not yet understood. However, it is generally believed, on the basis of strong physiological evidence, that the first stage is the absorption of the stimulus by light-sensitive elements in the retina. These light-sensitive elements, known as *cones*, form three classes, each having a different spectral sensitivity distribution. The exact spectral sensitivities are not known, but they are broad and overlap considerably. An estimate of the three classes of spectral sensitivity is given in Figure 3.1.

It is clear from this *trichromacy* of color vision that many different physical stimuli can evoke the same sensation of color. All that is required for two stimuli to be equivalent is that they each cause the same number of quanta to be absorbed by any given class of cone. In such cases, the neural impulses—and thus the color sensations—generated by the two stimuli will be the same. The visual system and the brain cannot differentiate between the two stimuli even though they are physically different. Such equivalent stimuli are known as *metamers* and the phenomenon as *metamerism*. Metamerism is fundamental to the science of colorimetry; without it color video reproduction as we know it could not exist. The stimulus produced by a video display is almost always a metamer of the original object and not a physical (spectral) match.

3.2.2 Color Matching

Because of the phenomenon of trichromacy, it is possible to match any color stimulus by a mixture of three primary stimuli. There is no unique set of primaries; any three stimuli will suffice as long as none of them can be matched by a mixture of the other two. In certain cases it is not possible to match a given stimulus with positive amounts of each of the three primaries, but a match is always possible if the primaries may be used in a negative sense.

Experimental measurements in color matching are carried out with an instrument called a *colorimeter*. This device provides a split visual field and a viewing eyepiece, as illustrated in Figure 3.2. The two halves of the visual field are split by a line and are arranged so that the mixture of three primary stimuli appears in one-half of the field. The amounts of the three primaries can be individually controlled so that a wide range of colors can be produced in this half of the field. The other half of the field accepts light from the sample to be matched. The amounts of the primaries are adjusted until the two halves of the field match. The amounts of the primaries are then recorded. For those cases where negative values of one or more of the primaries are needed to secure a match, the instrument is arranged to transfer any of the primaries to the other half of the field. The amount of a primary inserted in this manner is recorded as negative.

The operation of color matching may be expressed by the *match equation*:

$$\mathbf{C} \equiv R\mathbf{R} + G\mathbf{G} + B\mathbf{B}$$

This equation is read as follows: Stimulus $\mathbf{C}$ is matched by R units of primary stimulus $\mathbf{R}$ mixed with G units of primary stimulus $\mathbf{G}$ and B units of primary stimulus $\mathbf{B}$. The quantities R, G, and B are called *tristimulus values* and provide a convenient way of describing the stimulus $\mathbf{C}$. All the different physical stimuli that look the same as $\mathbf{C}$ will have the same three tristimulus values R, G, and B.

It is common practice (followed in this publication) to denote the primary stimuli by using boldface letters (usually $\mathbf{R}$, $\mathbf{G}$, and $\mathbf{B}$ or $\mathbf{X}$, $\mathbf{Y}$, and $\mathbf{Z}$) and the corresponding tristimulus values by italic letters R, G, and B or X, Y, and Z, respectively.

The case in which a negative amount of one of the primaries is required is represented by the match equation:

$$\mathbf{C} \equiv -R\mathbf{R} + G\mathbf{G} + B\mathbf{B}$$

This equation assumes that the red primary is required in a negative amount.

The extent to which negative values of the primaries are required depends upon the nature of the primaries, but no set of real physical primaries can eliminate the requirement en-

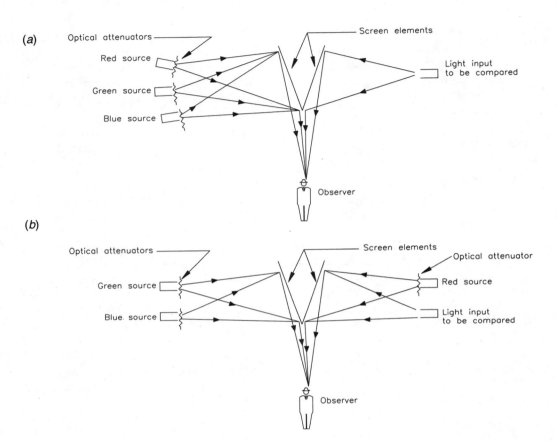

Figure 3.2 Tristimulus color matching instruments: (a) conventional colorimeter, (b) addition of a primary color to perform the match.

tirely. Experimental investigations have shown that in most practical situations, color matches obey the algebraic rules of additivity and linearity. These rules, as they apply to colorimetry, are known as *Grassmann's laws* [3]

To illustrate this point, assume two stimuli defined by the following match equations:

$$\mathbf{C}_1 \equiv R_1\mathbf{R} + G_1\mathbf{G} + B_1\mathbf{B}$$

$$\mathbf{C}_2 \equiv R_2\mathbf{R} + G_2\mathbf{G} + B_2\mathbf{B}$$

If $\mathbf{C}_1$ is added to $\mathbf{C}_2$ in one half of a colorimeter field, and the resultant mixture is matched with the same three primaries in the other half of the field, the amounts of the primaries will be given by the sums of the values in the individual equations. The match equation will be:

$$\mathbf{C}_1 + \mathbf{C}_2 \equiv (R_1 + R_2)\mathbf{R} + (G_1 + G_2)\mathbf{G} + (B_1 + B_2)\mathbf{B}$$

In this discussion, the symbols **R**, **G**, and **B** signify red, green, and blue as a set of primaries. The meaning of these color names must be specified exactly before the colorimetric expressions have precise scientific meaning. Such specification may be given in terms of three relative spectral-power-distribution curves, one for each primary. Similarly, the amounts of each primary must be specified in terms of some unit, such as watts or lumens.

The concept of matching the color of a stimulus by a mixture of three primary stimuli is, of course, the basis of color video reproduction. The three primaries are the three colored phosphors, and the additive mixture is performed in the observer's eye because of the eye's inability to resolve the small phosphor dots or stripes from one another.

3.2.3 Color-Matching Functions

In general, a color stimulus is composed of a mixture of radiations of different wavelengths in the visible spectrum. One important consequence of Grassmann's laws is that if the tristimulus values R, G, and B of a monochromatic (single-wavelength) stimulus of unit radiance are known at each wavelength, the tristimulus values of any stimulus can be calculated by summation. Thus, if the tristimulus values of the spectrum are denoted by

$$\bar{r}(\lambda), \bar{g}(\lambda), \bar{b}(\lambda)$$

per unit radiance, then the tristimulus values of a stimulus with a spectral concentration of radiance $L_e(\lambda)$ are given by:

$$R = \int_{380}^{780} L_e(\lambda) \bar{r}(\lambda) d\lambda$$

for the red value. For green:

$$G = \int_{380}^{780} L_e(\lambda) \bar{g}(\lambda) d\lambda$$

And for blue:

$$B = \int_{380}^{780} L_e(\lambda) \bar{b}(\lambda) d\lambda$$

If a set of primaries were selected and used with a colorimeter, all selections of color mixture could be set up and the appropriate matches made by an observer. A disadvantage of this method is that any selected observer can be expected to have color vision that differs from the "average vision" of many observers. The color matches made might not be satisfactory to the majority of individuals with normal vision.

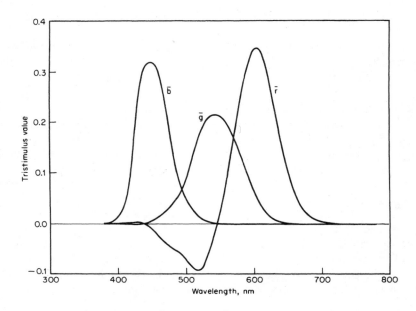

Figure 3.3 Color-matching functions of the CIE standard observer based on matching stimuli of wavelengths 700.0, 546.1, and 435.8 nm, with units adjusted to be equal for a match to an equienergy stimulus.

For this reason, it is desirable to adopt a set of universal data which is prepared by averaging the results of color-matching experiments made by a number of individuals who have normal vision. A spread of the readings taken in these color matches would indicate the variation to be expected among normal individuals. Averaging the results would give a reliable set of spectral tristimulus values. Many experimenters have conducted such psychophysical experiments. The results are in good agreement and have been used as the basis for industry standards.

The curves are now generally known as *color-matching functions*, although in older literature the terms *color-mixing functions* and *distribution coefficients* were sometimes used. The set of color-matching functions used by the CIE in 1931 as a basis for an international standard are shown in Figure 3.3 in terms of a particular set of real primaries, **R**, **G**, and **B**.

3.2.4 Luminance Relationships

Luminances are, by definition, additive quantities. Thus, the luminance of a stimulus with tristimulus values R, G, and B is given by:

$$L = L_R R + L_G G + L_B B$$

Where:

L_R = luminance unit amount of the **R** primary
L_G = luminance unit amount of the **G** primary
L_B = luminance unit amount of the **B** primary

In the special case of a monochromatic stimulus of unit radiance, the luminance is given by:

$$L = L_R \bar{r}(\lambda) + L_G \bar{g}(\lambda) + L_B \bar{b}(\lambda)$$

This luminance condition is also given by:

$$L = K_m V(\lambda)$$

Where:
K_m = 683 lm/W
$V(\lambda)$ = the spectral luminous efficiency function

It follows that $V(\lambda)$ must be a linear combination of the color-matching functions:

$$K_m V(\lambda) = L_R \bar{r}(\lambda) + L_G \bar{g}(\lambda) + L_B \bar{b}(\lambda)$$

3.2.5 Vision Abnormality

In the previous discussion of color matching, the term "normal" was deliberately used to exclude those individuals (about 8 percent of males and 0.5 percent of females) whose color vision differs from the majority of the population. These people are usually called *color-blind*, although very few (about 0.003 percent of the total population) can see no color at all. About 2.5 percent of males require only two primaries to make color matches. Most of these can distinguish yellows from blues but confuse reds and greens. The remaining 5.5 percent require three primaries, but their matches are different from the majority and their ability to detect small color differences is usually less [2, 4].

3.3 Color Representation

Tristimulus values provide a convenient method of measuring a stimulus. Any two stimuli with identical tristimulus values will appear identical under given viewing conditions. However, the actual appearance of the stimuli (whether they are, for example, red, blue, light, or dark) depends on a number of other factors, including:

- The size of the stimuli
- The nature of other stimuli in the field of view
- The nature of other stimuli viewed prior to the present ones

Color appearance cannot be predicted simply from the tristimulus values. Current knowledge of the human color-vision system is far from complete, and much remains to be learned before color appearance can be predicted adequately. However, the idea that the first stage is the absorption of radiation (light) by three classes of cone is accepted by most vision scientists and correlates well with the concept and experimental results of tristimulus colorimetry. Further, tristimulus values—and quantities derived from them—do provide a useful and orderly way of representing color stimuli and illustrating the relationships between them.

It is possible to describe the appearance of a color stimulus in words based upon a person's perception of it. The *trichromatic theory* of color leads to the expectation that this perception will have three dimensions or attributes. Everyday experience confirms this. One set of terms for these three attributes is:

- *Hue*—the attribute according to which an area appears to be similar to one, or to proportions of two, of the perceived colors red, orange, yellow, green, blue, and purple.
- *Brightness*—the attribute according to which an area appears to be emitting, transmitting, or reflecting more or less light.
- *Saturation*—the attribute according to which an area appears to exhibit more or less chromatic color.

A perceived color from a self-emitting object, therefore, is typically described by its hue, brightness, and saturation. For reflecting objects, two other attributes are often used:

- *Lightness*—the degree of brightness judged in proportion to the brightness of a similarly illuminated area that appears to be white.
- *Perceived chroma*—the degree of colorfulness judged in proportion to a similarly illuminated area that appears to be white.

Reflecting objects, thus, may be described by hue, brightness, and saturation or by hue, lightness, and perceived chroma. The perceptual color space formed by these attributes may be represented by a geometrical model, as illustrated in Figure 3.4. The *achromatic* colors (black, gray, white) are represented by points on the vertical axis, with lightness increasing along this axis. All colors of the same lightness lie on the same horizontal plane. Within such a plane, the various hues are arranged in a circle with a gradual progression from red through orange, yellow, green, blue, purple, and back to red. Saturation and perceived chroma both increase from the center of the circle outward along a radius but in different ways depending on the lightness. All colors of the same saturation lie on a conical surface, whereas all colors of the same perceived chroma lie on a cylindrical surface.

If two colors have equal saturation but different lightness, the darker one will have less perceived chroma because perceived chroma is judged relative to a white area. Conversely, if two colors have equal perceived chroma, the darker one will have greater saturation because saturation is judged relative to the brightness—or, in this case, lightness—of the color itself [4].

3.3.1 Munsell System

It is possible to construct a color chart based on the set of color attributes described in the previous section. One such chart, devised by A. H. Munsell, is known as the *Munsell sys-*

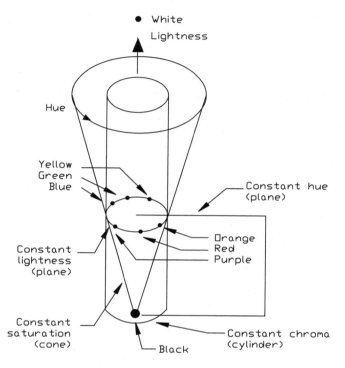

Figure 3.4 A geometrical model of perceptual color space for reflecting objects.

tem. The names used for the attributes by Munsell are *hue*, *chroma*, and *value* [5–7], corresponding, respectively, to the terms hue, perceived chroma, and lightness. In the Munsell system, the colors are arranged in a circle in the following order:

- Red (*R*)
- Red-purple (*RP*)
- Purple (*P*)
- Purple-blue (*PB*)
- Blue (*B*)
- Blue-green (*BG*)
- Green (*G*)
- Green-yellow (*GY*)
- Yellow (*Y*)
- Yellow-red (*YR*)

The Munsell system chart defines a hue circle having 10 hues. To give a finer hue division, each of the 10 hue intervals is further subdivided into 10 parts. For example, there are 10 red subhues, which are referred to as $1R$ to $10R$.

The attribute of chroma is described by having hue circles of various radii. The greater the radius, the greater is the chroma.

The attribute of *value* is divided into 10 steps, from zero (perfect black or zero reflectance factor) to 10 (perfect white or 100 percent reflectance factor). At each value level there is a set of hue circles of different chromas, the lightness of all the colors in the set being equal.

A color is specified in the Munsell system by stating in turn (1) hue, (2) value, and (3) chroma. Thus, the color specified as $6RP$ 4/8 has a red-purple hue of $6RP$, a value of 4, and a chroma of 8. Not all chromas or values can be duplicated with available pigments.

This arrangement of colors in steps of hue, value, and chroma was originally carried out by Munsell using his artistic eye as a judge of the correct classification. Later, a committee of the Optical Society of America [8] made extensive visual studies that resulted in slight modifications to Munsell's original arrangement. The committee's judgment is perpetuated in the form of a book of paper swatches colored with printer's ink and marked with the corresponding Munsell notation [9]. A set of chips arranged in the form of a color tree can also be obtained.

3.3.2 Other Color-Order Systems

In addition to the Munsell system, there are a number of other color-order systems [3]. The three scales of the various systems, and the spacing of samples along the scales, are chosen by different criteria. In some systems the scales and spacing are determined by a systematic mixture of dyes or pigments, or by systematic variation of parameters in a printing process. In others they are based on the rules of additive mixture, as in a tristimulus colorimeter. A third class of color-order systems (which includes the Munsell system) is based on visual perceptions. Within each class the exact rules by which the colors are ordered vary significantly from one to another.

3.3.3 Color Triangle

The *color triangle* is an alternative method of classifying and specifying colors. This method was originated by Newton and used extensively by Maxwell. It is a method of representing the matching and mixing of stimuli and is derived from the tristimulus values discussed previously.

The color triangle displays a given color stimulus in terms of the relative tristimulus values; that is, the relative amounts of three primaries needed to match it. Thus, the color triangle displays only the quality of the stimulus and not its quantity. In one form, the stimulus is represented by a point chosen so that the perpendicular distances to each of the three sides are proportional to the tristimulus values. The triangle need not be equilateral, although the triangles used by Newton and Maxwell were of this type. The method is illustrated in Figure 3.5.

This method of display is equivalent to the use of *trilinear coordinates*, which form a well-known coordinate system in analytical geometry. In this representation, the three pri-

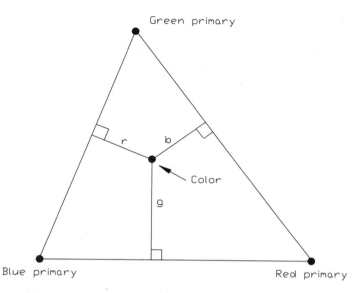

Figure 3.5 The color triangle, showing the use of trilinear coordinates. The amounts of the three primaries needed to match a given color are proportional to r, g, and b.

maries appear one at each of the three vertices of the triangle, because two of the trilinear coordinates vanish at each vertex.

It is more common, however, to use a right-angled triangle as shown in Figure 3.6. The quantities r, g, and b, called *chromaticity coordinates*, are calculated using the following equations:

$$r = \frac{R}{R+G+B}, \quad g = \frac{G}{R+G+B}, \quad b = \frac{B}{R+G+B}$$

The chromaticity coordinates are plotted with r as abscissa and g as ordinate. Because $r + g + b = 1$, it is not necessary to plot b because it can be derived by $b = 1 - r - g$. This (r, g) diagram, and others like it, are known as *chromaticity diagrams*.

A chromaticity diagram is a specification of a color stimulus, not appearance. A particular chromaticity may have many different appearances, depending on factors such as its luminance relative to a white reference in the field of view and the observer's *state of adaptation*. For example, there is a range of chromaticities that will appear white to an observer who is adapted to them. Nevertheless, for a rigidly defined set of observing conditions, it is sometimes useful to think of a chromaticity diagram in terms of appearance. Chromaticity coordinates are relative tristimulus values, and thus correlate with hue and saturation but not with brightness. For most observing conditions, stimuli near the middle of the diagram produce color appearances of low saturation, and stimuli at the edges produce high saturations. The hues are arranged in the usual order of red, orange, yellow, green, blue, purple, and

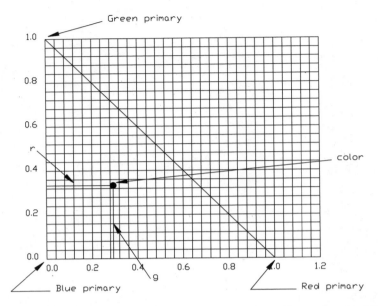

Figure 3.6 A chromaticity diagram. The amounts of the three primaries needed to match a given color are proportional to r, g, and b (= 1 − r − g).

black to red. Conventionally, the diagram is arranged so that this progression is counter-clockwise.

Center of Gravity Law

The chromaticity diagram is a useful way of representing additive color mixture. Consider two stimuli C_1 and C_2:

$$C_1 \equiv R_1 \mathbf{R} + G_1 \mathbf{G} + B_1 \mathbf{B}$$

$$C_2 \equiv R_2 \mathbf{R} + G_2 \mathbf{G} + B_2 \mathbf{B}$$

As explained previously, primary stimulus C_1 is matched by R_1 units of primary stimulus **R**, mixed with G_1 units of primary stimulus **G**, and B_1 units of primary stimulus **B**. In a similar manner, stimulus C_2 is matched by R_2, G_2, and B_2 units of the same primaries. It follows that the chromaticity coordinates are:

$$r_1 = \frac{R_1}{R_1 + G_1 + B_1}, \quad g_1 = \frac{G_1}{R_1 + G_1 + B_1}, \quad b_1 = \frac{B_1}{R_1 + G_1 + B_1}$$

The sum of $R_1 + G_1 + B_1 = T_1$, the total tristimulus value. Then:

$$r_1 = \frac{R_1}{T_1}, \quad g_1 = \frac{G_1}{T_1}, \quad b_1 = \frac{B_1}{T_1}$$

After reordering the equations it follows that:

$$R_1 = r_1 T_1, \quad G_1 = g_1 T_1, \quad B_1 = b_1 T_1$$

In a similar manner:

$$R_2 = r_2 T_2, \quad G_2 = g_2 T_2, \quad B_2 = b_2 T_2$$

In terms of chromaticity coordinates (r, g, and b), the equations for $\mathbf{C}_1$ and $\mathbf{C}_2$ may be written as:

$$\mathbf{CC}_1 = (r_1 T_1)\mathbf{R} + (g_1 T_1)\mathbf{G} + (b_1 T_1)\mathbf{B}$$

$$\mathbf{C}_2 = (r_2 T_2)\mathbf{R} + (g_2 T_2)\mathbf{G} + (b_2 T_2)\mathbf{B}$$

Thus, by Grassmann's laws, the stimulus $\mathbf{C}$ formed by mixing $\mathbf{C}_1$ and $\mathbf{C}_2$ is:

$$\mathbf{C} = R\mathbf{R} + G\mathbf{G} + B\mathbf{B}$$

Where:
$R = r_1 T_1 + r_2 T_2$
$G = g_1 T_1 + g_2 T_2$
$B = b_1 T_1 + b_2 T_2$

The chromaticity coordinates of the mixture, therefore, are:

$$r = \frac{r_1 T_1 + r_2 T_2}{T_1 + T_2}, \quad g = \frac{g_1 T_1 + g_2 T_2}{T_1 + T_2}, \quad b = \frac{b_1 T_1 + b_2 T_2}{T_1 + T_2}$$

The interpretation of this math in the chromaticity diagram is simply that the mixture lies on the straight lining joining the two components and divides it in the ratio T_2/T_1. This concept, illustrated in Figure 3.7, is known as the *center of gravity law* because of the analogy with the center of gravity of weights T_1 and T_2 placed at the points representing $\mathbf{C}_1$ and $\mathbf{C}_2$.

Alychne

As stated previously, the luminance of a stimulus with tristimulus values R, G, and B is

$$L = L_R R + L_G G + L_B B$$

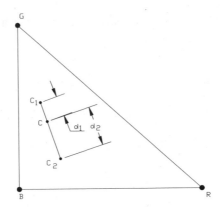

Figure 3.7 The center of gravity law in the chromaticity diagram. The additive mixture of color stimuli represented by C_1 and C_2 lies at C, whose location on the straight line C_1C_2 is given by $d_1T_1 = d_2T_2$, where T_1 and T_2 are the total tristimulus values of the component stimuli.

If this expression is divided by $R + G + B$ and if $L = 0$, then:

$$0 = L_R r + L_G g + L_B b$$

The foregoing is the equation of a straight line in the chromaticity diagram. It is the line along which colors of zero luminance would lie if they could exist. The line is called the *alychne*.

The alychne is illustrated in Figure 3.8, which is a chromaticity diagram based on monochromatic primaries of wavelengths 700.0, 546.1, and 435.8 nm with their units normalized so that equal amounts are required to match a stimulus in which the spectral concentration of radiant power per unit wavelength is constant throughout the visible spectrum (this stimulus is called the *equienergy stimulus*). The alychne lies wholly outside the triangle of primaries, as indeed it must, for no positive combination of real primaries can possibly have zero luminance.

Spectrum Locus

Because all color stimuli are mixtures of radiant energy of different wavelengths, it is interesting to plot, in a chromaticity diagram, the points representing monochromatic stimuli (stimuli consisting of a single wavelength). These can be calculated from the color-matching functions:

$$r(\lambda) = \frac{\bar{r}(\lambda)}{\bar{r}(\lambda) + \bar{g}(\lambda) + \bar{b}(\lambda)}, \quad g(\lambda) = \frac{\bar{g}(\lambda)}{\bar{r}(\lambda) + \bar{g}(\lambda) + \bar{b}(\lambda)}$$

Principles of Color Vision

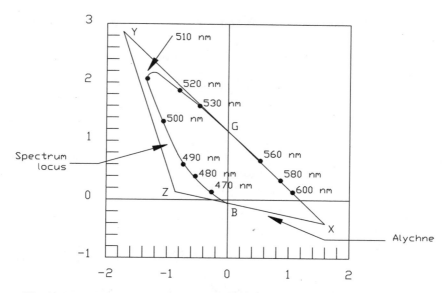

Figure 3.8 The spectrum locus and alychne of the CIE *1931 Standard Observer* plotted in a chromaticity diagram based on matching stimuli of wavelengths 700.0, 546.1, and 435.8 nm. The locations of the CIE primary stimuli *X*, *Y*, and *Z* are shown.

When these spectral chromaticity coordinates are plotted, as shown in Figure 3.8, they lie along a horseshoe-shaped curve called the *spectrum locus*. The extremities of the curve correspond to the extremities of the visible spectrum—approximately 380 nm for the blue end and 780 nm for the red end. The straight line joining the extremities is called the *purple boundary* and is the locus of the most saturated purples obtainable.

Because all color stimuli are combinations of spectral stimuli, it is apparent from the center of gravity law that all real color stimuli must lie on, or inside, the spectrum locus.

It is an experimental fact that the part of the spectrum locus lying between 560 and 780 nm is substantially a straight line. This means that broad-band colors in the yellow-orange-red region can give rise to colors of high saturation.

3.3.4 Subjective and Objective Quantities

It is important to distinguish clearly between perceptual (subjective) terms and psychophysical (objective) terms. Perceptual terms relate to attributes of sensations of light and color. They indicate subjective magnitudes of visual responses. Examples are hue, saturation, brightness, and lightness.

Psychophysical terms relate to objective measures of physical variables which identify stimuli that produce equal visual responses under specified viewing conditions. Examples include tristimulus values, luminance, and chromaticity coordinates.

Table 3.1 The Perceptual Terms and Their Psychophysical Correlates

Perceptual (Subjective)	Psychophysical (Objective)
Hue	Dominant wavelength
Saturation	Excitation purity
Brightness	Luminance
Lightness	Luminous reflectance or luminous transmittance

Psychophysical terms are usually chosen so that they correlate in an approximate way with particular perceptual terms. Examples of some of these correlations are given in Table 3.1.

3.4 The CIE System

In 1931, the International Commission on Illumination (known by the initials CIE for its French name, Commission Internationale de l'Eclairage) defined a set of color-matching functions and a coordinate system that have remained the predominant, international, standard method of specifying color to this day.

The CIE system deals with the three fundamental aspects of color experience:

- The object
- The light source that illuminates the object
- The observer

The light source is important because color appearance varies considerably in different lighting conditions. Generally, the *observer* is the person viewing the color, but it can also be a camera. Working from experiments conducted in the late 1920s, the CIE derived a set of color matching functions (the *Standard Observer*) that mathematically describe the sensitivity of the average human eye with normal color vision.

The color-matching functions were based on experimental data from many observers measured by Wright [10] and Guild [11]. Wright and Guild used different sets of primaries, but the results were transformed to a single set, namely, monochromatic stimuli of wavelengths 700.0, 546.1, and 435.8 nm. The units of the stimuli were chosen so that equal amounts were needed to match an equienergy stimulus (constant radiant power per unit wavelength throughout the visible spectrum). Figure 3.8 shows the spectrum locus in the (r, g) chromaticity diagram based on these color-matching functions.

At the same time it adopted these color-matching functions as a standard, the CIE also introduced and standardized a new set of primaries involving some ingenious concepts. The set of real physical primaries were replaced by a new set of imaginary nonphysical primaries with special characteristics. These new primaries are referred to as **X**, **Y**, and **Z**, and the corresponding tristimulus values as X, Y, and Z. The chromaticity coordinates of **X**, **Y**, and **Z** in the **RGB** system are shown in Figure 3.8. Primaries **X** and **Z** lie on the alychne and hence have zero luminance. All the luminance in a mixture of these three primaries is contributed by **Y**.

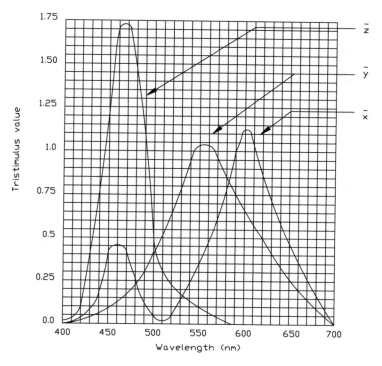

Figure 3.9 The CIE 1931 color-matching functions.

This convenient property depends only on the decision to locate **X** and **Z** on the alychne. It still leaves a wide choice of locations for all three primaries. The actual locations chosen by the CIE (illustrated in Figure 3.8) were based on the following additional considerations:

- The spectrum locus lies entirely within the triangle **XYZ**. This means that negative amounts of the primaries are never needed to match real colors. The color-matching functions , , and , shown in Figure 3.9, are therefore all positive at all wavelengths.

- The line $Z = 0$ (the line from **X** to **Y**) lies along the straight portion of the spectrum locus. Z is effectively zero for spectral colors with wavelengths greater than about 560 nm.

- The line $X = 0$ (the line from **Y** to **Z**) was chosen to minimize (approximately) the area of the **XYZ** triangle outside the spectrum locus. This choice led to a bimodal shape for the color-matching function because the spectrum locus curves away from the line $X = 0$ at low wavelengths. See Figure 3.8. A different choice of $X = 0$ (tangential to the spectrum locus at about 450 nm) would have eliminated the secondary lobe of but would have pushed **Y** much further from the spectrum locus.

- The units of **X**, **Y**, and **Z** were chosen so that the tristimulus values X, Y, and Z would be equal to each other for an equienergy stimulus.

Table 3.2 CIE Colorimetric Data (1931 Standard Observer)

Wave-length (mm)	Trichromatic Coefficients		Distribution Coefficients, Equal-energy Stimulus			Energy Distributions for Standard Illuminants			
	r	g	r̄	ḡ	b̄	E_A	E_B	E_C	E_{D65}
380	0.0272	−0.0115	0.0000	0.0000	0.0012	9.80	22.40	33.00	49.98
390	0.0263	−0.0114	0.0001	0.0000	0.0036	12.09	31.30	47.40	54.65
400	0.0247	−0.0112	0.0003	0.0001	0.0121	14.71	41.30	63.30	82.75
410	0.0225	−0.0109	0.0008	−0.0004	0.0371	17.68	52.10	80.60	91.49
420	0.0181	−0.0094	0.0021	−0.0011	0.1154	20.99	63.20	98.10	93.43
430	0.0088	−0.0048	0.0022	−0.0012	0.2477	24.67	73.10	112.40	86.68
440	−0.0084	0.0048	−0.0026	0.0015	0.3123	28.70	80.80	121.50	104.86
450	−0.0390	0.0218	0.0121	0.0068	0.3167	33.09	85.40	124.00	117.01
460	0.0909	0.0517	−0.0261	0.0149	0.2982	37.81	88.30	123.10	117.81
470	−0.1821	0.1175	−0.0393	0.0254	0.2299	42.87	92.00	123.80	114.86
480	−0.3667	0.2906	−0.0494	0.0391	0.1449	48.24	95.20	123.90	115.92
490	−0.7150	0.6996	−0.0581	0.0569	0.0826	53.91	96.50	120.70	108.81
500	1.1685	1.3905	0.0717	0.0854	0.0478	59.86	94.20	112.10	109.35
510	−1.3371	1.9318	−0.0890	0.1286	0.0270	66.06	90.70	102.30	107.80
520	−0.9830	1.8534	−0.0926	0.1747	0.0122	72.50	89.50	96.90	104.79
530	−0.5159	1.4761	0.0710	0.2032	0.0055	79.13	92.20	98.00	107.69
540	0.1707	1.1628	0.0315	0.2147	0.0015	85.95	96.90	102.10	104.41
550	0.0974	0.9051	0.0228	0.2118	−0.0006	92.91	101.00	105.20	104.05
560	0.3164	0.6881	0.0906	0.1970	−0.0013	100.00	102.80	105.30	100.00
570	0.4973	0.5067	0.1677	0.1709	−0.0014	107.18	102.60	102.30	96.33
580	0.6449	0.3579	0.2543	0.1361	0.0011	114.44	101.00	97.80	95.79
590	0.7617	0.2402	0.3093	0.0975	−0.0008	121.73	99.20	93.20	88.69
600	0.8475	0,1537	0.3443	0.0625	−0.0005	129.04	98.00	89.70	90.01
610	0.9059	0.0494	0.3397	0.0356	0.0003	136.35	98.50	88.40	89.60
620	0.9425	0.0580	0.2971	0.0183	−0.0002	143.62	99.70	88.10	87.70
630	0.9649	0.0354	0.2268	0.0083	−0.0001	150.84	101.00	88.00	83.29
640	0.9797	0.0205	0.1597	0.0033	0.000	157.98	102.20	87.80	83.70
650	0.9888	0.0113	0.1017	0.0012	0.0000	165.03	103.90	88.20	80.03
660	0.9940	0.0061	0.0593	0.0004	0.0000	171.96	105.00	87.90	80.21
670	0.9966	0.0035	0.0315	0.0001	0.0000	178.77	104.90	86.30	82.28
680	0.9984	0.0016	0.0169	0.0000	0.0000	185.43	103.90	84.00	78.28
690	0.9996	0.0004	0.0082	0.0000	0.0000	191.93	101.60	80.20	69.72
700	1.0000	0.0000	0.0041	0.0000	0.0000	198.26	99.10	76.30	71.61
710	1.0000	0.0000	0.0021	0.0000	0.0000	204.41	96.20	72.40	74.15
720	1.0000	0.0000	0.0011	0.0000	0.0000	210.36	92.90	68.30	61.60
730	1.0000	0.0000	0.0005	0.0000	0.0000	216.12	89.40	64.40	69.89
740	1.0000	0.0000	0.0003	0.0000	0.0000	221.67	86.90	61.50	75.09
750	1.0000	0.0000	0.0001	0.0000	0.0000	227.00	85.20	59.20	63.59
760	1.0000	0.0000	0.0001	0.0000	0.0000	232.12	84.70	58.10	46.42
770	1.0000	0.0000	0.0000	0.0000	0.0000	237.01	85.40	58.20	66.81
780	1.0000	0.0000	0.0000	0.0000	0.0000	241.68	87.00	59.10	63.38

This coordinate system and the set of color-matching functions that go with it are known as the CIE 1931 Standard Observer.

The color-matching data on which the 1931 Standard Observer is based were obtained with a visual field subtending 2° at the eye. Because of the slight nonuniformities of the retina, color-matching functions for larger fields are slightly different. In 1964, this prompted

Table 3.2 (continued)

Wave-length (mm)	Trichromatic Coefficients		Distribution Coefficients, Equal-energy Stimulus			Distribution Coefficients Weighted by Illuminant C		
	x	y	$\bar{x}$	$\bar{y}$	$\bar{z}$	$E_c\bar{x}$	$E_c\bar{y}$	$E_c\bar{z}$
380	0.1741	0.0050	0.0014	0.0000	0.0065	0.0036	0.0000	0.0164
390	0.1738	0.0049	0.0042	0.0001	0.0201	0.0183	0.0004	0.0870
400	0.1733	0,0048	0.0143	0.0004	0.0679	0.0841	0.0021	0.3992
410	0.1726	0.0048	0.0435	0.0012	0.2074	0.3180	0.0087	1.5159
420	0.1714	0.0051	0.1344	0.0040	0.6456	1.2623	0.0378	6.0646
430	0.1689	0.0069	0.2839	0.0116	1.3856	2.9913	0.1225	14.6019
440	0.1644	0.0109	0.3483	0.0230	1.7471	3.9741	0.2613	19.9357
450	0.1566	0.0177	0.3362	0.0380	1.7721	3.9191	0.4432	20.6551
460	0.1440	0.0297	0.2908	0.0600	1.6692	3.3668	0.6920	19.3235
470	0.1241	0.0578	0.1954	0.0910	1.2876	2.2878	1.0605	15.0550
480	0.0913	0.1327	0.0956	0.1390	0.8130	1.1038	1.6129	9.4220
490	0.0454	0.2950	0.0320	0.2080	0.4652	0.3639	2.3591	5.2789
500	0.0082	0.5384	0.0049	0.3230	0.2720	0.0511	3.4077	2.8717
510	0.0139	0.7502	0.0093	0.5030	0.1582	0.0898	4.8412	1.5181
520	0.0743	0.8338	0.0633	0.7100	0.0782	0.5752	6.4491	0.7140
530	0.1547	0.8059	0.1655	0.8620	0.0422	1.5206	7.9357	0.3871
540	0.2296	0.7543	0.2904	0.9540	0.0203	2.7858	9.1470	0.1956
550	0.3016	0.6923	0.4334	0.9950	0.0087	4.2833	9.8343	0.0860
560	0.3731	0.6245	0.5945	0.9950	0.0039	5.8782	9.8387	0.0381
570	0.4441	0.5547	0.7621	0.9520	0.0021	7.3230	9.1476	0.0202
580	0.5125	0.4866	0.9163	0.8700	0.0017	8.4141	7.9897	0.0147
590	0.5752	0.4242	1.0263	0.7570	0.0011	8.9878	6.6283	0.0101
600	0.6270	0.3725	1.0622	0.6310	0.0008	8.9536	5.3157	0.0067
610	0.6658	0.3340	1.0026	0.5030	0.0003	8.3294	4.1788	0.0029
620	0.6915	0.3083	0.8544	0.3810	0.0002	7.0604	3.1485	0.0012
630	0.7079	0.2920	0.6424	0.2650	0.0000	5,3212	2.1948	0.0000
640	0,7190	0.2809	0.4479	0.1750	0.0000	3.6882	1.4411	0.0000
650	0.7260	0.2740	0.2835	0.1070	0.0000	2.3531	0.8876	0.0000
660	0.7300	0.2700	0.1649	0,0610	0.0000	1.3589	0.5028	0.0000
670	0.7320	0.2680	0.0874	0.0320	0.0000	0.7113	0.2606	0.0000
680	0.7334	0.2666	0.0468	0.0170	0.0000	0.3657	0.1329	0.0000
690	0.7344	0.2656	0.0227	0.0082	0.0000	0.1721	0.0621	0.0000
700	0.7347	0.2653	0.0114	0.0041	0.0000	0.0806	0.0290	0.0000
710	0.7347	0,2653	0.0058	0.0021	0.0000	0.0398	0.0143	0.0000
720	0.7347	0.2653	0.0029	0.0010	0.0000	0.0183	0.0064	0.0000
730	0.7347	0.2653	0.0014	0.0005	0.0000	0.0085	0.0030	0.0000
740	0.7347	0.2653	0.0007	0.0003	0.0000	0.0040	0.0017	0.0000
750	0.7347	0.2653	0.0003	0.0001	0.0000	0.0017	0.0006	0.0000
760	0.7347	0.2653	0.0002	0.0001	0.0000	0.0008	0.0003	0.0000
770	0.7347	0.2653	0.0001	0.0000	0.0000	0.0003	0.0000	0.0000
780	0.7347	0.2653	0.0000	0.0000	0.0000	0.0000	0.000	0.0000

the CIE to recommend a second Standard Observer, known as the CIE 1964 Supplementary Standard Observer, for use in colorimetric calculations when the field size is greater than 4°.

3.4.1 Color-Matching Functions

The color-matching functions of the CIE Standard Observer, shown in Figure 3.9, are listed in Table 3.2. They are used to calculate tristimulus specifications of color stimuli and to determine whether two physically different stimuli will match each other. Such calculated matches represent the results of the average of many observers, but may not represent an exact match for any single real observer. For most purposes, this restriction is not important; the match of an average observer is all that is required.

3.4.2 Tristimulus Values and Chromaticity Coordinates

By exact analogy with the calculation of the tristimulus values R, G, B, the tristimulus values X, Y, Z of a stimulus $L_e(\lambda)$ are calculated by:

$$X = \int_{380}^{780} L_e(\lambda)\bar{x}(\lambda)d\lambda$$

$$Y = \int_{380}^{780} L_e(\lambda)\bar{y}(\lambda)d\lambda$$

$$Z = \int_{380}^{780} L_e(\lambda)\bar{z}(\lambda)d\lambda$$

The chromaticity coordinates x, y are then calculated by:

$$x = \frac{X}{X+Y+Z}, y = \frac{Y}{X+Y+Z}$$

The chromaticity coordinates x, y are plotted as rectangular coordinates to form the CIE 1931 chromaticity diagram, as shown in Figure 3.10.

It is important to remember that the CIE chromaticity diagram is not intended to illustrate appearance. The CIE system tells only whether two stimuli match in color, not what they look like. Appearance depends on many factors not taken into account in the chromaticity diagram. Nevertheless, it is often useful to know approximately where colors lie on the diagram. Figure 3.11 gives some color names for various parts of the diagram based on observations of self-luminous areas against a dark background.

3.4.3 Conversion Between Two Systems of Primaries

To transform tristimulus specifications from one system of primaries to another, it is necessary and sufficient to know—in one system—the tristimulus values of the primaries of the other system. For example, consider two systems **R**, **G**, **B** (in which tristimulus values are represented by R, G, B and chromaticity coordinates by r, g and **R′ G′ B′** (in which tristimulus values are represented by R' G' B' and chromaticity coordinates by r' g'). If one system is defined in terms of the other by the match equations:

$$\mathbf{R'} \equiv a_{11}\mathbf{R} + a_{21}\mathbf{G} + a_{31}\mathbf{B}$$

$$G' \equiv a_{12}R + a_{22}G + a_{32}B$$

$$B' \equiv a_{13}R + a_{23}G + a_{33}B$$

Then, the following equations can be derived to relate the tristimulus values and chromaticity coordinates of a color stimulus measured in one system to those of the same color stimulus measured in the other system:

$$R = a_{11}R' + a_{12}G' + a_{13}B'$$

$$G = a_{21}R' + a_{22}G' + a_{23}B'$$

$$B = a_{31}R' + a_{32}G' + a_{33}B'$$

$$R' = b_{11}R + b_{12}G + b_{13}B$$

$$G' = b_{21}R + b_{22}G + b_{23}B$$

$$B' = b_{31}R + b_{32}G + b_{33}B$$

$$r = \frac{\alpha_{11}r' + \alpha_{12}g' + \alpha_{13}}{t}$$

$$g = \frac{\alpha_{21}r' + \alpha_{22}g' + \alpha_{23}}{t}$$

$$t = \alpha_{31}r' + \alpha_{32}g' + \alpha_{33}$$

$$r' = \frac{\beta_{11}r + \beta_{12}g + \beta_{13}}{t'}$$

$$g' = \frac{\beta_{21}r + \beta_{22}g + \beta_{23}}{t'}$$

$$t' = \beta_{31}r + \beta_{32}g + \beta_{33}$$

These equations are known as *projective transformations*, which have the property of retaining straight lines as straight lines. In other words, a straight line in the r, g diagram will transform to a straight line in the r', g' diagram. Another important property is that the center of gravity law continues to apply. The derivation of the CIE **XYZ** color-matching functions and chromaticity diagram from the corresponding data in the **RGB** system is an example of the use of the transformation equations previously given.

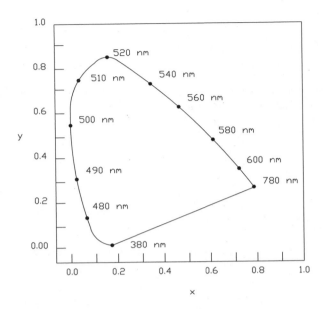

Figure 3.10 The CIE 1931 chromaticity diagram showing spectrum locus and wavelengths in nanometers.

3.4.4 Luminance Contribution of Primaries

Because **X** and **Z** were chosen to be on the alychne, their luminances are zero. Thus, all the luminance of a mixture of **X**, **Y**, and **Z** primaries is contributed by **Y**. This means that the Y tristimulus value is proportional to the luminance of the stimulus.

3.4.5 Standard Illuminants

The CIE has recommended a number of standard illuminants $E(\lambda)$ for use in evaluating the tristimulus values of reflecting and transmitting objects. Originally, in 1931, it recommended three—known as A, B, and C. These illuminants are specified by tables of relative spectral distribution and were chosen so that they could be reproduced by real physical sources. (CIE terminology distinguishes between *illuminants*, which are tables of numbers, and *sources*, which are physical emitters of light.) The sources are defined as follows:

- **Source A**. A tungsten filament lamp operating at a color temperature of about 2856K. Its chromaticity coordinates are $x = 0.4476$, and $y = 0.4074$. Source A represents incandescent light.

- **Source B**. A source with a composite filter made of two liquid filters of specified chemical composition [3]. The chromaticity coordinates of source B are $x = 0.3484$ and $y = 0.3516$. Source B represents noon sunlight.

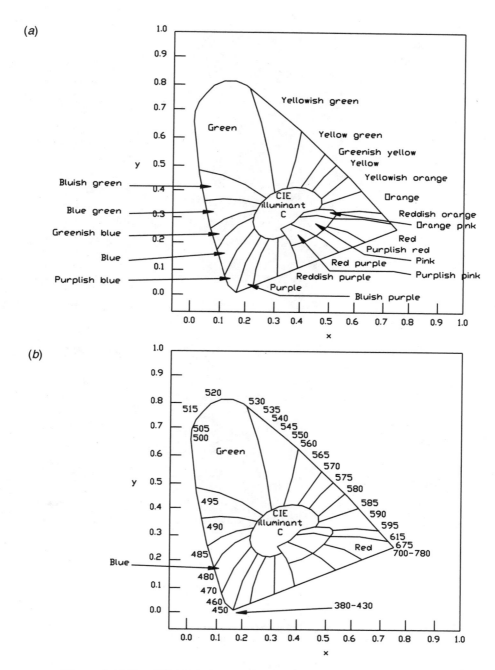

Figure 3.11 The CIE 1931 chromaticity diagram: (a) regions corresponding to various color names derived from observations of self-luminous areas against a dark background, (b) equivalent wavelengths of major divisions, (c, next page) illustration combining all elements of the chromaticity diagram into a single drawing. (*After* [12].)

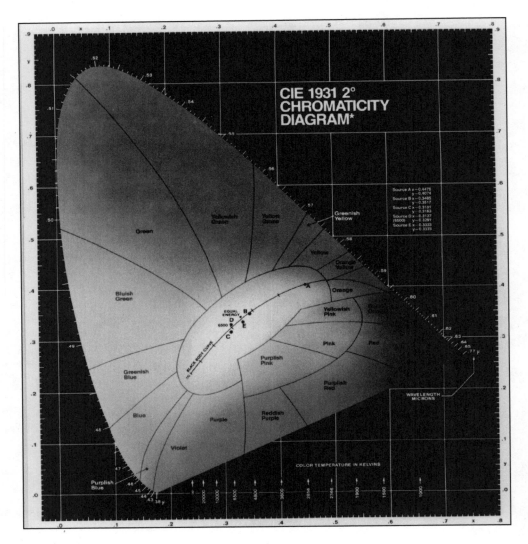

Figure 3.11(c) *Courtesy of Photo Research.*

- **Source C.** This source is also produced by source A with two liquid filters [3]. Its chromaticity coordinates are $x = 0.3101$ and $y = 0.3162$. Source C represents average daylight according to information available in 1931.

In 1971, the CIE introduced a new series of standard illuminants that represented daylight more accurately than illuminants B and C [13]. The improvement is particularly marked in the ultraviolet part of the spectrum, which is important for fluorescent samples.

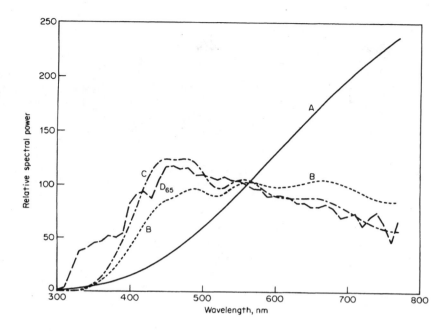

Figure 3.12 The relative spectral power distributions of CIE standard illuminants A, B, C, and D_{65}.

The most important of the D illuminants is D_{65} (sometimes written D6500), which has chromaticity coordinates of $x = 0.3127$ and $y = 0.3290$.

The relative spectral power distributions of illuminants A, B, C, and D_{65} are given in Figure 3.12

3.4.6 Gamut of Reproducible Colors

In a system that seeks to match or reproduce colors with a set of three primaries, only those colors can be reproduced that lie inside the triangle of primaries. Colors outside the triangle cannot be reproduced because they would require negative amounts of one or two of the primaries.

In a color-reproducing system, it is important to have a triangle of primaries that is sufficiently large to permit a satisfactory gamut of colors to be reproduced. To illustrate the kinds of requirements that must be met, Figure 3.13 shows the maximum color gamut for real surface colors and the triangle of typical color television receiver phosphors as standardized by the European Broadcasting Union (EBU). These are shown in the CIE 1976 u', v' chromaticity diagram in which the perceptual spacing of colors is more uniform than in the x, y diagram. High-purity blue-green and purple colors cannot be reproduced by these phosphors, whereas the blue phosphor is actually of slightly higher purity than any real surface colors.

96 Video Display Engineering

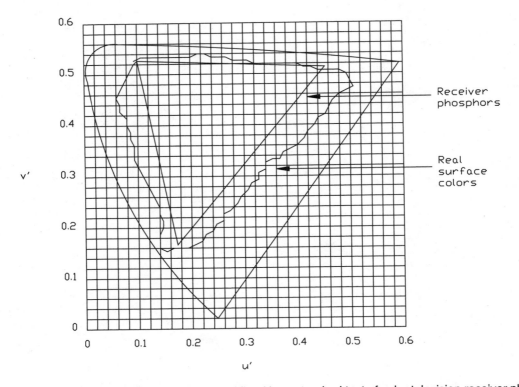

Figure 3.13 The color triangle defined by a standard test of color television receiver phosphors compared with the maximum real color gamut on a u', v' chromaticity diagram. (*After* [14].)

3.4.7 Vector Representation

In the preceding sections the representation of color has been reduced to two dimensions by eliminating consideration of quantity (luminance) and discussing only chromaticity. Because color requires three numbers to specify it fully, a three-dimensional representation can be made, taking the tristimulus values as vectors. For the sake of simplicity, this discussion will be confined to CIE tristimulus values and to a rectangular framework of coordinate axes.

Tristimulus values have been treated as scalar quantities. They may be transformed into vector quantities by multiplying by unit vectors $\mathbf{i}$, $\mathbf{j}$, and $\mathbf{k}$ in the x, y, and z directions. As a result, X, Y, and Z will become vector quantities $\mathbf{i}X$, $\mathbf{j}Y$, and $\mathbf{k}Z$. A color can then be represented by three vectors $\mathbf{i}X$, $\mathbf{j}Y$, and $\mathbf{k}Z$ along the x, y, and z coordinate axes, respectively. Combining these vectors gives a single resultant vector $\mathbf{V}$, represented by the *vector equation*:

$$\mathbf{V} = \mathbf{i}X + \mathbf{j}Y + \mathbf{k}Z$$

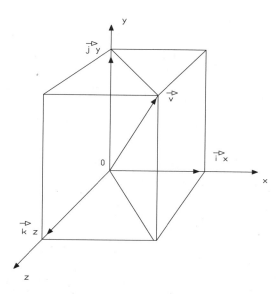

Figure 3.14 The combination of vectors.

The resultant is obtained by the usual vector methods, which are shown in Figure 3.14.

Some implications of this form of color representations are illustrated in Figure 3.15. The diagram shows a vector **V** representing a color, a set of coordinate axes, and the plane $x + y + z = 1$ passing through the points $L(1,0,0)$, $M(0,1,0)$, and $N(0,0,1)$. The vector passes through point Q in this plane, having coordinates x, y, and z. Point P is the projection of point Q into the xy plane, and therefore has coordinates (x, y). From the geometry of the figure it can be seen that:

$$\frac{X}{x} = \frac{Y}{y} = \frac{Z}{z} = \frac{X+Y+Z}{x+y+z}$$

Because Q is on the plane $x + y + z = 1$, it follows:

$$\frac{X}{x} = \frac{Y}{y} = \frac{Z}{z} = X+Y+Z$$

Therefore:

$$x = \frac{X}{X+Y+Z}, \quad y = \frac{Y}{X+Y+Z}$$

Thus, x and y are the CIE chromaticity coordinates of the color. Therefore, the xy plane in color space represents the CIE chromaticity diagram, when the vectors have magnitudes equal to the CIE tristimulus values and the axes are rectangular. The triangle LMN is a

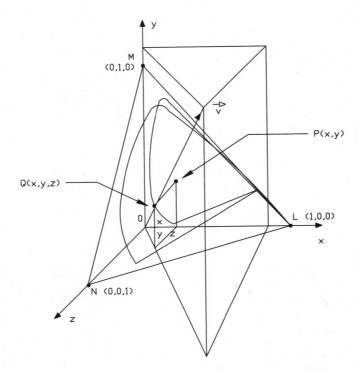

Figure 3.15 The relationship between color space and the CIE chromaticity diagram.

Maxwell triangle for CIE primaries. The spectrum locus in the *xy* plane can be thought of as defining a cylinder with generators parallel to *z*; this cylinder intersects the plane of the triangle *LMN* in the spectrum locus for the Maxwell triangle.

3.5 Refinements to the 1931 CIE Model

The CIE XYZ tristimulus values form the basis for all CIE numerical color descriptors.[2] However, the *XYZ* values are not intuitive for most people. As a result, the CIE continued to develop and refine other approaches to color specifications. Just as geographers use maps to represent geographic coordinates and other information, color scientists have developed two- and three-dimensional diagrams and graphic models to represent color information. Hence, the newer models are often also referred to as *color spaces*. A three-dimensional representation of chromaticity space is shown in Figure 3.16. The point at

---

2 This section was adapted from: "Colorimetry and Television Camera Color Measurement," application note 21W-7165, Tektronix, Beaverton, OR, 1992.

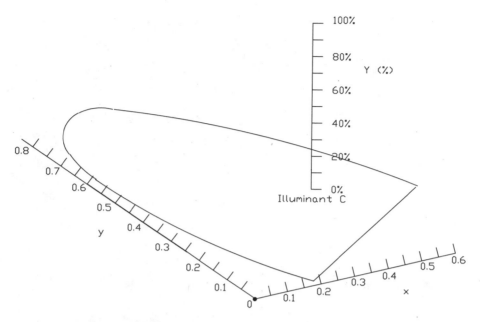

Figure 3.16 A drawing of the 1931 CIE color standard illustrating all three dimensions, x, y, and Y.

which the luminance axis (Y) touches the x, y plane is dependent upon the chromaticity coordinates of the illuminant or *white point* being referenced.

3.5.1 Improved Visual Uniformity

The 1931 chromaticity diagram was developed primarily for color specification and was not intended to provide information on color appearance. Consequently, the system do͏͞ not display perceptual uniformity. That is, colors do not appear to be equally sp͏͞ ally. For example, colors that consist of the same visually perceived h͏͞ (those associated with a specific wavelength) do not follow str͏͞ gram, but are curved instead. This non-uniformity is similar t͏͞ tion world map distorts what is truly represented on a globe. diagram is that black, or the absence of color, does not hav͏͞

The search for a chromaticity diagram with greater visu͏͞ CIE u', v' uniform chromaticity scales (UCS) diagram͏͞ v' chromaticity coordinates for any real color are locate͏͞ locus and the line of purples that joins the spectrum er͏͞ the 1931 diagram, the u' and v' coordinates do not͏͞ they contain no information on its inherent lightness. ͏͞ denoted by the tristimulus value Y, which represents the͏͞ tion is perpendicular to the u', v' plane, extending up f͏͞

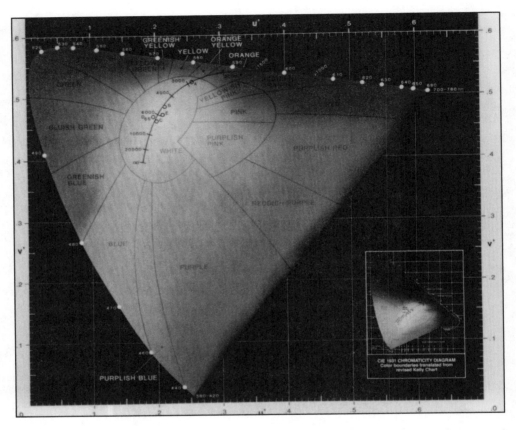

Figure 3.17 The 1976 CIE UCS diagram. The u', v' chromaticity coordinates for any real color are located within the bounds the horse-shoe-shaped spectrum locus and the line of purples that joins the spectrum ends. (*Courtesy of Photo Research.*)

The u', v' coordinates of the 1976 UCS diagram can be derived from a simple transformation (defined by the CIE) of the 1931 x, y coordinates, or, more directly, from a transformation of the XYZ tristimulus values:

$$u' = \frac{4x}{-2x+12y+3}, \quad v' = \frac{9y}{-2x+12y+3}$$

These quantities can also be expressed as:

$$u' = \frac{4X}{X+15y+3Z}, \quad v' = \frac{9Y}{X+15y+3Z}$$

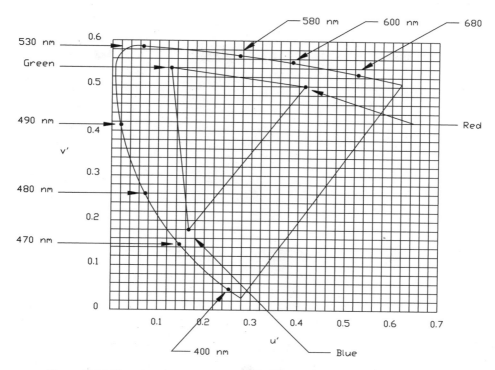

Figure 3.18 The triangle representing the range of chromaticities generally achievable using the additive mixture of typical red, green, and blue phosphors on a CRT display.

The 1976 UCS diagram displays marked improvement over the 1931 x, y diagram in overall visual uniformity. It is not without flaws, but it enjoys wide acceptance. For video applications, the CIE 1976 UCS diagram is particularly useful because the region of the space delineated by the chromaticity limits of typical phosphor primaries falls within the most uniform region of the diagram, as illustrated in Figure 3.18. The most severe visual nonuniformities occur outside of the phosphor triangles and at the extreme limits of the chromaticity diagram.

3.5.2 CIELUV

In an ideal color space, the numerical magnitude of a color difference should bear a direct relationship to the difference in color appearance. While the 1976 UCS diagram provides better visual uniformity than earlier approaches, it does not meet the critical goal of full visual uniformity, because uniform differences in chromaticity do not necessarily correspond to equivalent visual differences in Y or luminance. In addition, the diagram lacks a built-in capability to incorporate the important aspect of white reference, which significantly affects color appearance.

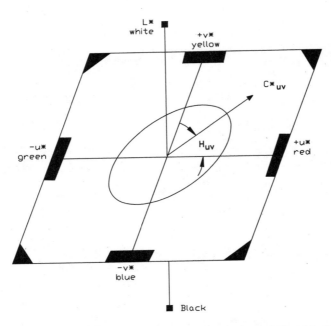

Figure 3.19 The CIELUV color space illustrating the relationship between the opponent color axes and the axis representing 1976 CIE metric lightness.

The CIE 1976 L^*, u^*, v^* (CIELUV) color space addresses these concerns. CIELUV is another tristimulus color space that uses three hues to describe a color. It integrates the CIE 1976 u', v' parameters with the 1976 metric lightness function, L^*.

CIELUV is an *opponent-type* color space. Opponent-color theory is based on the assumption that the eye and brain encode all colors into light-dark, red-green, and yellow-blue signals. In a system of this type, colors are mutually exclusive in that a color cannot be red and green at the same time, or yellow and blue at the same time. A color such as purple can be described as red and blue, however, because these are not opponent colors.

In the CIELUV system, the u^* and v^* coordinate axes describe the chromatic attributes of color. The u^* axis represents the red-green coordinate, while the v^* axis represents yellow-blue. Positive values of u^* denote red colors, while negative values denote green. Similarly, positive values of v^* represent yellows and negative values signify blues. The L^* axis denotes variations in lightness or darkness and lies perpendicular to the u^*, v^* plane. The achromatic or neutral colors (black, gray, white) lie on the L^* axis (the point where u^* and v^* intersect, $u^* = 0$, $v^* = 0$). Figure 3.19 shows the basic layout of the axes with respect to one another.

Calculating CIELUV coordinates requires a full chromaticity specification of the reference white point. For calculating CIELUV values in this section, all specifications are in the

1976 u', v', Y format. The following equations are standard u', v', Y to CIELUV transforms defined by the CIE:

$$L_s^* = \sqrt[3]{116\left(\frac{Y_s}{Y_w}\right) - 16}$$

for $(Y_s/Y_w) > 0.008856$

The transform can also be described as:

$$L_s^* = 903.29 \frac{Y_s}{Y_w}$$

for $(Y_s/Y_w) \leq 0.008856$

$$u_s^* = 13L^*\left(u_s^* - u_w^*\right) \quad v_s^* = 13L^*\left(v_s^* - v_w^*\right)$$

Note that the subscript s (Y_s) refers to color coordinates of the sample color, while coordinates with the subscript w (Y_w) are reference white.

Because each of the opponent coordinates is a function of metric lightness, the CIELUV color space has a unique location for black. As a result, when L^* is equal to zero, the coordinates u^* and v^* are also zero; thus, black lies at a single point on the L^* (neutral) axis.

The total CIELUV color space is represented by an irregularly shaped spheroid, illustrated in Figure 3.20. There are two reasons why the space approaches absolute limits at each end of the L^* scale. At the lower end of the color space L^* approaches zero. Thus, by definition, so do u^* and v^* because they are functions of L^*. This is what ensures a unique, single-point definition for black. The top portion of the CIELUV space converges for a different reason. Remember that in the CIE u', v' color standard, as Y increases, the limits on chromaticity become severe and less chromatic variation is possible. This results in a unique white point. The irregularity of the CIELUV color solid reflects the fact that certain colors are inherently capable of greater dynamic range than others.

In addition to the basic parameters of color, the CIE has also developed several *psychometric coordinates* designed to equate more with how color is perceived. In the CIELUV system, the two most often used are *psychometric hue angle* and *psychometric chroma*.

Psychometric hue angle (h_{uv}) represents an angle in the u^*, v^* plane that is correlated to a color family name or hue. The metric is defined as:

$$h_{uv} = \tan^{-1} \frac{u^*}{v^*}$$

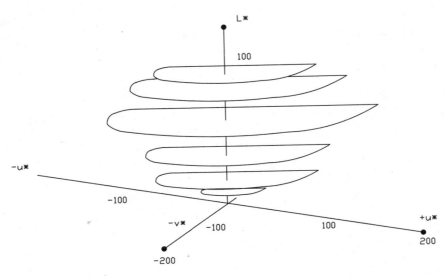

Figure 3.20 The CIELUV object-color solid showing constant lightness planes, $L^* = 10.0$, 20.0, 40.0, 60.0, 80.0, and 90.0.

Psychometric chroma is a quantity that represents overall vibrancy or colorfulness. It is defined by:

$$C_{uv}^* = \sqrt{u^{*2} + v^{*2}}$$

Currently, the CIELUV system and its related quantities are the best visually-uniform color space for additive colors accepted for international use. The deviations from uniformity found in the 1976 CIE u', v' diagram are more obvious in certain regions of color space than in others, particularly towards the limits of the spectrum locus. Fortunately, the gamut of achievable colors on a CRT lie in the most uniform region of the u', v' diagram. Furthermore, because the CIELUV system is based on the CIE u', v' parameters, it provides the same advantages to video-based implementations.

3.5.3 Colorimetry and Color Targets

Color targets serve as a standardized method for describing specific colors. These targets generally provide reproduction of the resident colors within some well-controlled tolerance. Colors, when expressed scientifically, are described relative to a particular lighting condition. Changing the illuminant can dramatically change the color's composite spectral curve and, hence, the appearance of the color. These changes occur in different areas of the visible spectrum, depending on the character of the illuminant used. As a result, certain colors can be more affected by a shift in illuminant than others.

The colors in a color target are subject to the same laws of physics as any other colors. For example, the Macbeth color checker is widely used as a color reproduction standard in many industries and across applications, most notably as a means of assessing the color balance of photographic films. Like other color targets, however, this chart is standardized for only one specific viewing situation. The widely published color coordinate data for this chart is accurate only when the lighting environment approximates CIE Illuminant C ("average" daylight). The results will be erroneous if the target is used in other lighting environments. Furthermore, it is not enough to simply take the origin (Illuminant C-based) data and transpose it relative to a new illuminant. If the target is used in a different lighting environment, the only way to truly represent the appearance of the color target under the new condition is to obtain data that is representative of each chart color in the new environment.

3.6 Color Models

A color model is a specification of a three-dimensional color coordinate system. The model describes a visible subset in the system within which all colors in a particular color gamut lie. The RGB color model, for example, is the *unit cube* subset of the 3D Cartesian coordinate system. The purpose of a model is to permit convenient specification of colors within a given gamut, such as that for CRT monitors. The color gamut is a subset of all visible chromaticities. The model, therefore, can be used to specify all visible colors on a given display.

The primary color models of interest for display technology are:

- *RGB model*, used with color CRT devices
- *YIQ model*, used with conventional video (NTSC) systems
- *CMY model*, used in printing applications
- *HSV model*, used to describe color independent of a given hardware implementation

The RGB, YIQ, and CMY models are hardware-oriented systems. The HSV model, and variations of the model, are not based on a particular hardware system. Instead, they relate to the intuitive notions of hue, saturation, and brightness.

Conversion algorithms permit translation of one color model to another to facilitate comparison of color specifications.

3.6.1 RGB Color Model

The RGB (red, green, blue) color model is used to specify color CRT monitors. As illustrated in Figure 3.21, it is a unit cube subset of the 3D Cartesian coordinate system. As discussed previously, the RGB color model is additive. Combination of the three primary colors in the proper amounts yields white. As the figure shows, the monochrome (gray scale) vector stretches from black (0, 0, 0 primaries) to white (1, 1, 1 primaries). The color gamut covered by the RGB model is defined by the chromaticities of the CRT phosphors. Therefore, it follows that devices using different phosphors will have different color gamuts.

To convert colors specified in the gamut of one CRT to the gamut of another device,

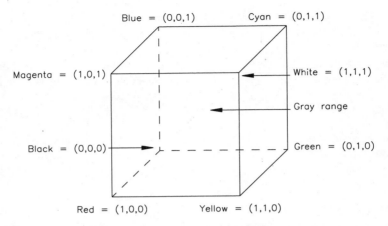

Figure 3.21 The RGB color model cube. (*After* [15].)

transformations M_1 and M_2 are used from the RGB color space of each monitor to the (X, Y, Z) color space. The form of each transformation is [15]:

$$\begin{Bmatrix} X \\ Y \\ Z \end{Bmatrix} = \begin{Bmatrix} X_r & X_g & X_b \\ Y_r & Y_g & Y_b \\ Z_r & Z_g & Z_b \end{Bmatrix} \begin{Bmatrix} R \\ G \\ B \end{Bmatrix}$$

Where:
X_n = the weights applied to the monitor RGB colors to find X, Y, and Z

If M is defined as the 3 × 3 matrix of color-matching coefficients, the preceding equation can be written:

$$\begin{Bmatrix} X \\ Y \\ Z \end{Bmatrix} = M \begin{Bmatrix} R \\ G \\ B \end{Bmatrix}$$

Given that M_1 and M_2 are matrices that convert from each of the CRT color gamuts to CIE, then ($M_2^{-1} M_1$) converts the RGB specification of monitor 1 to that of monitor 2. As long as the color in question lies in the gamut of both monitors, this matrix product is accurate.

3.6.2 YIQ Model

The YIQ model describes the color television broadcasting system used in the U.S. and elsewhere. Known as NTSC, the system is optimized for efficient transmission of color information within a limited terrestrial bandwidth. A primary criteria for NTSC is compatibility with monochrome television receivers. The components of YIQ are:

- Y—the luminance component
- I and Q—the encoded chrominance components

The Y signal contains sufficient information to display a black-and-white picture of the encoded color signal.

The YIQ color model uses a 3D Cartesian coordinate system. RGB-to-YIQ mapping is defined by the following [15]:

$$\begin{Bmatrix} Y \\ I \\ Q \end{Bmatrix} = \begin{bmatrix} 0.299 & 0.587 & 0.114 \\ 0.596 & -0.275 & -0.321 \\ 0.212 & -0.528 & 0.311 \end{bmatrix} \begin{Bmatrix} R \\ G \\ B \end{Bmatrix}$$

The quantities in the first row of the equation reflect the relative importance of green and red in producing brightness, and the smaller role that blue plays. The equation assumes that the RGB color specification is based on the standard NTSC RGB phosphor set. The CIE coordinates of the set are:

	Red	Green	Blue
x	0.67	0.21	0.14
y	0.33	0.71	0.08

The YIQ model capitalizes on two important properties of the human visual system:

- The eye is more sensitive to changes in luminance than to changes in hue or saturation
- Objects that cover a small part of the field of view produce a limited color sensation

These properties form the basis upon which the NTSC television color system was developed.

3.6.3 CMY Model

Cyan, magenta, and yellow are complements of red, green, and blue, respectively. When used to subtract color from white light, they are referred to as *subtractive primaries*. The subset of the Cartesian coordinate system for the CMY model is identical to RGB except that white is the origin, rather than black.

In the CMY model, colors are specified by what is removed (subtracted) from white light, rather than what is added to a black screen. The CMY system is commonly used in printing applications, with the white light being that light reflected from paper. For example, when a portion of paper is coated with cyan ink, no red light is reflected from the surface. Cyan subtracts red from the reflected white light. The color relationship is illustrated in Figure 3.22. As shown, the subtractive primaries may be combined to produce a spectrum of colors. For example, cyan and yellow combine to produce green.

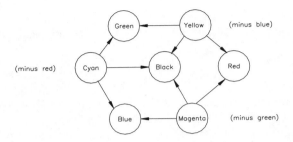

Figure 3.22 The CMY color model primaries and their mixtures. (*After* [15].)

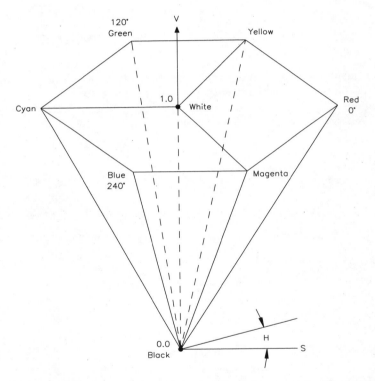

Figure 3.23 The single-hexcone HSV color model. The *V*=1 plane contains the RGB model *R* =1, *G* = 1, and *B* = 1 planes in the regions illustrated. (*After* [16].)

CMYK is a variation of the CMY model used for some color output devices and most color printing presses. *K* refers to the black component of the image. The addition of black to the process is particularly useful for printing applications, where text is almost always printed as black.

3.6.4 HSV Model

The HSV model [16] utilizes the intuitive elements of hue, saturation, and value to describe color. The coordinate system is cylindrical. The subset of the space within which the model is defined is a *hexcone* (six-sided pyramid), as illustrated in Figure 3.23. The top of the hexcone corresponds to $V = 1$, which contains the relatively bright colors. Hue (H) is measured by the angle around the vertical axis, with red at 0°, green at 120°, and blue at 240°. The complementary colors in the HSV hexcone are 180° opposite one another. The value of S is a ratio ranging from 0 on the center line (V axis) to 1 on the triangular sides of the hexcone. Saturation is measured relative to the color gamut represented by the model.

3.7 References

1. Smith, V. C., and J. Pokorny: "Spectral Sensitivity of the Foveal Cone Pigments Between 400 and 500 nm," *Vision Res.* vol. 15, pp. 161–171, 1975.
2. Boynton, R.M.: *Human Color Vision*, Holt, New York, N.Y., p. 404, 1979.
3. Judd, D. B., and G. Wyszencki: *Color in Business, Science, and Industry*,. 3rd ed., Wiley, New York, N.Y., pp. 44-45, 1975.
4. Wyszecki, G., and W. S. Stiles: *Color Science*, 2nd ed., Wiley, New York, N.Y., 1982.
5. Nickerson, D.: "History of the Munsell Color System, Company and Foundation, I," *Color Res. Appl.* vol. 1, pp. 7–10, 1976.
6. Nickerson, D.: "History of the Munsell Color System, Company and Foundation, II: Its Scientific Application," *Color Res. Appl.* vol. 1, pp. 69–77, 1976.
7. Nickerson, D.: "History of the Munsell Color System, Company and Foundation, III," *Color Res. Appl.* vol. 1, pp. 121–130, 1976.
8. Newhall, S. M., D. Nickerson, and D. B. Judd: "Final Report of the OSA Subcommittee on the Spacing of the Munsell Colors," *Journal of the Optical Society of America*, vol. 33, pp. 385–418, 1943.
9. *Munsell Book of Color*. Munsell Color Co., 2441 No. Calvert Street, Baltimore, MD 21218.
10. Wright, W. D.: "A Redetermination of the Trichromatic Coefficients of the Spectral Colours," *Trans. Opt. Soc.*, vol. 30, pp. 141–164, 1928–1929.
11. Guild, J.: "The Colorimetric Properties of the Spectrum," *Phil. Trans. Roy. Soc. A.*, vol. 230, pp. 149–187, 1931.
12. Kelly, K. L.: "Color Designations for Lights," *J. Opt. Soc. Am.*, vol. 33, pp. 627–632, 1943.
13. "Colorimetry," Publication no. 15, CIE, Paris, 1971.
14. Pointer, M. R.: "The Gamut of Real Surface Colours," *Color Res. Appl.*, vol. 5, pp. 145–155, 1980.
15. Foley, James D., et al.: *Computer Graphics: Principles and Practice*, 2nd ed., Addison-Wesley, Reading, Mass., pp. 584–592, 1991.
16. Smith, A. R.: "Color Gamut Transform Pairs," *SIGGRAPH 78*, 12–19, 1978.

3.8 Bibliography

"Colorimetry," Publication no. 15, Commission Internationale de l'Eclairage, Paris, 1971.

Kaufman, J. E. (ed.): *IES Lighting Handbook-1981 Reference Volume*, Illuminating Engineering Society of North America, New York, N.Y., 1981.

Tektronix application note #21W-7165: "Colorimetry and Television Camera Color Measurement," Tektronix, Beaverton, Ore., 1992.

Wright, W. D.: *The Measurement of Colour*, 4th ed., Adam Hilger, London, 1969.

Chapter 4

Application of Visual Properties

4.1 Introduction

Advanced display systems improve on earlier techniques primarily by better utilizing the resources of human vision. The primary objective of an advanced display is to enhance the visual field occupied by the video image. In many applications this has called for larger, wider pictures that are intended to be viewed more closely than conventional video. In other applications this has called for miniature displays that serve specialized purposes. To satisfy the viewer at this closer inspection, the displayed image must possess proportionately finer detail and sharper outlines.

4.2 The Television System

Terrestrial broadcast television is the basis of all video display systems. It was the first electronic system to convert an image into electrical signals, encode them for transmission, and display a representative image of the original at a remote location.

The technology of television is based on the conversion of light rays from still or moving scenes and pictures into electronic signals for transmission or storage, and subsequent reconversion into visual images on a screen. A similar function is provided in the production of motion picture film. However, where film records the brightness variations of a complete scene on a single frame in a short exposure no longer than a fraction of a second, the elements of a television picture must be scanned one piece at a time. In the television system, a scene is dissected into a *frame* composed of a mosaic of *picture elements* (*pixels*). A pixel is defined as the smallest area of a television image that can be transmitted within the parameters of the system. This process is accomplished by:

- Analyzing the image with a photoelectric device in a sequence of *horizontal scans* from the top to the bottom of the image to produce an electric signal in which the brightness and color values of the individual picture elements are represented as voltage levels of a video waveform.

- Transmitting the values of the picture elements in sequence as voltage levels of a video signal.

- Reproducing the image of the original scene in a video signal display of parallel scanning lines on a viewing screen.

4.2.1 Scanning Lines and Fields

The image pattern of electrical charges on a camera pickup element (corresponding to the brightness levels of a scene) are converted to a video signal in a sequential order of picture elements in the scanning process. At the end of each horizontal line sweep, the video signal is *blanked* while the beam returns rapidly to the left side of the scene to start scanning the next line. This process continues until the image has been scanned from top to bottom to complete one *field scan*.

After completion of this first field scan, at the midpoint of the last line, the beam again is blanked as it returns to the top center of the target where the process is repeated to provide a second field scan. The spot size of the beam as it impinges upon the target must be sufficiently fine to leave unscanned areas between lines for the second scan. The pattern of scanning lines covering the area of the target, or the screen of a picture display, is called a *raster*.

4.2.2 Interlaced Scanning Fields

Because of the half-line offset for the start of the beam return to the top of the raster and for the start of the second field, the lines of the second field lie in between the lines of the first field. Thus, the lines of the two are *interlaced*. The two interlaced fields constitute a single television *frame*. Figure 4.1 shows a frame scan with interlacing of the lines of two fields.

Reproduction of the camera image on a cathode ray tube (CRT) is accomplished by an identical operation, with the scanning beam modulated in density by the video signal applied to an element of the electron gun. This control voltage to the CRT varies the brightness of each picture element on the phosphor screen.

Blanking of the scanning beam during the return trace (*retrace*) is provided for in the video signal by a "blacker-than-black" pulse waveform. In addition, in most receivers and monitors another blanking pulse is generated from the horizontal and vertical scanning circuits and applied to the CRT electron gun to ensure a black screen during scanning retrace.

The interlaced scanning format, standardized for monochrome and compatible color, was chosen primarily for two partially related and equally important reasons:

- To eliminate viewer perception of the intermittent presentation of images, known as *flicker*.
- To reduce the video bandwidth requirements for an acceptable flicker threshold level.

The standards adopted by the Federal Communications Commission (FCC) for monochrome television in the United States specified a system of 525 lines per frame, transmitted at a frame rate of 30 Hz, with each frame composed of two interlaced fields of horizontal lines. Initially in the development of television transmission standards, the 60 Hz power line waveform was chosen as a convenient reference for vertical scan. Furthermore, in the event of coupling of power line hum into the video signal or scanning/deflection circuits, the visible effects would be stationary and less objectionable than moving *hum bars* or distortion of

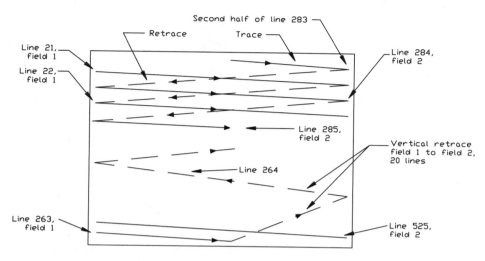

Figure 4.1 The interlace scanning pattern (raster) of the television image.

horizontal-scanning geometry. In the United Kingdom and much of Europe, a 50 Hz interlaced system was chosen for many of the same reasons. With improvements in television receivers, the power line reference was replaced with a stable crystal oscillator.

The initial 525-line monochrome standard was retained for color in the recommendations of the National Television System Committee (NTSC) for compatible color television in the early 1950s. The NTSC system, adopted in 1953 by the FCC, specifies a scanning system of 525 horizontal lines per frame, with each frame consisting of two interlaced fields of 262.5 lines at a field rate of 59.94 Hz. Forty-two of the 525 lines in each frame are blanked as black picture signals and reserved for transmission of the vertical scanning synchronizing signal. This results in 483 visible lines of picture information.

4.2.3 Synchronizing Video Signals

In monochrome television transmission, two basic synchronizing signals are provided to control the timing of picture-scanning deflection:

- Horizontal sync pulses at the line rate.

- Vertical sync pulses at the field rate in the form of an interval of wide horizontal sync pulses at the field rate. Included in the interval are *equalizing pulses* at twice the line rate to preserve interlace in each frame between the even and odd fields (offset by line).

In color transmissions, a third synchronizing signal is added during horizontal scan blanking to provide a frequency and phase reference for signal encoding circuits in cameras and decoding circuits in receivers. These synchronizing and reference signals are combined with the picture video signal to form a *composite video waveform*.

The receiver scanning and color-decoding circuits must follow the frequency and phase of the synchronizing signals to produce a stable and geometrically accurate image of the proper color hue and saturation. Any change in timing of successive vertical scans can impair the interlace of even and odd fields in a frame. Small errors in horizontal scan timing of lines in a field can result in a loss of resolution in vertical line structures. Periodic errors over several lines that may be out of the range of the horizontal-scan automatic frequency control circuit in the receiver will be evident as jagged vertical lines.

4.2.4 Television Industry Standards

There are three primary standard-definition color transmission standards in use today:

- *NTSC* (National Television Systems Committee)—used in the United States, Canada, Central America, some of South America, and Japan. In addition, NTSC is used in various countries or possessions heavily influenced by the United States.

- *PAL* (Phase Alternate each Line)—used in the United Kingdom, most countries and possessions influenced by England, most European countries, and China. Variation exists in PAL systems.

- *SECAM* (SEquential Color with [Avec] Memory)—used in France, countries and possessions influenced by France, the USSR (generally the former Soviet Bloc nations, including East Germany) and other areas influenced by Russia.

The three standards are incompatible for a variety of reasons.

4.2.5 Composite Video

The term *composite* is used to denote a video signal that contains:

- Picture luminance and chrominance information

- Timing information for synchronization of scanning and color signal processing circuits

The negative-going portion of the composite waveform, shown in Figure 4.2, is used to transmit information for synchronization of scanning circuits. The positive-going portion of the amplitude range is used to transmit luminance information representing brightness and—for color pictures—chrominance.

At the completion of each line scan in a receiver or monitor, a horizontal synchronizing (*H-sync*) pulse in the composite video signal triggers the scanning circuits to return the beam rapidly to the left of the screen for the start of the next line scan. During the return time, a horizontal blanking signal at a level lower than that corresponding to the blackest portion of the scene is added to avoid the visibility of the retrace lines. In a similar manner, after completion of each field, a vertical blanking signal blanks out the retrace portion of the scanning beam as it returns to the top of the picture to start the scan of the next field. The small-level difference between video reference black and the blanking level is called *setup*. Setup is used in the NTSC system as a guard band to ensure separation of the synchronizing and video-information functions, and to ensure adequate blanking of the scanning retrace lines on receivers.

Application of Visual Properties 115

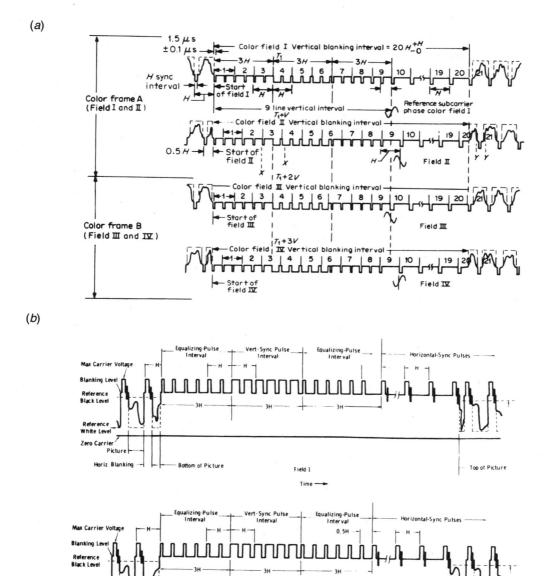

Figure 4.2 The NTSC color television waveform: (*a*) principle components, (*b*) detail of picture elements. (*Source: Electronic Industries Association*.)

116   Video Display Engineering

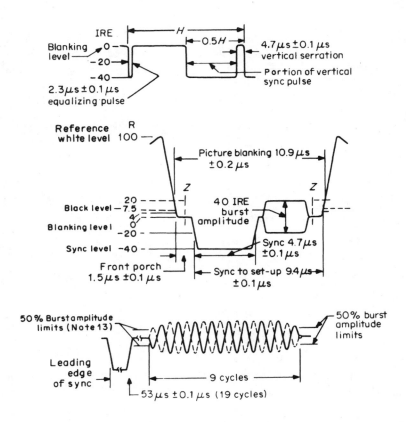

Figure 4.3 Detail of sync and color subcarrier pulse widths for the NTSC system. (*Source: Electronic Industries Association.*)

Table 4.1 Video and Sync Levels in IRE Units

Signal Level	RE Level
Reference white	100
Color burst sine wave peak	+20 to –20
Reference black	7.5
Blanking	0
Sync level	–40

The waveforms of Figure 4.3 show the various reference levels of video and sync in the composite signal. The unit of measurement for video level was specified initially by the Institute of Radio Engineers (IRE). These *IRE units* are still used to quantify video signal levels. Primary IRE values are given in Table 4.1.

4.2.6 Color Signal Encoding

To facilitate an orderly introduction of color television broadcasting in the United States and other countries with existing monochrome services, it was essential that new transmissions be compatible. In other words, color pictures would provide acceptable quality on unmodified monochrome receivers. In addition, because of the limited availability of RF spectrum, another related requirement was fitting approximately 2 MHz bandwidth of color information into the 4.2 MHz video bandwidth of the then existing 6 MHz broadcasting channels with little or no modification of existing transmitters. This is accomplished using the band-sharing color system developed by the NTSC, and by taking advantage of the fundamental characteristics of the eye regarding color sensitivity and resolution.

The video-signal spectrum generated by scanning an image consists of energy concentrated near harmonics of the 15,734 Hz line scanning frequency. Additional lower amplitude sideband components exist at multiples of 59.94 Hz (the field scan frequency) from each line scan harmonic. Substantially no energy exists halfway between the line scan harmonics, that is, at odd harmonics of one-half the line frequency. Thus, these blank spaces in the spectrum are available for the transmission of a signal for carrying color information and its sideband. In addition, a signal modulated with color information injected at this frequency is of relatively low visibility in the reproduced image because the odd harmonics are of opposite phase on successive scanning lines and in successive frames, requiring four fields to repeat. Furthermore, the visibility of the color video signal is reduced further by the use of a subcarrier frequency near the cutoff of the video bandpass.

In the NTSC system, color is conveyed using two elements:

- A luminance signal
- A chrominance signal

The luminance signal is derived from components of the three primary colors, red, green, and blue in the proportions for *reference white*, E_y, as follows:

$$E_y = 0.3E_R + 0.59E_G + 0.11E_B$$

Where:
E_R = the red chrominance component
E_G = green chrominance component
E_B = blue chrominance component

These transmitted values equal unity for white and thus result in the reproduction of colors on monochrome receivers at the proper luminance level (the *constant-luminance* principle).

The color signal consists of two chrominance components, I and Q, transmitted as amplitude modulated sidebands of two 3.579545 MHz subcarriers in quadrature (differing in phase by 90°). The subcarriers are suppressed, leaving only the sidebands in the color signal. Suppression of the carriers permits demodulation of the components as two separate color signals in a receiver by reinsertion of a carrier of the phase corresponding to the desired

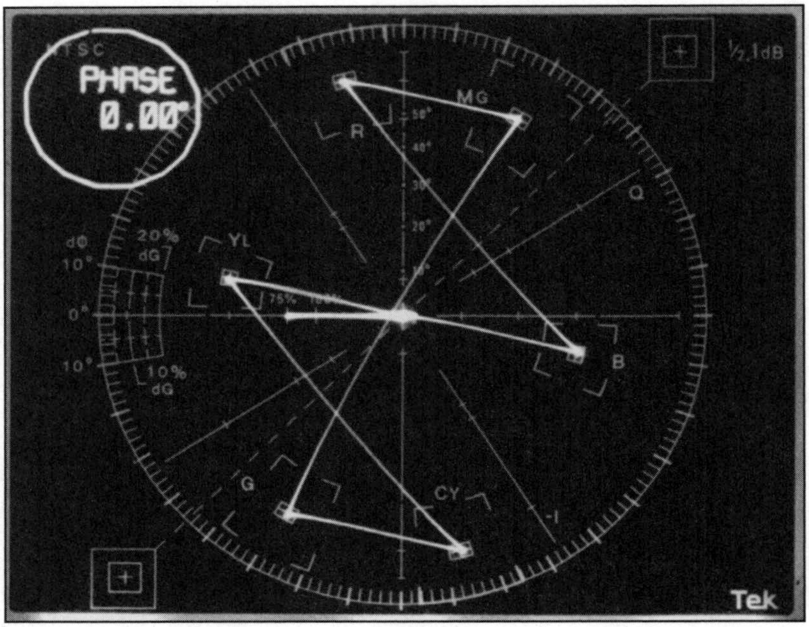

118 Video Display Engineering

Figure 4.4 Vectorscope representation of vector and chroma amplitude relationships in the NTSC system for a color bars signal. (*Courtesy of Tektronix.*)

color signal. This system for recovery of the color signals is called *synchronous demodulation*.

I and Q signals are composed of red, green, and blue primary color components produced by color cameras and other signal generators. The phase relationship among the I and Q signals, the derived primary and complementary colors, and the color synchronizing burst can be shown graphically on a *vectorscope* display. The horizontal and vertical sweep signals on a vectorscope are produced from $R - Y$ and $B - Y$ subcarrier sine waves in quadrature, producing a circular display. The chrominance signal controls the intensity of the display. A vectorscope display of an Electronic Industries Association (EIA) color bar signal is shown in Figure 4.4.

4.2.7 Color Signal Decoding

Each of the two chrominance signal carriers can be recovered individually by means of synchronous detection. A reference subcarrier of the same phase as the desired chroma signal is applied as a gate to a balanced demodulator. Only the modulation of the signal in the same phase as the reference will be present in the output. A lowpass filter can be added to remove second harmonic components of the chroma signal generated in the process.

4.2.8 Deficiencies of Conventional Video Signals

The composite transmission of luminance and chrominance in a single channel is achieved in the NTSC system by choosing the chrominance subcarrier to be an odd multiple of one-half the line-scanning frequency. This causes the component frequencies of chrominance to be interleaved with those of luminance. The intent of this arrangement is to make it possible to easily separate the two sets of components at the receiver, thus avoiding interference prior to the recovery of the primary color signals for display.

In practice, this process has been fraught with difficulty. The result has been a substantial limitation on the horizontal resolution available in consumer receivers. Signal intermodulation arising in the bands occupied by the chrominance subcarrier signal produces degradations in the image known as *cross color* and *cross luminance*. Cross color causes a display of false colors to be superimposed on repetitive patterns in the luminance image. Cross luminance causes a crawling dot pattern that is primarily visible around colored edges.

These effects have been sufficiently prominent that manufacturers of NTSC receivers have tended to limit the luminance bandwidth to less than 3 MHz (below the 3.58 MHz subcarrier frequency). This is far short of the 4.2 MHz maximum potential of the broadcast signal. The end result is that the horizontal resolution in such receivers is confined to about 250 lines. The filtering typically employed to remove chrominance from luminance is a simple notch filter tuned to the subcarrier frequency.

Comb Filtering

It is clear that the quality of conventional receivers would improve if the signal mixture between luminance and chrominance could be substantially reduced, if not eliminated. This has become possible with the development and widespread use of the *comb filter*. A common version of the comb filter consists of a glass fiber connected between two transducers—a video-to-acoustic transducer and an acoustic-to-video transducer. The composite signal fed to the input transducer produces an acoustic analog version of the signal, which reaches the far end of the fiber after an acoustic delay equal to one line-scan interval (63.555 μs for NTSC and 64 μs for PAL). The delayed electrical output version of the signal is then removed.

When the delayed output signal is added to the input, the sum represents luminance nearly devoid of chrominance content. This process is illustrated in Figure 4.5. Conversely, when the delayed output is subtracted from the input, the sum represents chrominance similarly devoid of luminance. When these signals are used to recover the primary-color information, cross-color and cross-luminance effects are largely removed. Newer versions of comb filters use charge-coupled devices that perform the same function without acoustical treatment.

Similar improvements in video quality can be realized through improved encoding techniques. Comb filters of advanced design have been produced that adapt their characteristics to changes in the image content. By these means, luminance/chrominance component separation is greatly increased. Filters using two or more line and/or field delays have been used. The greater the number of delays, the sharper the cutoff of the filter passband. Figure 4.6 shows a simplified diagram of a 2-*H* comb filter, and its luminance passband.

(a)

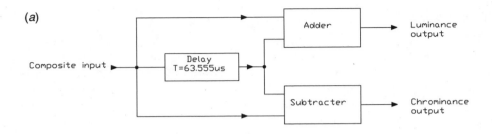

(b)

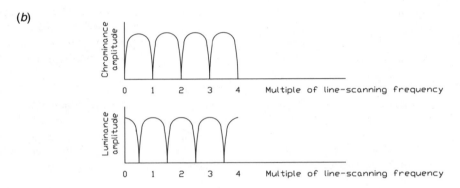

Figure 4.5 Comb filtering: (*a*) circuit introducing a one scan-line delay, (*b*) the luminance and chrominance passband.

4.3 Video Colorimetry

A video display can be regarded as a series of small visual colorimeters. In each picture element, a colorimetric match is made to an element of the original scene. The primaries are the red, green, and blue phosphors. The mixing occurs inside the eye of the observer because the eye cannot resolve the individual phosphor dots; they are too closely spaced. The outputs (R, G, B) of the three phosphors may be regarded as tristimulus values. Coefficients in the related equations depend on the chromaticity coordinates of the phosphors and on the luminous outputs of each phosphor for unit electrical input. Usually the gains of each of the three channels are set so that equal electrical inputs to the three produce a standard displayed white such as CIE illuminant D_{65}.

Therefore, a video camera must produce, for each picture element, three electrical signals representative of the three tristimulus values (R, G, B) of the required display. To accomplish this the system must have three optical channels with spectral sensitivities equal to the color-matching functions , , and , corresponding to the three primaries of the display.

Thus, the information to be conveyed by the electronic circuits comprising the camera, transmission system, and receiver/display is the amount of each of the three primaries (phosphors) required to match the input color. This information is based on the following items:

(a)

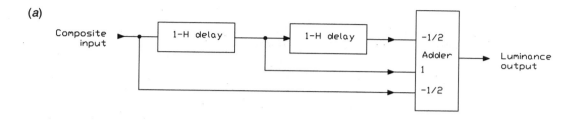

(b)

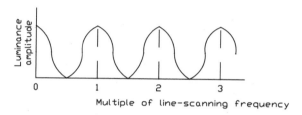

Figure 4.6 Comb filtering: (*a*) circuit introducing a two scan-line delay; (*b*) the luminance passband.

- An agreement concerning the chromaticities of the three primaries to be used.
- The representation of the amounts of these three primaries by electrical signals suitably related to them.
- The specification (typically) that the electrical signals shall be equal at some specified chromaticity.

The electrical signal voltages are then representative of the tristimulus values of the original scene. They obey all the laws to which tristimulus values conform, including being transformable to represent the amounts of primaries of other chromaticities than those for which the signals were originally composed. Such transformations can be arranged by forming three sets of linear combinations of the original signals.

Unfortunately the simple objective of producing an exact colorimetric match between each picture element in the display and the corresponding element of the original scene is difficult to fulfill, and, in any case, may not achieve the ultimate objective of equality of appearance between the display and the original scene. There are several reasons for this, including:

- It may be difficult to achieve the luminance of the original scene because of the limitation of the maximum luminance that can be generated by the reproducing system.
- The adaptation of the eye may be different for the reproduction than it is for the original scene because the surrounding conditions are different.

122 Video Display Engineering

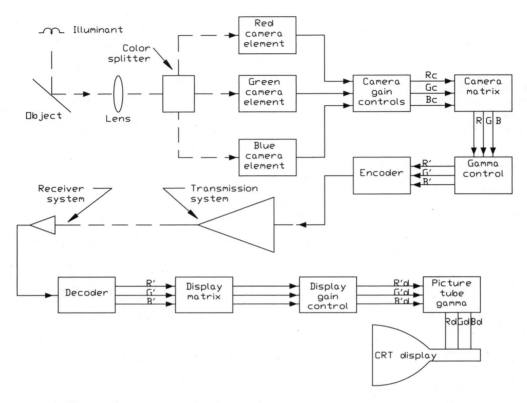

Figure 4.7 Block diagram of a simplified color television system.

- Ambient light complicates viewing the reproduced picture and changes its effective contrast ratio.
- The angle subtended by the reproduced picture may be different from that of the original scene.

Although an oversimplification, it is often considered that adequate reproduction is achieved when the chromaticity is accurately reproduced, while the luminance is reproduced *proportionally* to the luminance of the original scene. Even though the adequacy of this approach is somewhat questionable, it provides a starting point for system designers and enables the establishment of targets for system performance.

A block diagram of a simplified color television system is shown in Figure 4.7. Light reflected (or transmitted) by an object is split into three elements so that a portion strikes each camera tube, producing outputs proportional to the tristimulus values R, G, and B. Gain controls are provided so that the three signals can be made equal when the camera is viewing a standard white object. For transmission, the signals are encoded into three different waveforms and then decoded back to R, G, and B at the receiver. The primary purpose of encoding is to enable the signal to be transmitted within a limited bandwidth and to maintain compatibility with monochrome reception systems. The decoded R, G, and B signals are applied to the picture tube to excite the three phosphors. Gain controls are provided so equal inputs of R, G, and B will produce a standard white on the display.

4.3.1 Gamma

So far in this discussion, a linear relationship has been assumed between corresponding electrical and optical quantities in both the camera and the receiver. In practice, however, the *transfer function* is often not linear. For example, over the useful operating range of a typical color receiver, the light output of each phosphor follows a power-law relationship to the video voltage applied to the grid or cathode of the CRT. The light output (L) is proportional to the video-driving voltage (E_v) raised to the power γ:

$$L = KE_v^\gamma$$

Where γ is typically about 2.5 for a color CRT. This produces black compression and white expansion. Compensation for these three nonlinear transfer functions is accomplished by three electronic *gamma correctors* in the color camera video processing amplifiers. Thus, the three signals that are encoded, transmitted, and decoded are not in fact R, G, and B, but rather R', G', and B', given by:

$$R' = R^{1/\gamma}, \quad G' = G^{1/\gamma}, \quad B' = B^{1/\gamma}$$

If the rest of the system is linear, application of these signals to the color picture tube causes light outputs that are linearly related to the R, G, and B tristimulus inputs to the color camera, and so the correct reproduction is achieved.

4.3.2 Display White

The NTSC signal specifications were designed so that equal signals ($R = G = B$) would produce a display white of the chromaticity of illuminant C. For many years, most home receivers (but not studio monitors) were set so that equal signals produced a much bluer white. The correlated color temperature was about 9300K and the chromaticity was usually slightly on the green side of the Planckian locus. The goal of this practice was to achieve satisfactorily high brightness and avoid excessive red/green current ratios with available phosphors. With modern phosphors, high brightness and red/green current equality can be achieved for a white at the chromaticity of D_{65} so that both monitors and receivers are now usually balanced to D_{65}. Because D_{65} is close to illuminant C, the color rendition is generally better than with the bluer balance of older display systems.

4.3.3 Scene White

When the original scene is illuminated by daylight (of which D_{65} is representative), it is a clearly reasonable aim to reproduce the chromaticities of each object exactly in the final display. However, many video images are taken in a studio with incandescent illumination of about 3000K. In viewing the original scene, the eye adapts to a great extent so that most objects have similar appearance in both daylight and incandescent light. In particular, whites appear white under both types of illumination. However, if the chromaticities were to be reproduced exactly, studio whites would appear much yellower than the outdoor whites. This is because the viewer's adaptation is controlled more by the ambient viewing illuminant than by the scene illuminant and, therefore, does not correct fully for the change of scene illuminant. Because of this property, exact reproduction of chromaticities is not necessarily a good objective.

The ideal objective is for the reproduction to have exactly the same appearance as the original scene, but not enough is yet known about the chromatic adaptation of the human eye to define what this means in terms of chromaticity. A simpler criterion is to aim to reproduce objects with the same chromaticity they would have if the original scene were illuminated by D_{65}. This can be achieved by placing an optical filter (colored glass) in front of the camera with the spectral transmittance of the filter being equal to the ratio of the spectral power distributions of D_{65} and the actual studio illumination. As far as the camera is concerned, this has exactly the same effect as putting the same filter over every light source.

This solution has disadvantages because a different correction filter (in effect, a different set of camera sensitivities) is required for every scene illuminant. For example, every phase of daylight requires a special filter. In addition, insertion of a filter increases light scattering and can slightly degrade the contrast, resolution, and S/N.

An alternative method involves the adjustment of gain controls in the camera so that a white object produces equal signals in the three channels irrespective of the actual chromaticity of the illuminant. For colors other than white this solution does not produce exactly the same effect as a correction filter, but in practice it is satisfactory. Typically, the camera operator focuses a white reference of the scene inside a cursor on the monitor, and a microprocessor in a the camera performs white balancing automatically.

4.3.4 Phosphor Chromaticities

The phosphor chromaticities specified by the NTSC in 1953 were based on phosphors in common use for color television displays at that time. Since then, different phosphors have been introduced, mainly to increase the brightness of displays. These modern phosphors, especially the green ones, have different chromaticities so that the gamut of reproducible chromaticities has been reduced. However, because of the increased brightness, the overall effect on color rendition has been beneficial. Figure 4.8 shows two sets of modern phosphors plotted on the CIE chromaticity diagram.

4.4 Video System Characteristics

The central objective of any video service is to offer the viewer a sense of presence in the scene, and of participation in the events portrayed.[1] To meet this objective, the video image

---

[1] Portions of this section were contributed by Lawrence J. Thorpe, Sony Corporation of America.

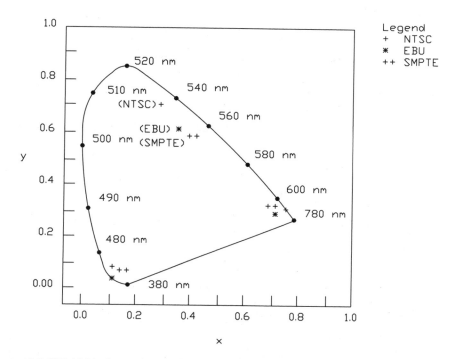

Figure 4.8 CIE 1931 chromaticity diagram showing three sets of phosphors used in color television displays.

should convey as much of the spatial and temporal content of the scene as is economically and technically feasible. Experience in the motion picture industry has demonstrated that a larger, wider picture, viewed closely, contributes greatly to the viewer's sense of presence and participation.

Current development of HDTV services for consumer applications is directed toward this same end. From the visual point of view, the term "high-definition" is, to some extent, a misnomer as the primary visual objective of the system is to provide an image that occupies a larger part of the visual field. Higher definition is secondary; it need be no higher than is adequate for the closer scrutiny of the image.

4.4.1 Foveal and Peripheral Vision

There are two areas of the retina of the eye that must be satisfied by video images: the *fovea* and the areas peripheral to it. Foveal vision extends over approximately one degree of the visual field, whereas the total field to the periphery of vision extends about 160° horizontally and 80° vertically. Motions of the eye and head are necessary to assure that the fovea is positioned on that part of the retinal image where the detailed structure of the scene is to be discerned.

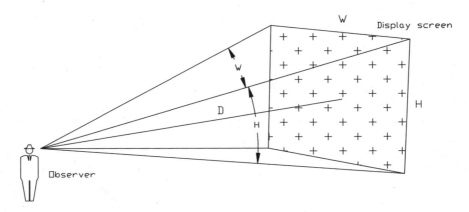

Figure 4.9 Geometry of the field of view occupied by a television image.

The portion of the visual field outside the foveal region provides the remaining visual information. Thus, a large part of visual reality is conveyed by this extra-foveal region. The vital perceptions of extra-foveal vision, notably motion and flicker, have received only secondary attention in the development of video engineering. Attention has first been paid to satisfying the needs of foveal vision. This is true because designing a system capable of resolving fine detail presents the major technical challenge, requiring a transmission channel that offers essentially no discrimination in the amplitude or time delay among the signals carried over a wide band of frequencies.

The properties of peripheral vision impose a number of constraints. Peripheral vision has great sensitivity to even modest changes in brightness and position. Thus, the bright portions of a wide image viewed closely are more apt to flicker at the left and right edges than the narrower image of conventional systems.

Figure 4.9 illustrates the geometry of the field occupied by the video image. The viewing distance D determines the angle h subtended by the picture height. This angle is usually measured as the ratio of the viewing distance to the picture height (D/H). The smaller the ratio, the more fully the image fills the field of view.

The useful limit to the viewing range is that distance at which the eye can just perceive the finest details in the image. Closer viewing serves no purpose in resolving detail, while more distant viewing prevents the eye from resolving all the detailed content of the image. The preferred viewing distance, expressed in picture heights, is the *optimal viewing ratio*, commonly referred to as the *optimal viewing distance*. It defines the distance at which a viewer with normal vision would prefer to see the image, when pictorial clarity is the criterion.

The optimal viewing ratio is not a fixed value; it varies with subject matter, viewing conditions, and visual acuity of the viewer. It does serve, however, as a convenient basis for comparing the performance of conventional and advanced display systems.

Computation of the optimal viewing ratio depends upon the degree of detail offered in the vertical dimension of the image, without reference to its pictorial content. The discernible detail is limited by the number of scanning lines presented to the eye, and by the ability of those lines to present the image details separately.

Ideally, each detail of a given scene would be reproduced by one pixel. That is, each scanning line would be available for one picture element along any vertical line in the image. In practice, however, some of the details in the scene inevitably fall between scanning lines. Two or more lines, therefore, are required for such picture elements. Some vertical resolution is lost in the process. Measurements of this effect show that only about 70 percent of the vertical detail is presented by the scanning lines. This ratio is known as the *Kell factor*; it applies irrespective of the manner of scanning, whether the lines follow each other sequentially (progressive scan) or alternately (interlaced scan).

When interlaced scanning is used, the 70 percent figure applies only when the image is fully stationary and the line of sight from the viewer does not move vertically. In practice these conditions are seldom met, so an additional loss of vertical resolution, the *interlace factor*, occurs under typical conditions.

This additional loss depends upon many aspects of the subject matter and viewer attention. Under favorable conditions, the additional loss reduces the effective value of vertical resolution to not more than 50 percent; that is, no more than half the scanning lines display the vertical detail of an interlaced image. Under unfavorable conditions, a larger loss can occur. The effective loss also increases with image brightness because the scanning beam becomes larger.

Because interlacing imposes this additional detail loss, it was decided by some system designers early in the development of advanced systems to abandon interlaced scanning for image display. Progressive-scanned displays can be derived from interlaced transmissions by digital image storage techniques. Such scanning conversion improves the vertical resolution by about 40 percent.

4.4.2 Horizontal Detail and Picture Width

Because the fovea is approximately circular in shape, its vertical and horizontal resolutions are nearly the same. This would indicate that the horizontal resolution of a display should be equal to its vertical resolution. Such equality is the usual basis of video system design, but it is not a firm requirement. Considerable variation in the shape of the picture element produces only a minor degradation in the sharpness of the image, provided that its area is unchanged. This being the case, image sharpness depends primarily on the *product* of the resolutions, that is, the total number of picture elements in the image.

The ability to extend the horizontal dimensions of the picture elements has been applied in wide-screen motion pictures. For example, the Fox CinemaScope system uses anamorphic optical projection to enlarge the image in the horizontal direction. Because the emulsion of the film has equal vertical and horizontal resolution, the enlargement lowers the horizontal resolution of the image.

When wide-screen motion pictures are displayed at the conventional aspect ratio, the full width of the film cannot be shown. This requires that the viewed area be moved laterally when scanned to keep the center of interest within the area displayed by the picture monitor. The interrelation of the aspect ratios for various services is shown in Figure 4.10.

The picture width-to-height ratio chosen for HDTV (widescreen) service is 1.777. This is a compromise with some of the wider aspect ratios of the film industry, imposed by constraints on the bandwidth of the HDTV channel. Other factors being equal, the video baseband increases in direct proportion to the picture width.

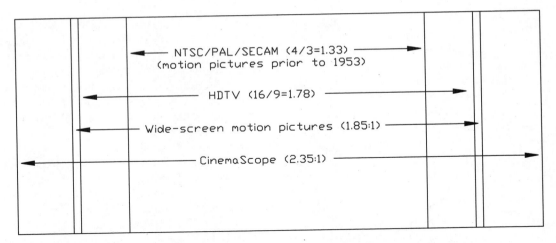

Figure 4.10 Comparison of the aspect ratios of television and motion pictures.

The 16:9 aspect ratio of HDTV offers an optimal viewing distance of 3.3 times the picture height, and a viewing angle of 39°, which is about 20 percent of the total horizontal visual field. While this is a small percentage, it covers that portion of the field within which the majority of visual information is conveyed.

Perception of Depth

Perception of depth depends on the angular separation of the images received by the eyes of the viewer. Successful binocular video systems have been produced, but their cost and inconvenience have restricted large scale adoption as of this writing. A considerable degree of depth perception is inferred in the flat image of video from the following:

- The perspective appearance of the subject matter.
- Camera techniques through the choice of the focal length of lenses and by changes in depth of focus.

Continuous adjustment of focal length by the zoom lens provides the viewer with an experience in depth perception wholly beyond the scope of natural vision. No special steps have been taken in the design of advanced visual services to offer depth clues not previously available. However, the wide field of view offered by HDTV and similar displays significantly improves depth perception, compared with that of conventional systems.

4.4.3 Contrast and Tonal Range

The range of brightness accommodated by video displays is limited compared to natural vision. This limitation of video has not been overcome in advanced display systems. Moreover, the display brightness of wide-screen images is restricted by the need to spread the available light over a large area.

Within the upper and lower limits of display brightness, the quality of the image depends on the relationship between changes in brightness in the scene and corresponding changes in the image. It is an accepted rule of image reproduction that the brightness be directly proportional to the source; the curve relating input and output brightness should be a straight line. Because the output versus input curves of cameras and displays are, in many cases, not straight lines, intermediate adjustment (*gamma correction*) is required. Gamma correction in wide-screen displays requires particular attention to avoid excessive noise and other artifacts evident at close scrutiny.

4.4.4 Chrominance Properties

Two aspects of color science are of particular interest with regard to advanced display systems. The first is the total range of colors perceived by the camera and offered by the display. The second is the ability of the eye to distinguish details presented in different colors. The *color gamut* defines the realism of the display, while attention to color acuity is essential to avoid presenting more chromatic detail than the eye can resolve.

In the earliest work on color television, equal bandwidth was devoted to each primary color, although it had long been known by color scientists that the color acuity of the normal eye is greatest for green light, less for red, and much less for blue. The NTSC system made use of this fact by devoting much less bandwidth to the red-minus-luminance channel, and still less to blue-minus-luminance. All color television services employ this device in one form or another. Properly applied, it offers economy in the use of bandwidth, without loss of resolution or color quality in the displayed image.

4.4.5 Temporal Factors in Vision

The response of the eye over time is essentially continuous. The cones of the retina convey nerve impulses to the visual cortex of the brain in pulses recurring about 1000 times per second. If video were required to follow this design, a channel several kilohertz wide would be required in HDTV (in an 1125/60—SMPTE 240M—system) for each of nearly a million picture elements. This would be equivalent to a video baseband exceeding 1 GHz.

Fortunately, video display systems can take advantage of a temporal property of the eye, persistence of vision. The special properties of this phenomenon must be carefully observed to assure that motion of the scene is free from the effects of flicker in all parts of the image. A common conflict encountered in the design of advanced imaging systems is the need to reduce the rate at which some of the spatial aspects of the image are reproduced, thus conserving bandwidth while maintaining temporal integrity, particularly the position and shape of objects in motion. To resolve such conflicts, it has become standard practice to view the video signal and its processing simultaneously in three dimensions: width, height, and time.

Basic to video and motion pictures is the presentation of a rapid succession of slightly different frames. The human visual system retains the image during the dark interval. Under specific conditions, the image appears to be continuously present. Any difference in the position of an object from one frame to the next is interpreted as motion of that object. For this process to represent visual reality, two conditions must be met:

- The rate of repetition of the images must be sufficiently high so that motion is depicted smoothly, with no sudden jumps from frame to frame.

- The repetition rate must be sufficiently high so that the persistence of vision extends over the interval between flashes. Here an idiosyncrasy of natural vision appears: the brighter the flash, the shorter the persistence of vision.

4.4.6 Continuity of Motion

Early in the development of motion pictures it was found that movement could be depicted smoothly at any frame rate faster than about 15 frames/s. To accurately portray rapid motion, a worldwide standard of 24 frames/s was selected for the motion picture industry. This frame rate does not solve all the problems of reproducing fast action. Restrictive production techniques (such as limited camera angles and restricted rates of panning) must be observed when rapid motion is encountered.

The acuity of the eye when viewing objects in motion is impaired because the temporal response of the fovea is slower than the surrounding regions of the retina. Thus, a loss of sharpness in the edges of moving objects is an inevitable aspect of natural vision. This property represents an important component of video system design.

Much greater losses of sharpness and detail, under the general term *smear*, occur whenever the image moves across the sensitive surface of the camera. Each element in the surface then receives light not from one detail, but from a succession of them. The signal generated by the camera is the sum of the passing light, not that of a single detail, and smear results. As in photography, this effect can be reduced by using a short exposure, provided there is sufficient light relative to the sensitivity of the camera. Electronic shutters can be used to limit the exposure to 1/1000 s or less when sufficient light is available.

Another source of smear occurs if the camera response carries over from one frame scan to the next. The retained signal elements from the previous scan are then added to the current scan, and any changes in their relative position causes a misalignment and consequent loss of detail. Such *carry-over smear* occurs when the exposure occupies the full scan time. A similar carry-over smear can occur in the display, when the light given off by one line persists long enough to be appreciably present during the successive scan of that line. Such carry-over also helps reduce flicker in the display; there is room for compromise between flicker reduction and loss of detail in moving objects.

4.4.7 Flicker Effects

The process of interlaced scanning is designed to eliminate flicker in the displayed image. Perception of flicker is primarily dependent upon two conditions:

- The brightness level of an image

- The relative area of an image in a picture

The 30 Hz transmission rate for a full 525 line conventional video frame is comparable to the highly successful 24-frame-per-second rate of motion-picture film. However, at the higher brightness levels produced on video screens, if all 483 lines (525 less blanking) of an image were to be presented sequentially as single frames, viewers would observe a disturbing flicker in picture areas of high brightness. For comparison, motion-picture theaters, on average, produce a screen brightness of 10 to 25 ft·L (footlambert), whereas a direct-view CRT may have a highlight brightness of 50 to 80 ft·L.

Through the use of interlaced scanning, single field images with one-half the vertical resolution capability of the 525 line system are provided at the high flicker-perception threshold rate of 60 Hz. Higher resolution of the full 490 lines (525 less vertical blanking) of vertical detail is provided at the lower flicker-perception threshold rate of 30 Hz. The result is a relatively flickerless picture display at a screen brightness of well over 50 to 75 ft·L, more than double that of motion-picture film projection. Both 60 Hz fields and 30 Hz frames have the same horizontal resolution capability.

The second advantage of interlaced scanning, compared to progressive scanning, is a reduction in video bandwidth for an equivalent flicker threshold level. Progressive scanning of 525 lines would have to be completed in 1/60 s to achieve an equivalent level of flicker perception. This would require a line scan to be completed in half the time of an interlaced scan. The bandwidth would then double for an equivalent number of pixels per line.

Interlace scanning, however, is the source of several degradations of image quality. While the total area of the image flashes at the rate of the field scan, twice that of the frame scan, the individual lines repeat at the slower frame rate. This gives rise to several degradations associated with the effect known as *interline flicker*. This causes small areas of the image, particularly when they are aligned horizontally, to display a shimmering or blinking that is visible at the usual viewing distance. A related effect is unsteadiness in extended horizontal edges of objects, as the edge is portrayed by a particular line in one field and by another line in the next. These effects become more pronounced as the vertical resolution provided by the camera and its image enhancement circuits is increased.

Interlacing also produces aberrations in the vertical and diagonal outlines of moving objects. This occurs because vertically-adjacent picture elements appear at different times in successive fields. An element on one line appears 1/60 s (actually 1/59.95 s) later than the vertically-adjacent element on the preceding field. If the objects in the scene are stationary, no adverse effects arise from this time difference. If an object is in rapid motion, however, the time delay causes the elements of the second field to be displaced to the right of, instead of vertically or diagonally adjacent to, those of the first field. Close inspection of such moving images shows that their vertical and diagonal edges are not sharp, but actually a series of step-wide serrations, usually coarser than the basic resolution of the image. Because the eye loses some acuity as it follows objects in motion, these serrations are often overlooked. They represent, however, an important impairment compared with motion picture images. All of the picture elements in a film frame are exposed and displayed simultaneously so the impairments resulting from interlacing do not occur.

As previously noted, the defects of interlacing have been an important target in advanced display system development. To avoid them, scanning at the camera must be progressive, using only one set of adjacent lines per frame. At the receiver, the display scan must match the camera.

4.4.8 Video Bandwidth

The time consumed in scanning fields and frames in a conventional video (NTSC/PAL systems) is measured in hundredths of a second. Much less time, less than one ten-millionth of a second, is available to scan a picture element in many advanced imaging systems, such as 1125/60 HDTV. The visual basis lies in the large number of picture elements (600,000 to 900,000) that are required to satisfy the eye when the image is under the close scrutiny offered by a wide-screen display. The eye further requires that signals representative of each of these elements be transmitted during the short time persistence of vision allows for flicker-free images, not more than 1/24 s. It follows that the rate of scanning depends upon the picture elements, ranging up to 22.5 million per second and beyond. This requirement is directly derived from the properties of the eye. When translated into engineering terms, the rates of scanning picture elements are stated in video frequencies. At best, one cycle of uncompressed video can represent only two horizontally adjacent picture elements, so the scanning rate of 22.5 million picture elements/s requires video frequencies of up to about 11 MHz for each channel.

4.5 Bibliography

Baldwin, M., Jr.: "The Subjective Sharpness of Simulated Television Images," *Proceedings of the IRE*, vol. 28, July 1940.

Belton, J.: "The Development of the CinemaScope by Twentieth Century Fox," *SMPTE Journal*, vol. 97, SMPTE, White Plains, N.Y., September 1988.

Benson, K. B., and D. G. Fink: *HDTV: Advanced Television for the 1990s*, McGraw-Hill, New York, N.Y., 1990.

Benson, K. B., and J. C. Whitaker (eds.): *Television Engineering Handbook*, revised ed., McGraw-Hill, New York, N.Y., 1991.

Bingley, F. J.: "Colorimetry in Color Television—Pt. I," *Proc. IRE*, vol. 41, pp. 838–851, 1953.

Bingley, F. J.: "Colorimetry in Color Television—Pts. II and III," *Proc. IRE*, vol. 42, pp. 48–57, 1954.

Bingley, F. J.: "The Application of Projective Geometry to the Theory of Color Mixture," *Proc. IRE*, vol. 36, pp. 709–723, 1948.

DeMarsh, L. E.: "Colorimetric Standards in US Color Television," *J. SMPTE*, vol. 83, pp. 1–5, 1974.

Epstein, D. W.: "Colorimetric Analysis of RCA Color Television System," *RCA Review*, vol. 14, pp. 227–258, 1953.

Fink, D. G.: "Perspectives on Television: The Role Played by the Two NTSCs in Preparing Television Service for the American Public," *Proceedings of the IEEE*, vol. 64, IEEE, New York, N.Y., September 1976.

Fink, D. G.: *Color Television Standards*, McGraw-Hill, New York, N.Y., 1955.

Fink, D. G., et. al.: "The Future of High Definition Television," *SMPTE Journal*, vol. 89, SMPTE, White Plains, N.Y., February/March 1980.

Fink, D. G.: *Color Television Standards*, McGraw-Hill, New York, N.Y., 1986.

Fujio, T., J. Ishida, T. Komoto and T. Nishizawa: "High Definition Television Systems—Signal Standards and Transmission," *SMPTE Journal*, vol. 89, SMPTE, White Plains, N.Y., August 1980.

Herman, S.: "The Design of Television Color Rendition," *J. SMPTE*, SMPTE, White Plains, N.Y., vol. 84, pp. 267–273, 1975.

Hubel, David H.: *Eye, Brain and Vision*, Scientific American Library, New York, N.Y., 1988.

Hunt, R. W. G.: *The Reproduction of Colour*, 3d ed., Fountain Press, England, 1975.

Judd, D. B.: "The 1931 C.I.E. Standard Observer and Coordinate System for Colorimetry," *Journal of the Optical Society of America*, vol. 23, 1933.

Kelly, K. L.: "Color Designation of Lights," *Journal of the Optical Society of America*, vol. 33, 1943.

Kelly, R. D., A. V. Bedbord and M. Trainer: "Scanning Sequence and Repetition of Television Images," *Proceedings of the IRE*, vol. 24, April 1936.

Miller, Howard: "Options in Advanced Television Broadcasting in North America," *Proceedings of the ITS*, International Television Symposium, Montreux, Switzerland, 1991.

Morizono, M.: "Technological Trends in High-Resolution Displays Including HDTV," *SID International Symposium Digest*, paper 3.1, May 1990.

Neal, C. B.: "Television Colorimetry for Receiver Engineers," *IEEE Trans. BTR*, vol. 19, pp. 149–162, 1973.

Pearson, M. (ed.): *Proc. ISCC Conf. on Optimum Reproduction of Color*, Williamsburg, Va., 1971, Graphic Arts Research Center, Rochester, N.Y., 1971.

Pitts, K. and N. Hurst: "How Much Do People Prefer Widescreen (16 × 9) to Standard NTSC (4 × 3)?," *IEEE Transactions on Consumer Electronics*, IEEE, New York, N.Y., August 1989.

Pointer, R. M.: "The Gamut of Real Surface Colors, *Color Res. App.*, vol. 5, 1945.

Pritchard, D. H.: "US Color Television Fundamentals—A Review," *IEEE Trans. CE*, vol. 23, pp. 467–478, 1977.

Sproson, W. N.: *Colour Science in Television and Display Systems*, Adam Hilger, Bristol, England, 1983.

Uba, T., K. Omae, R. Ashiya, and K. Saita: "16:9 Aspect Ratio 38V-High Resolution Trinitron for HDTV," *IEEE Transactions on Consumer Electronics*, IEEE, New York, N.Y., February 1988.

van Raalte, John A.: "CRT Technologies for HDTV Applications," *1991 HDTV World Conference Proceedings*, National Association of Broadcasters, Washington, D.C., April 1991.

Wentworth, J. W.: *Color Television Engineering*, McGraw-Hill, New York, N.Y., 1955.

Wintringham, W. T.: "Color Television and Colorimetry," *Proc. IRE*, vol. 39, pp. 1135–1172, 1951.

Whitaker, Jerry C., and K. Blair Benson (eds.): *Standard Handbook of Video and Television Engineering,* 3rd ed., McGraw-Hill, New York, N.Y., 1999.

Chapter 5

Measuring Display Parameters

5.1 Introduction

New applications for electronic displays are pushing vendors to produce systems that provide higher resolution, larger screen size, and better accuracy. More stringent performance demands improved quality assessment techniques in the manufacture, installation, and maintenance of display systems.

5.1.1 Visual Acuity

Visual acuity is the ability of the eye to distinguish between small objects (and hence, to resolve the details of an image) and is expressed in reciprocal minutes of the angle subtended at the eye by two objects that can be separately identified. When objects and background are displayed in black and white (as in monochrome television), at 100 percent contrast, the range of visual acuity extends from 0.2 to about 2.5 reciprocal minutes (5 to 0.4 minutes of arc, respectively). An acuity of 1 reciprocal minute is usually taken as the basis of television system design. At this value, stationary white points on two scanning lines separated by an intervening line (the remainder of the scanning pattern being dark) can be resolved at a distance of about 20 times the picture height. Adjacent scanning lines, properly interlaced, cannot ordinarily be distinguished at distances greater than six or seven times the picture height.

Visual acuity varies markedly as a function of the following:

- Luminance of the background
- Contrast of the image
- Luminance of the area surrounding the image
- Luminosity to which the eye is adapted

The acuity is approximately proportional to the logarithm of the background luminance, increasing from about 1 reciprocal minute at 1 ft·L (3.4 cd/m^2) background brightness and 100 percent contrast, to about 2 reciprocal minutes at 100 ft·L (340 cd/m^2). Figure 5.1 shows experimental data on visual acuity as a function of luminance, when the contrast is nearly 100 percent, using incandescent lamps as the illumination. These data apply to de-

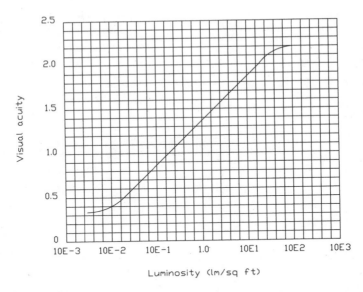

Figure 5.1 Visual acuity (the ability to resolve details of an image) as a function of the luminosity to which the eye is adapted. (*After R. J. Lythgoe.*)

tails viewed in the center of the field of view (by cone vision near the fovea of the retina). The acuity of rod vision, outside the foveal region, is poorer by a factor of about five times.

Visual acuity falls off rapidly as the contrast of the image decreases. Under typical conditions (1 ft·L background luminance) acuity increases from about 0.2 reciprocal minute at 10 percent contrast to about 1.0 reciprocal minute at 100 percent contrast, the acuity being roughly proportional to the percent contrast. Figure 5.2 shows experimental measurements of this effect.

When colored images are viewed, visual acuity depends markedly on the color. Acuity is higher for objects illuminated by monochromatic light than those illuminated by a source of the same color having an extended spectrum. This improvement results from the lack of chromatic aberration in the eye. In a colored image reproduced by primary colors, acuity is highest for the green primary and lowest for the blue primary. This is partly explained by the relative luminances of the primaries. The acuity for blue and red images is approximately two-thirds that for a white image of the same luminance. For the green primary, acuity is about 90 percent that of a white image of the same luminance. When the effect of the relative contributions to luminance of the standard FCC/NTSC color primaries (approximately, green:red:blue = 6:3:1) is taken into account, acuities for the primary images are in the approximate ratio green:red:blue = 8:3:1.

Visual acuity is also affected by *glare*, that is, regions in or near the field of view whose luminance is substantially greater than the object viewed. When viewing a video image, acuity fails rapidly if the area surrounding the image is brighter than the background luminance of the image. Acuity also decreases slightly if the image is viewed in darkness, that is, if the surround luminance is substantially lower than the average image luminance. Acuity is

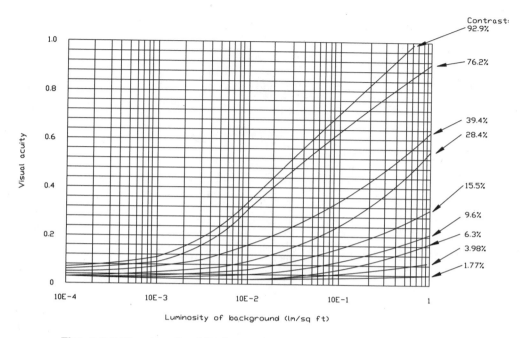

Figure 5.2 Visual acuity of the human eye as a function of luminosity and contrast (experimental data). (*After J. P. Conner and R. E. Ganoung.*)

not adversely affected if the surround luminance has a value in the range from 0.1 to 1.0 times the average image luminance.

5.1.2 Contrast Sensitivity

The ability of the eye to distinguish between the luminances of adjacent areas is known as *contrast sensitivity*, expressed as the ratio of the luminance to which the eye is adapted (usually the same as the background luminance) to the least perceptible luminance difference between the background luminance of the scene and the object luminance.

Two forms of contrast vision must be distinguished. In rod vision, which occurs at a background luminance of less than about 0.01 ft·L, the contrast sensitivity ranges from 2 to 5 (the average luminance is two to five times the least perceptible luminance difference), increasing slowly as the background luminance increases from 0.0001 to 0.01 ft·L. In cone vision, which takes place above approximately 0.01 ft·L, the contrast sensitivity increases proportionately to background luminance from a value of 30 at 0.01 ft·L to 150 at 10 ft·L. Contrast sensitivity does not vary markedly with the color of the light. It is somewhat higher for blue light than for red at low background luminance. The opposite case applies at high background luminance levels. Contrast sensitivity, like visual acuity, is reduced if the surround is brighter than the background. Figures 5.3 and 5.4 illustrate experimental data on contrast sensitivity.

138 Video Display Engineering

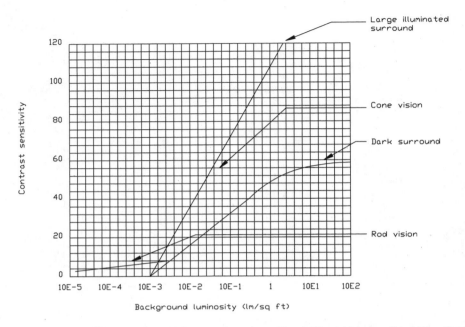

Figure 5.3 Contrast sensitivity as a function of background luminosity. (*After P. H. Moon.*)

5.1.3 Flicker

The perceptibility of flicker varies so widely with viewing conditions that it is difficult to describe quantitatively. In one respect, however, flicker phenomena can be readily compared. When the viewing situation is constant (no change in the image or surround other than a proportional change in the luminance of all their parts, and no change in the conditions of observation), the luminance at which flicker just becomes perceptible varies logarithmically with the luminance (the *Ferry-Porter law*). Numerically, a positive increment in the flicker frequency of 12.6 Hz raises the luminance flicker threshold 10 times. In video (NTSC) scanning, the applicable flicker frequency is the field frequency (the rate at which the area of the image is successively illuminated). Table 5.1 lists the flicker-threshold luminance for various flicker frequencies, based on 180 ft·L for 60 fields per second. The following factors affect the flicker threshold:

- The luminance of the flickering area
- Color of the area
- Solid angle subtended by the area at the eye
- Absolute size of the area
- Luminance of the surround
- Variation of luminance with time and position within the flickering area

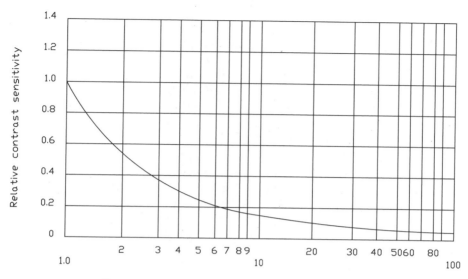

Figure 5.4 The effects of surround luminance on contrast sensitivity of the human eye. (*After P. H. Moon.*)

Table 5.1 Relative Flicker Threshold for Various Luminances
(The luminances tabulated have only relative significance; they are based on a value of 180 ft·L for a flicker frequency of 60 Hz, which is typical performance under NTSC.)

Flicker Frequency (Hz)	System	Frames/s	Flicker Threshold Luminance (ft·L)
48	Motion pictures	24	20
50	Television scanning	25	29
60	Television scanning	30	180

- Adaptation and training of the observer

5.2 Video Signal Spectrum

The spectrum of the video signal arising from the scanning process in a television camera extends from a lower limit determined by the time rate of change of the average luminance of the scene to an upper limit determined by the time during which the scanning spots cross the sharpest vertical boundary in the scene as focused within the camera. This concept is illustrated in Figure 5.5. The distribution of spectral components within these limits is determined by the following:

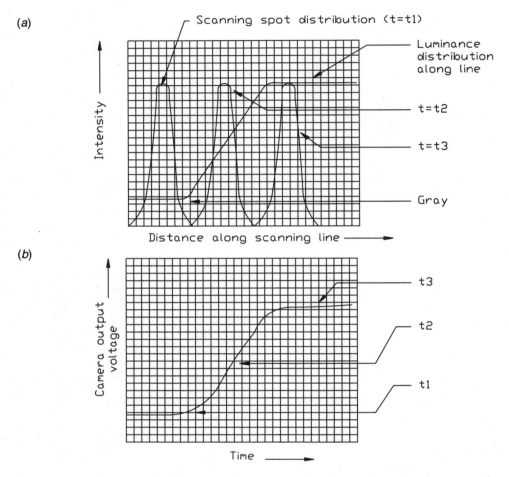

Figure 5.5 Video signal spectra: (*a*) camera scanning spot, shown with a guassian distribution, passing over a luminance boundary on the scanning line, (*b*) corresponding camera output signal resulting from convolution of the spot and luminance distributions.

- The distribution of energy in the camera scanning spots
- Number of lines scanned per second
- Percentage of line-scan time consumed by horizontal blanking
- Number of fields scanned per second
- Rates at which the luminances and chrominances of the scene change in size, position, and boundary sharpness

To the extent that the contents and dynamic properties of the scene cannot be predicted, the spectrum limits and energy distribution are not defined. However, the spectra associated with certain static and dynamic test charts and tapes may be used as the basis for video system design. Among the configurations of interest are:

- Flat fields of uniform luminance and/or chrominance
- Fields divided into two or more segments of different luminance by sharp vertical, horizontal, or oblique boundaries

The latter case includes the horizontal and vertical wedges of test charts and the concentric circles of *zone plate* (*zone pattern*) charts, illustrated in Figure 5.6. The reproductions of such patterns typically display diffuse boundaries and other degradations that may be introduced by the camera scanning process, the amplitude and phase responses of the transmission system, the receiver scanning spots, and other artifacts associated with scanning.

The upper limit of the video spectrum actually employed in reproducing a particular image is most often determined by the amplitude-vs.-frequency and phase-vs.-frequency responses of the receiving system. These responses are selected as a compromise between the image sharpness demanded by typical viewers and the deleterious effects of noise, interference, and incomplete separation of the luminance and chrominance signals in the receiver.

5.2.1 Minimum Video Frequency

To reproduce a uniform value of luminance from top to bottom of an image scanned in the conventional interlaced fashion, the video signal spectrum must extend downward to include the field-scanning frequency. This frequency represents the lower limit of the spectrum arising from scanning an image whose luminance does not change. Changes in the average luminance are reproduced by extending the video spectrum to a lower frequency equal to the reciprocal of the duration of the luminance change. Because a given average luminance may persist for many minutes, the spectrum extends sensibly to zero frequency (dc). Various techniques of preserving or restoring the dc component are employed to extend the spectrum from the field frequency down to zero frequency.

5.2.2 Maximum Video Frequency

In the analysis of maximum operating frequency for a video system, three values must be distinguished:

- Maximum output signal frequency generated by the camera or other pickup/generating device
- Maximum modulating frequency corresponding to (1) the fully transmitted radiated sideband, or (2) the system used to convey the video signal from the source to the display
- Maximum video frequency present at the picture tube (display) control electrodes

The maximum camera frequency is determined by the size and current distribution of the camera scanning spots and the time the spots take to move over the sharpest vertical boundary in the scene as focused on the scanned surfaces of the camera. The resulting transient

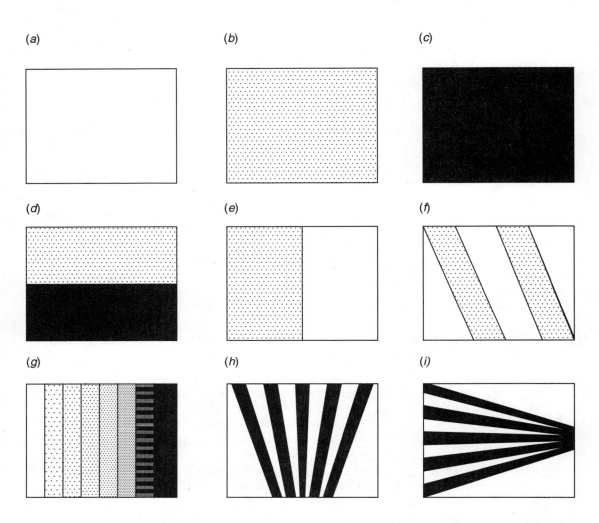

Figure 5.6 Scanning patterns of interest in analyzing conventional video signals: (*a*), (*b*), (*c*) flat fields useful for determining color purity and transfer gradient (gamma); (*d*) horizontal half-field pattern for measuring low-frequency performance; (*e*) vertical half field for examining high-frequency transient performance; (*f*) display of oblique bars; (*g*) in monochrome, a tonal wedge for determining contrast and luminance transfer characteristics; in color, a display used for hue measurements and adjustments; (*h*) wedge for measuring horizontal resolution; (*i*) wedge for measuring vertical resolution.

video signal current is represented by the convolution integral of the spot distributions and the brightness distribution of the boundary, both measured along the scanning line (Figure 5.5). The spectrum of the video signal transient includes significant amplitudes up to a frequency equal to the inverse of the time of rise of the resultant transient. Higher frequencies are also present, but have a relatively small effect on the shape of the reproduced transient after transmission of the video signal. Conventional television cameras and scanners are usually designed to generate maximum camera frequencies of 8 to 10 MHz, that is, approximately twice the maximum modulating frequency in the NTSC system.

The maximum modulating frequency is determined by the extent of the video channel reserved for the fully transmitted sideband. The channel width, in turn, is chosen to provide a value of horizontal resolution approximately equal to the vertical resolution implicit in the scanning pattern.

The maximum receiver video frequency as applied to the picture tube depends on the bandwidth of the picture IF amplifier, picture second detector, and video amplifiers. The value selected for conventional television is typically 2.8 MHz (a value substantially lower than the maximum modulating frequency of 4.2 MHz in the NTSC system).

Until the advent of the comb filter, it was not possible to separate the luminance signal completely from the chrominance signal in the vicinity of the 3.58 MHz color subcarrier. As discussed in the previous chapter, when strong luminance signal components are produced at or near that frequency, cross-color and cross-luminance are produced. Cutoff of the luminance signal at or about 2.8 MHz attenuates these components.

5.2.3 Horizontal Resolution

The *horizontal resolution* factor is the proportionality factor between horizontal resolution and video frequency. It may be expressed as:

$$H_r = \frac{R_h}{\alpha} \times \iota$$

Where:
H_r = horizontal resolution factor in lines per megahertz
R_h = lines of horizontal resolution per hertz of the video waveform
α = aspect ratio of the display
ι = active line period in microseconds

For NTSC, the horizontal resolution factor is:

$$78.7 = \frac{2}{4/3} \times 52.5$$

5.2.4 Video Frequencies Arising from Scanning

The signal spectrum arising from scanning comprises a number of discrete components at multiples of the scanning frequencies. Each spectrum component is identified by two numbers, *m* and *n*, which describe the pattern that would be produced if that component alone were present in the signal. The value of *m* represents the number of sinusoidal cycles of brightness measured horizontally (in the width of the picture) and *n* the number of cycles measured vertically (in the picture height). The 0, 0 pattern is the dc component of the signal, the 0, 1 pattern is produced by the field-scanning frequency, and the 1, 0 pattern by

144 Video Display Engineering

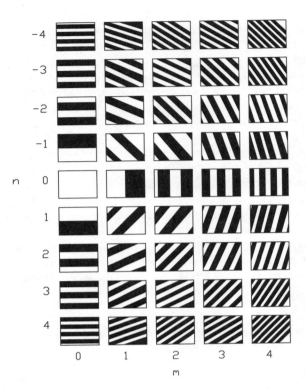

Figure 5.7 An array of image patterns corresponding to indicated values of *m* and *n*. (*After* [1].)

the line-scanning frequency. Typical patterns for various values of *m* and *n* are shown in Figure 5.7. By combining a number of such patterns (including *m* and *n* values up to several hundred), in the appropriate amplitudes and phases, any image capable of being represented by the scanning pattern may be built up. This is a two-dimensional form of the Fourier series.

The amplitudes of the spectrum components decrease as the values of *m* and *n* increase. Because *m* represents the order of the harmonic of the line-scanning frequency, the corresponding amplitudes are those of the left-to-right variations in brightness. A typical spectrum resulting from scanning a static scene is shown in Figure 5.8. The components of major magnitude include:

- The dc component
- Field-frequency component
- Components of the line frequency and its harmonics

Surrounding each line-frequency harmonic is a cluster of sideband components, each separated from the next by an interval equal to the field-scanning frequency.

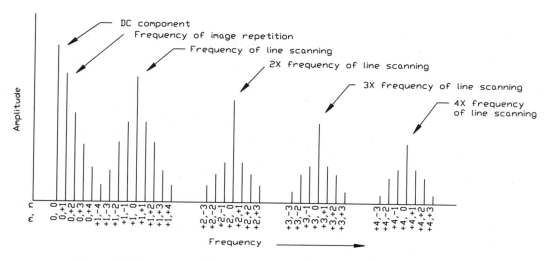

Figure 5.8 The typical spectrum of a video signal, showing the harmonics of the line-scanning frequency surrounded by clusters of components separated at intervals equal to the field-scanning frequency. (*After* [1].)

It is possible for the clusters surrounding adjacent line-frequency harmonics to overlap one another. This corresponds to two patterns in Figure 5.7 situated on adjacent vertical columns, which produce the same value of video frequency when scanned. Such *intercomponent confusion* of spectral energy is fundamental to the scanning process. Its effects are visible when a heavily striated pattern (such as that of a fabric having an accented weave) is scanned with the striations approximately parallel to the scanning lines. In the NTSC and PAL color systems, in which the luminance and chrominance signals occupy the same spectral region (one being interlaced in frequency with the other) such intercomponent confusion may produce prominent color fringes. Precise filters, which sharply separate the luminance and chrominance signals (comb filters), can remove this effect.

In static and slowly moving scenes, the clusters surrounding each line-frequency harmonic are compact, seldom extending further than 1 or 2 kHz on either side of the line-harmonic frequency. The space remaining in the signal spectrum is unoccupied and may be used to accommodate the spectral components of another signal having the same structure and frequency spacing. For scenes where motion is sufficiently slow for the eye to perceive the detail of the moving objects, it may be safely assumed that less than half the spectral space between line-frequency harmonics is occupied by energy of significant magnitude. It is on this principle that the NTSC and PAL compatible color television systems are based.

5.3 Measurement of Color Displays

The chromaticity and luminance of a portion of a color display device may be measured in several ways. The most fundamental approach involves a complete spectroradiometric

146   Video Display Engineering

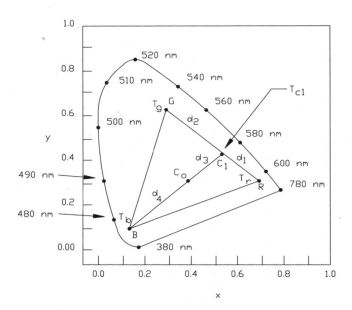

Figure 5.9 The CIE 1931 chromaticity diagram illustrating use of the center of gravity law (i.e., $T_r d_1 = T_g d_2$, $T_{c1} = T_r + T_g$, $T_{c1} d_3 = T_b d_4$).

measurement followed by computation using tables of color-matching functions. Portable spectroradiometers with built-in computers are available for this purpose. Another method, somewhat faster but less accurate, involves the use of a photoelectric colorimeter. These devices have spectral sensitivities approximately equal to the CIE color-matching functions, and thus provide direct readings of tristimulus values.

For setting up the reference white, it is often simplest to use a split-field visual comparator and to adjust the display device until it matches the reference field (usually D_{65}) of the comparator. However, because there is usually a large spectral difference (large metamerism) between the display and the reference, different observers will often make different settings by this method. Thus settings by one observer—or a group of observers—with normal color vision are often used simply to provide a reference point for subsequent photoelectric measurements.

An alternative method of determining the luminance and chromaticity coordinates of any area of a display involves measurement of the output of each phosphor separately and combining them using the center of gravity law in which the total tristimulus output of each phosphor is considered as an equivalent weight located at the chromaticity coordinates of the phosphor.

Consider the CIE chromaticity diagram shown in Figure 5.9 to be a uniform flat surface positioned in a horizontal plane. For the case illustrated, the center of gravity of the three weights (T_r, T_g, T_b), or the *balance point*, will be at the point C_o. This point determines the chromaticity of the mixture color. The luminance of the color C_o will be the linear sum of the luminance outputs of the red, green, and blue phosphors. The chromaticity coordinates of the display primaries may be obtained from the manufacturer. The total tristimulus output of one phosphor may be determined by turning off the other two CRT guns, measuring the luminance of the specified area, and dividing this value by the y chromaticity coordinate of the energized phosphor. This procedure is then repeated for the other two phosphors. From these data the color resulting from given excitations of the three phosphors may be calculated as follows:

- Chromaticity coordinates of red phosphor = x_r, y_r
- Chromaticity coordinates of green phosphor = x_g, y_g
- Chromaticity coordinates of blue phosphor = x_b, y_b
- Luminance of red phosphor = Y_r
- Luminance of green phosphor = Y_g
- Luminance of blue phosphor = Y_b

$$\text{Total tristimulus value of red phosphor} = X_r + Y_r + Z_r = \frac{Y_r}{y_r} = T_r$$

$$\text{Total tristimulus value of green phosphor} = X_g + Y_g + Z_g = \frac{Y_g}{y_g} = T_g$$

$$\text{Total tristimulus value of blue phosphor} = X_b + Y_b + Z_b = \frac{Y_b}{y_b} = T_b$$

Consider T_r as a weight located at the chromaticity coordinates of the red phosphor and T_g as a weight located at the chromaticity coordinates of the green phosphor. The location of the chromaticity coordinates of color C_1 (blue gun of color CRT turned off) can be determined by taking moments along line RG to determine the center of gravity of weights T_r and T_g:

$$T_r \times d_1 = T_g \times d_2$$

The total tristimulus value of C_1 is equal to $T_r + T_g = T_{c1}$. Taking moments along line C_1B will locate the chromaticity coordinates of the mixture color C_o:

$$T_{c1} \times d_3 = T_b \times d_4$$

The luminance of the color C_o is equal to $Y_r + Y_g + Y_b$.

5.3.1 Assessment of Color Reproduction

A number of factors contribute to poor color rendition in a display system. To assess their effects, it is necessary to define system objectives, and then establish a method of measuring departures from the objectives. Visual image display may be categorized as follows:

- *Spectral color reproduction*—the exact reproduction of the spectral power distributions of the original stimuli. Clearly, this is not possible in a video system with three primaries.

- *Exact color reproduction*—the exact reproduction of tristimulus values. The reproduction is then a metameric match to the original. Exact color reproduction will result in equality of appearance only if the viewing conditions for the picture and the original scene are identical. These conditions include the angular subtense of the picture, the luminance and chromaticity of the surround, and glare. In practice, exact color reproduction often cannot be achieved because of limitations on the maximum luminance that can be produced on a color monitor.

- *Colorimetric color reproduction*—a variant of exact color reproduction in which the tristimulus values are proportional to those in the original scene. In other words, the chromaticity coordinates are reproduced exactly, but the luminances are all reduced by a constant factor. Traditionally, color video systems have been designed and evaluated for colorimetric color reproduction. If the original and the reproduced reference whites have the same chromaticity, if the viewing conditions are the same, and if the system has an overall gamma of unity, colorimetric color reproduction is indeed a useful criterion. However, these conditions often do not hold and then colorimetric color reproduction is inadequate.

- *Equivalent color reproduction*—the reproduction of the original color appearance. This might be considered as the ultimate objective but cannot be achieved because of the limited luminance that can be generated in a display system.

- *Corresponding color reproduction*—a compromise in which colors in the reproduction have the same appearance as the colors in the original if illuminated to produce the same average luminance level and the same reference white chromaticity as that of the reproduction. For most purposes, corresponding color reproduction is the most suitable objective of a color video system.

- *Preferred color reproduction*—a departure from the preceding categories that recognizes the preferences of the viewer. It is sometimes argued that corresponding color reproduction is not the ultimate aim for some display systems, such as color television, but that account should be taken of the fact that people prefer some colors to be different from their actual appearance. For example, sun-tanned skin color is preferred to average real skin color, and sky is preferred bluer and foliage greener than they really are.

Even if corresponding color reproduction is accepted as the target, it is important to remember that some colors are more important than others. For example, flesh tones must be acceptable—not obviously reddish, greenish, purplish, or otherwise incorrectly rendered.

Similarly the sky must be blue and the clouds white, within the viewer's range of acceptance. Similar conditions apply to other well-known colors of common experience.

5.3.2 Chromatic Adaptation and White Balance

With properly adjusted cameras and displays, whites and neutral grays are reproduced with the chromaticity of D_{65}. Tests have shown that such whites (and grays) appear satisfactory in home viewing situations even if the ambient light is a different color temperature. Problems occur, however, when the white balance is slightly different from one camera to the next, or when the scene shifts from studio to daylight or vice versa. In the first case, unwanted shifts of the displayed white occur, whereas in the other, no shift occurs even though the viewer subconsciously expects a shift.

By always reproducing a white surface with the same chromaticity, the system is mimicking the human visual system, which adapts so that white surfaces always appear the same whatever the chromaticity of the illuminant (at least within the range of common light sources). The effect on other colors, however, is more complicated. In video cameras the white balance adjustment is usually made by gain controls on the R, G, and B channels. This is similar to the von Kries model of human chromatic adaptation, although the R, G, and B primaries of the model are not the same as the video primaries. It is known that the von Kries model does not accurately account for the appearance of colors after chromatic adaptation, and it follows that simple gain changes in a video camera is not the ideal approach. Nevertheless, this approach seems to work well in practice, and the viewer does not object to the fact—for example—that the relative increase in the luminances of reddish objects in tungsten light is lost.

5.3.3 Overall Gamma Requirements

Colorimetric color reproduction requires that the overall gamma of the system, including the camera, the display, and any gamma-adjusting electronics, be unity. This simple criterion is the one most often used in the design of a video color rendition system. However, the more sophisticated criterion of corresponding color reproduction takes into account the effect of the viewing conditions. In particular, several studies have shown that the luminance of the surround is important. For example, a dim surround requires a gamma of about 1.2, and a dark surround requires a gamma of about 1.5 for optimum color reproduction. It is important to note that the compatibility of a color image on a monochrome display is materially compromised because the display gamma of the monochrome tube is incorrect.

5.3.4 Perception of Color Differences

The CIE 1931 chromaticity diagram does not map chromaticity on a uniform-perceptibility basis. A just-perceptible change of chromaticity is not represented by the same distance in different parts of the diagram. Many investigators have explored the manner in which perceptibility varies over the diagram. The most often quoted study is that of MacAdam [2] who identified a set of ellipses that are contours of equal perceptibility about a given color, as shown in Figure 5.10.

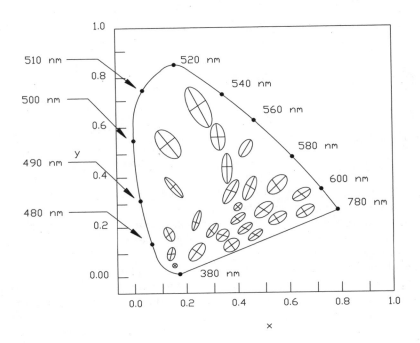

Figure 5.10 Ellipses of equally perceptible color differences. (*After* [2].)

From this and similar studies it is apparent, for example, that large distances represent relatively small perceptible changes in the green sector of the diagram. In the blue region, much smaller changes in the chromaticity coordinates are readily perceived.

Furthermore, viewing identical images on dissimilar displays can result in observed differences in the appearance of the image [3]. There are several factors that affect the appearance of the image:

- Physical factors, including display gamut limitations, illumination level, and black point
- Psychophysical factors, including chromatic induction and color constancy

Each of these factors interact in such a way that predicting the appearance of an image on a given display becomes difficult. System designers have experimented with colorimetry to facilitate the sharing of image data among display devices that vary in manufacture, calibration, and location. Of particular interest is the application of colorimetry to imaging in a networked window system environment, where it is often necessary to assure that an image displayed remotely looks like the image displayed locally.

Studies have indicated that image context and image content are also factors that affect color appearance. While it is popular to use highly chromatic backgrounds in a windowed display system, this will impact the appearance of the colors in the foreground.

5.3.5 Display Resolution and Pixel Format

The pixel represents the smallest resolvable element of a display. The size of the pixel varies from one display type to another. In a monochrome CRT, pixel size is determined primarily by the following factors:

- Spot size of the electron beam (the current density distribution)
- Phosphor particle size
- Thickness of the phosphor layer

The term *pixel* was developed in the era of monochrome television, and the definition was—at that time—straightforward. With the advent of color triad-based CRTs and solid state display systems, the definition is not nearly so clear.

For a color CRT, a single triad of red, green, and blue phosphor dots constitutes a single pixel. This definition assumes that the mechanical and electrical parameters of the CRT will permit each triad to be addressed without illuminating other elements on the face of the tube. Most display systems, however, will not meet this criterion. Depending on the design, a number of triads may constitute a single pixel in a CRT display. A more all-inclusive definition for the pixel is *the smallest spatial-information element as seen by the viewer* [4].

Dot pitch is one of the principle mechanical criteria of a CRT that determines—to a large extent—the resolution of the display. Dot pitch is defined as the center-to-center distance between adjacent green phosphor dots of the red, green, blue triad.

The *pixel format* is the arrangement of pixels into horizontal rows and vertical columns. For example, an arrangement of 640 horizontal pixels by 480 vertical pixels results in a 640 × 480 pixel format. This description is not a resolution parameter by itself, simply the arrangement of pixel elements on the screen. Resolution is the measure of the ability to delineate picture detail; it is the smallest discernible and measurable detail in a visual presentation [5].

Pixel density is a parameter closely related to resolution, stated in terms of pixels per linear distance. Pixel density specifies how closely the pixel elements are spaced on a given display. It follows that a display with a given pixel format will not provide the same pixel density (resolution) on a large size screen—such as 27-in diagonal—as on a small size screen—such as 12-in diagonal.

Television lines is another term used to describe resolution. The term refers to the number of discernible lines on a standard test chart. As before, the specification of television lines is not by itself a description of display resolution. A 525-line display on a 17-in monitor will appear to have greater resolution to a viewer than a 525-line display on a 30-inch monitor. Pixel density is the preferred resolution parameter.

Contrast Ratio

The purpose of a video display is to convey information by controlling the illumination of phosphor dots on a screen, or by controlling the reflectance or transmittance of a light source. The *contrast ratio* specifies the observable difference between a pixel that is switched *on* and its corresponding *off* state:

$$C_r = \frac{L_{on}}{L_{off}}$$

Where:
C_r = contrast ratio of the display
L_{on} = luminance of a pixel in the *on* state
L_{off} = luminance of a pixel in the *off* state

The area encompassed by the contrast ratio is an important parameter when considering the performance of a display. Two contrast ratio divisions are typically specified:

- *Small area*—comparison of the *on* and *off* states of a pixel-sized area
- *Large area*—comparison of the *on* and *off* states of a group of pixels

For most display applications, the small area contrast ratio is the more critical parameter.

5.4 Applications of the Zone Pattern Signal

The increased information content of advanced, high-definition display systems requires sophisticated processing to make recording and transmission practical.[1] This processing often uses some form of bandwidth compression, scan rate changes, motion detection and prediction algorithms, or other new techniques. Zone patterns are well suited to exercising a complex video system in the three dimensions of its signal spectrum: horizontal, vertical, and temporal. Zone pattern signals, unlike most conventional test signals, can be complex and dynamic. Because of this, they are capable of simulating much of the detail and movement of actual picture video, exercising the system under test with signals representative of the intended application. These digitally generated and controlled signals also have other important characteristics needed in test signals.

A signal intended for meaningful testing of a video system must be carefully controlled, so that any departure from a known parameter of the signal is attributable to a distortion or other change in the system under test. Also, the test signal must be predictable, so it can be accurately reproduced at other times or places. These constraints have usually led to test signals that are electronically generated. In a few special cases, a standardized picture has been produced by a camera or monoscope—usually for a subjective, but more detailed evaluation of system performance. A zone plate [6] is a specific optical pattern. Electronic signal generators are now capable of readily producing zone-type patterns of various characteristics. The label "zone pattern", thus, is applied to a wide variety of signals created by video test instruments.

Conventional test signals, for the most part limited by the practical considerations of electronic generation, have represented relatively simple images. Each signal is capable of testing a narrow range of possible distortions; several test signals are needed for a more complete evaluation. Even with several signals, this method may not reveal all possible dis-

---

[1] Portions of this section were adapted from: "Broadening the Applications of Zone Plate Generators," Application Note 20W7056, Tektronix, Beaverton, Ore., 1992.

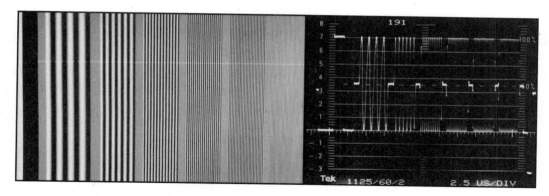

Figure 5.11 Multiburst video test waveform: (*a*, left) picture display, (*b*, right) multiburst signal as viewed on a waveform monitor (1-*H*). (*Courtesy of Tektronix*.)

tortions, nor does it allow study of all pertinent characteristics. This is especially true in video systems employing various forms of sophisticated signal processing.

5.4.1 Simple Zone Plate Patterns

Historically, the basic testing of a video communication channel has involved the application of several single frequencies—in effect, spot checking the spectrum of interest. A well-known, and quite practical adaptation of this idea is the *multiburst* signal, as shown in Figure 5.11. This test waveform has been used since the earliest days of video. The multiburst signal provides several discrete frequencies along a TV line.

The *frequency sweep* signal is an improvement on multiburst. While harder to implement in earlier generators, it was easier to use. The frequency sweep signal (illustrated in Figure 5.12) varies the applied signal frequency continuously along the TV line.[2] In some cases, the signal is swept as a function of the vertical position (field time). Even in these cases, the signal being swept is appropriate for testing the spectrum of the horizontal dimension of the picture.

Figure 5.13 shows the output of a zone plate generator configured to produce a horizontal single frequency output. Figure 5.14 shows a zone plate generator configured to produce a frequency sweep signal. Electronic test patterns, such as these, may be used to evaluate the following system characteristics:

- Channel frequency response
- Horizontal resolution

---

2 Figure 5.12(*a*) and other photographs in this section show the "beat" effects introduced by the screening process used for photographic printing. This is largely unavoidable. The screening process is similar to the scanning or sampling of a television image. The patterns are designed to identify this type of problem.

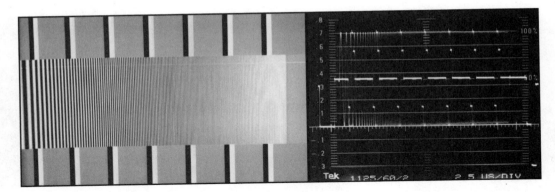

Figure 5.12 Conventional sweep frequency test waveform: (*a*, left) picture display, (*b*, right) waveform monitor display, with markers (1-*H*). (*Courtesy of Tektronix.*)

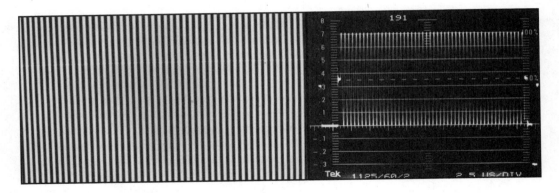

Figure 5.13 Single horizontal frequency test signal from a zone plate generator: (*a*, left) picture display, (*b*, right) waveform monitor display (1-*H*). (*Courtesy of Tektronix.*)

- Moiré effects in recorders and displays
- Other impairments

Patterns that test vertical (field) response have—traditionally—been less frequently used. As new technologies implement conversion from progressive to interlaced scan, line doubling display techniques, vertical anti-aliasing filters, scan conversion, motion detection, or other processes that combine information from line to line, vertical testing patterns are more in demand.

In the vertical dimension, as well as the horizontal, tests may be done at a single frequency or with a frequency sweep signal. Figure 5.15 illustrates a magnified vertical rate waveform display. Each "dash" in the photo represents one horizontal scan line. Sampling of vertical frequencies is inherent in the scanning process, and the photo shows the effects

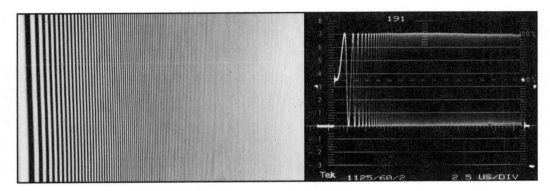

Figure 5.14 Horizontal frequency sweep test signal from a zone plate generator: (*a*, left) picture display, (*b*, right) waveform monitor display (1-*H*). (*Courtesy of Tektronix.*)

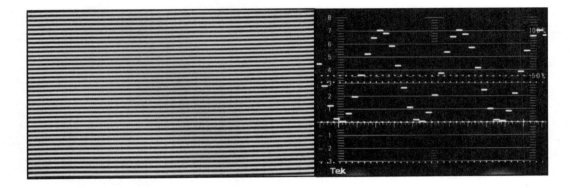

Figure 5.15 Single vertical frequency test signal: (*a*, left) picture display, (*b*, right) magnified vertical rate waveform, showing the effects of scan sampling. (*Courtesy of Tektronix.*)

on the signal waveform. Also, note that the signal voltage remains constant during each line, and changes only from line to line in accord with the vertical dimension sine function of the signal. Figure 5.16 shows a vertical frequency sweep picture display.

While the horizontal and vertical sinewaves and sweeps are quite useful, they do not use the full potential of a zone plate signal source.

5.4.2 Producing the Zone Plate Signal

A zone plate signal is created in real time by the test signal generator. The value of the signal at any instant is represented by a number in the digital hardware. This number is incremented as the scan progresses through the three dimensions that define a point in the video image; horizontal position, vertical position, and time.

Figure 5.16 Vertical frequency sweep picture display. (*Courtesy of Tektronix.*)

Figure 5.17 Combined horizontal and vertical frequency sweep picture display. (*Courtesy of Tektronix.*)

The exact method in which these dimensions alter the number is controlled by a set of coefficients. These coefficients determine the initial value of the number and control the size of the increments as the scan progresses along each horizontal line, from line to line vertically, and from field to field temporally. A set of coefficients uniquely determines a pattern, or a sequence of patterns, when the time dimension is active.

This process produces a sawtooth waveform; overflow in the accumulator holding the signal number effectively resets the value to zero at the end of each cycle of the waveform. Usually it is desirable to minimize the harmonic energy content of the output signal; in this case, the actual output is a sine function of the number generated by the incrementing process.

Complex Patterns.

A pattern of sinewaves or sweeps in multiple dimensions may be produced, using the unique architecture of the zone plate generator. The pattern shown in Figure 5.17, for example is a signal sweeping both horizontally and vertically. Figure 5.18 shows the waveform of a single selected line (line 263 in the 1125/60/2 HDTV system). Note that the horizontal waveform is identical to Figure 5.12(*b*), even though the vertical dimension sweep is now also active. Actually, different lines will give slightly different waveforms. The hor-

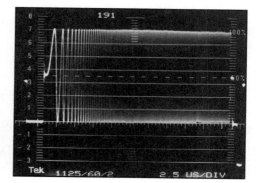

Figure 5.18 Combined horizontal and vertical frequency sweeps, selected line waveform display (1-*H*). This figure shows the maintenance of horizontal structure in the presence of vertical sweep. (*Courtesy of Tektronix.*)

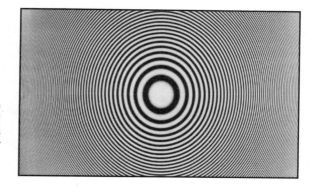

Figure 5.19 The best-known zone plate pattern, combined horizontal and vertical frequency sweeps with zero frequency in the center screen. (*Courtesy of Tektronix.*)

izontal frequency and sweep characteristics will be identical, but the starting phase must be different from line to line to construct the vertical signal.

Figure 5.19 shows a two-axis sweep pattern that is most often identified with zone plate generators; perhaps because it quite closely resembles the original optical pattern. In this *circle* pattern, both horizontal and vertical frequencies start high, sweep to zero (in the center of the screen), and sweep up again to the end of their respective scans. The concept of two-axis sweeps is actually more powerful than it might, at first, appear. In addition to purely horizontal or vertical effects, there are possible distortions or artifacts that are only apparent with simultaneous excitation in both axes. In other words, the response of a system to diagonal detail may not be predictable from information taken from the horizontal and vertical responses.

Consider an example from NTSC. A comb filter will suppress cross-color effects from a horizontal luminance frequency signal near the subcarrier frequency (such as the higher frequency packets in a NTSC multiburst signal). If, however, the right vertical component is added, creating a specific diagonal luminance pattern, even the most complex decoders will interpret the pattern as colored. In this case, the two-axis sweep shows very different effects than the same sweeps shown individually. A multi-dimensional sweep is a powerful tool for identifying analogous effects in other complex signal-processing systems.

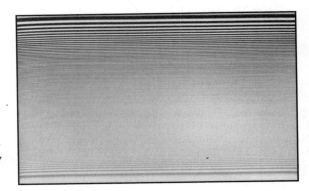

Figure 5.20 Vertical frequency sweep picture display. (*Courtesy of Tektronix.*)

5.4.3 The Time (Motion) Dimension

Incrementing the number in the accumulator of the zone pattern generator from frame to frame (or field to field in an interlaced system) creates a predictably different pattern for each vertical scan. This, in turn, creates apparent motion and exercises the signal spectrum in the temporal dimension. Analogous to the single frequency and frequency sweep examples previously given, the appropriate setting of time related coefficients of the generator will create constant motion or motion sweep (acceleration).

Specific motion detection and interpolation algorithms in a system under test may be exercised by determining the coefficients of a critical sequence of patterns. These patterns may then be saved for subsequent testing during development or adjustment. In an operational environment, appropriate response to a critical sequence could ensure expected operation of the equipment or facilitate fault detection.

While motion artifacts are difficult to portray in the still image constraints of a printed book, the following example provides some idea of the value of a versatile generator. In Figure 5.20 the vertical sweep maximum frequency has been increased to the point where it is zero-beating with the scan at the bottom of the screen. (The cycles per picture height of the pattern matches the lines per picture height per field of the scan.) Actually, in direct viewing, there is another noticeable artifact in the vertical center of the screen; a harmonic beat related to the gamma of the display CRT. Because of interlace, this beat flickers at the field rate. The photograph integrates the interfield flicker and thereby hides the artifact, which is readily apparent when viewed in real time.

Figure 5.21 is identical to the previous photo, except for one important difference—upward motion of ½ cycle per field has been added to the pattern. Now the sweep pattern itself is integrated out, as is the first order beat at the bottom. The harmonic effects in center screen no longer flicker, because the change of scan vertical position from field to field is compensated by a change in position of the image. The resulting beat pattern does not flicker and is easily photographed or scanned to determine depth of modulation.

A change in coefficients produces hyperbolic, rather than circular two-axis patterns (as shown in Figure 5.22). Another interesting pattern, which has been suggested for checking complex codecs, is shown in Figure 5.23. This is also a moving pattern, which was slightly altered to freeze some aspects of the movement to take the photograph.

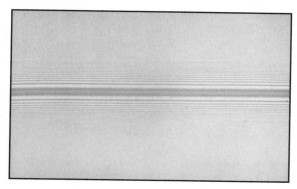

Figure 5.21 The same vertical sweep as Figure 5.20 except that appropriate pattern motion has been added to "freeze" the beat pattern in the center screen for photography or other analysis. (*Courtesy of Tektronix.*)

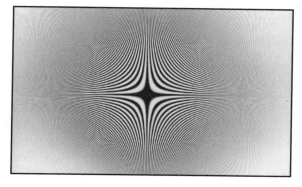

Figure 5.22 A hyperbolic variation of the two-axis zone plate frequency sweep. (*Courtesy of Tektronix.*)

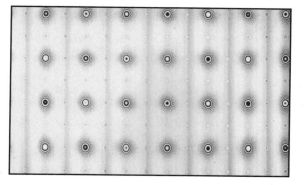

Figure 5.23 A two-axis frequency sweep in which the range of frequencies is swept several times in each axis. Complex patterns such

5.5 References

1. Mertz, P.: "Television-The Scanning Process," *Proc. IRE*, vol. 29, pp. 529–537, October 1941.
2. MacAdam, D. L.: "Visual Sensitivities to Color Differences in Daylight," *J. Opt. Soc. Am.*, vol. 32, pp. 247–274, 1942.

3. Bender, Walter, and Alan Blount: "The Role of Colorimetry and Context in Color Displays," *Human Vision, Visual Processing, and Digital Display III*, Bernice E. Rogowitz ed., Proc. SPIE 1666, SPIE, Bellingham, Wash., pp. 343–348, 1992.
4. Tannas, Lawrence E., Jr.: *Flat Panel Displays and CRTs*," Van Nostrand Reinhold, New York, N.Y., pg. 18, 1985.
5. Standards and Definitions Committee, Society for Information Display, San Jose, Calif.
6. M. Weston: "The Zone Plate: Its Principles and Applications," *EBU Review—Technical*, no. 195, October 1982.

5.6 Bibliography

Allen, E. S.: *Six-Place Tables*, 7th ed., McGraw-Hill, New York, N.Y., 1947.

Baldwin, M. W., Jr.: "Subjective Sharpness of Additive Color Pictures," *Proc. IRE*, vol. 39, pp. 1173–1176, October 1951.

Baldwin, M. W., Jr.: "The Subjective Sharpness of Simulated Television Images," *Proc. IRE*, vol. 28, p. 458, October 1940.

Bartleson, C. J., and E. J. Breneman: "Brightness Reproduction in the Photographic Process," *Photog. Sci. Eng.*, vol. 11, pp. 254–262, 1967.

Benson, K. B.: "Report on Sources of Variability in Color Reproduction as Viewed on the Home Television Receiver," *IEEE Trans. BTR*, vol. 19, pp. 269–275, 1973.

Benson, K. B., and J. C. Whitaker: *Television and Audio Handbook for Engineers and Technicians*, McGraw-Hill, New York, N.Y., 1989.

Bingley, F. J.: "The Application of Projective Geometry to the Theory of Color Mixture," *Proc. IRE*, vol. 36, pp. 709–723, 1948.

Brodeur, R., K. R. Field, and D. H. McRae: "Measurement of Color Rendition in Color Television," in M. Pearson (ed.), *Proc. ISCC Conf. Optimum Reproduction of Color*, Williamsburg, Va., 1971, Graphic Arts Research Center, Rochester, N.Y., 1971.

Castellano, Joseph A.: *Display Systems*, Academic Press, New York, N.Y., 1992.

DeMarsh, L. E.: "Color Rendition in Television," *IEEE Trans. CE*, vol. 23, pp. 149–157, 1977.

DeMarsh, L. E.: "Colorimetric Standards in US Color Television," *J. SMPTE*, SMPTE, White Plains, N.Y., vol. 83, pp. 1–5, 1974.

Dwight, H. B.: *Tables of Integrals and Other Mathematical Data*, 4th ed., Macmillan, New York, N.Y., 1961.

Ekstrand, R.: "A Flesh-Tone Correction Circuit," *IEEE Trans. BTR*, vol. 17, pp. 182–189, 1971.

Epstein, D. W.: "Colorimetric Analysis of RCA Color Television System," *RCA Review*, vol. 14, pp. 227–258, 1953.

Eshbach, O. W.: *Handbook of Engineering Fundamentals*, 2d ed., Wiley, New York, N.Y., 1936.

Fink, D. G.: *Color Television Standards*, McGraw-Hill, New York, N.Y., 1955.

Fink, Donald G., and Donald Christiansen (eds.): *Electronics Engineers' Handbook*, 2d ed., McGraw-Hill, New York, N.Y., 1982.

Fink, Donald G., and H. Wayne Beaty (eds.): *Standard Handbook for Electrical Engineers*, 11th ed., McGraw-Hill, New York, N.Y., 1978.

Fowle, F. E.: *Smithsonian Physical Tables*, 9th ed., Smithsonian Institution, Washington, D.C., 1954.

Green, Marc: "Temporal Sampling Requirements for Stereoscopic Displays, *Human Vision, Visual Processing, and Digital Display III*, Bernice E. Rogowitz ed., Proc. SPIE 1666, SPIE, Bellingham, Wash., pp. 101–111, 1992.

Harwood, L. A.: "A Chrominance Demodulator IC with Dynamic Flesh Correction," *IEEE Trans. CE*, vol. 22, pp. 111–118, 1976.

Henney, Keith, (ed.): *Radio Engineering Handbook*, 5th ed., McGraw-Hill, New York, N.Y., 1959.

Henney, Keith, and Beverly Dudley: *Handbook of Photography*, McGraw-Hill, New York, N.Y., 1939.

Herman, S.: "The Design of Television Color Rendition," *J. SMPTE*, SMPTE, White Plains, N.Y., vol. 84, pp. 267–273, 1975.

Hudson, R. G.: *The Engineer's Manual*, Wiley, New York, N.Y., 1939.

Hunt, R. W. G.: *The Reproduction of Colour*, 3d ed., Fountain Press, England, 1975.

Langford-Smith, F., (ed.): *Radiotron Designer's Handbook*, 4th ed., Radio Corporation of America, Harrison, N.J., 1953.

Mertz, P., and F. Gray: "A Theory of Scanning and Its Relation to the Characteristics of the Transmitted Signal in Telephotography and Television," *Bell System Tech. J.*, vol. 13, pp. 464–515, July 1934.

Naiman, Avi C., and Walter Makous: "Spatial Non-linearities of Grayscale CRT pixels," *Human Vision, Visual Processing, and Digital Display III*, Bernice E. Rogowitz ed., Proc. SPIE 1666, SPIE, Bellingham, Wash., pp. 41–56, 1992.

Neal, C. B.: "Television Colorimetry for Receiver Engineers," *IEEE Trans. BTR*, vol. 19, pp. 149–162, 1973.

Novick, S. B.: "Tone Reproduction from Colour Telecine Systems," *Br. Kin. Sound TV*, vol. 51, pp. 342–347, 1969.

Pearson, M. (ed.): *Proc. ISCC Conf. on Optimum Reproduction of Color*, Williamsburg, Va., 1971, Graphic Arts Research Center, Rochester, N.Y., 1971.

Pritchard, D. H.: "US Color Television Fundamentals—A Review," *IEEE Trans. CE*, vol. 23, pp. 467–478, 1977.

Reinhart, William F.: "Gray-scale Requirements for Anti-aliasing of Stereoscopic Graphic Imagery," *Human Vision, Visual Processing, and Digital Display III*, Bernice E. Rogowitz ed., Proc. SPIE 1666, SPIE, Bellingham, Wash., pp. 90–100, 1992.

Robertson, A. R.: "Colour Differences," *Die Farbe*, vol. 29, pp. 273–296, 1981.

"Setting Chromaticity and Luminance of White for Color Television Monitors Using Shadow Mask Picture Tubes," SMPTE Recommended Practice 71-1977 SMPTE, White Plains, N.Y., 1977.

Smith, P. F., and W. R. Longley: *Mathematical Tables and Formulas*, Wiley, New York, N.Y., 1929.

Sproson, W. N.: *Colour Science in Television and Display Systems*, Adam Hilger, Bristol, England, 1983.

Sullivan, James R., and Lawrence A. Ray: "Secondary Quantization of Gray-Level Images for Minimum Visual Distortion," *Human Vision, Visual Processing, and Digital Display III*, Bernice E. Rogowitz ed., Proc. SPIE 1666, SPIE, Bellingham, Wash., pp. 27–40, 1992.

Terman, F. E.: *Radio Engineers' Handbook*, McGraw-Hill, New York, N.Y., 1943.

Thomas, G. A.: "An Improved Zone Plate Test Signal Generator," *Proceedings, Eleventh International Broadcasting Conference*, Brighton, UK, pp. 358–361, 1986.

Thomas, Woodlief, Jr. (ed.): *SPSE Handbook for Photographic Science and Engineering*, Wiley, New York, N.Y., 1973.

Wentworth, J. W.: *Color Television Engineering*, McGraw-Hill, New York, N.Y., 1955.

Whitaker, Jerry C., and K. Blair Benson (eds): *Standard Handbook of Video and Television Engineering*, McGraw-Hill, New York, N.Y., 1999.

Wintringham, W. T.: "Color Television and Colorimetry," *Proc. IRE*, vol. 39, pp. 1135–1172, 1951.

Wyszecki, G.: "Proposal for a New Color-Difference Formula," *J. Opt. Soc. Am.*, vol. 53, pp. 1318–1319, 1963.

Chapter 6

Cathode Ray Tube Fundamentals

6.1 Introduction

The cathode ray tube (CRT) remains the dominant display technology for a wide range of applications—both consumer and professional. As requirements for greater resolution and color purity have increased, improvements have also been made in the design and manufacture of CRT devices and signal-driving circuits. Within the last 10 years, improvements to the basic monochrome and/or color CRT have been pushed by the explosion of the personal computer industry and the increased resolution demanded by end-users. Display size has also been a key element in CRT development. Consumer interest in large-screen home television has been strong within the last decade, and the acceptance of high-definition digital television will accelerate this trend.

6.1.1 Basic Operating System

The CRT produces visible or ultraviolet radiation by bombardment of a thin layer of phosphor material by an energetic beam of electrons. Nearly all commercial applications involve the use of a sharply-focused electron beam directed time-sequentially toward relevant locations on the phosphor layer by means of externally controlled electrostatic or electromagnetic fields. In addition, the current in the electron beam can be controlled or modulated in response to an externally applied varying electric signal. A generalized CRT consists of the following elements:

- An electron beam-forming system
- Electron-beam deflecting system (electrostatic or electromagnetic)
- Phosphor screen
- Evacuated envelope

Figure 6.1 shows the basic design of a monochrome CRT. The electron beam is formed in the electron gun, where it is modulated and focused. The beam then travels through the *deflection region*, where it is directed toward a specific spot or sequence of spots on the phosphor screen. At the phosphor screen the electron beam gives up some of the energy of the

164 Video Display Engineering

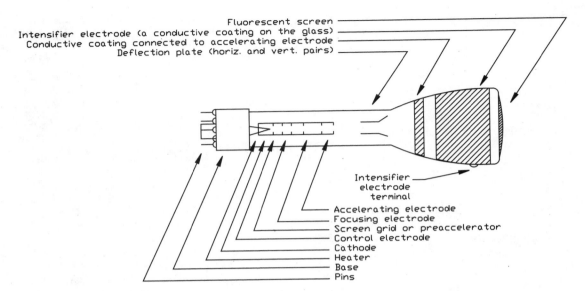

Figure 6.1 A generalized schematic of a cathode ray tube using electrostatic deflection.

electrons in producing light or other radiation, some in generating secondary electrons, and the remainder in producing heat.

6.1.2 Classification of CRT Devices

Tubes may be classified in terms of bulb parameters and screen/gun geometry. Principle categories that separate one class of device from another include:

- **Tube size**. Conventionally, tube size is measured as the screen diagonal dimension in rounded inch or centimeter units. This number is usually included in tube-type numbers.
- **Neck diameter** (OD). The gun, yoke, neck hardware, and socketing are affected by this dimension (typically given in millimeters). Common neck sizes are 36.5, 29, and 22.5 mm.
- **Deflection angle**. This parameter is calculated from the rated full-screen diagonal and glassware drawings, using the yoke reference plane as an assumed center of deflection. Angles in common use include 90°, 100°, and 110°. Higher deflection angles enable shorter tubes but entail other tradeoffs.
- Other characteristics, including gun type (delta or in-line) and screen structure (stripes or dots) for color CRTs.

6.1.3 The CRT Envelope

The cathode ray tube envelope consists of the faceplate, bulb, funnel, neck, base press, base, faceplate safety panels, shielding, and potting. (Not all CRTs incorporate each of these components.) The faceplate is the most critical component of the envelope because the display on the phosphor must be viewed through it. Most faceplates are pressed in molds from molten glass and trimmed and annealed before further processing. Some specialized CRTs for photographic recording or flying-spot scanning use optical-quality glass faceplates sealed to the bulb section in such a way as to produce minimum distortion.

To minimize return scattering of ambient light from the white phosphor, many CRT types use a neutral-gray-tinted faceplate. While the display information is attenuated as it makes a single pass through this glass, ambient light will be attenuated both going in and coming out, thus squaring the attenuation ratio and increasing contrast.

Certain specialized CRTs have faceplates made wholly or partially of fiber optics, which may have extraordinary characteristics, such as high ultraviolet transmission. A fiber-optic region in the faceplate permits direct-contact exposure of photographic or other sensitive film without the necessity for external lenses or space for optical projection.

The bulb section of the CRT is the transition element necessary to enclose the full deflection volume of the electron beam between the deflection region and the phosphor screen on the faceplate. In most CRTs, this is a roughly cone-shaped molded-glass component, as illustrated in previous figure. Instead of an ordinary glass bulb, many large CRTs include metal cone sections made of a glass-sealing iron alloy. The metal cones are generally lighter in weight than the corresponding glass sections.

The junction region of the bulb of a CRT with the neck section is critical to the geometry of the device. Tubes made with these separate sections are intended for electromagnetic deflection, and this region is where the extended deflection yoke is located.

The neck diameter of a CRT depends to a great extent upon the type of deflection used and the intended application of the CRT. In general, electrostatic deflection CRTs have large neck diameters, while the electromagnetic-deflection devices have small diameters.

6.1.4 Arc Protection

Because of the high operating voltages applied to a CRT, internal arcing is possible. Arcs or flashovers can cause tube damage and/or circuit component failures. Corrective or protective methods include:

- Internal cleanliness. Manufacturing processes are designed to prevent sharp edges or points on electrodes.

- Subjecting the finished tube to a programmed high-voltage conditioning process called *spot knocking*.

- Inclusion of spark gaps on the tube base.

- Inclusion of an internal resistive coating between the bulb anode button and the gun anode.

- Discrete internal resistors incorporated into specific electrodes.

- Use of an arc shield in the gun assembly.

Table 6.1 lists typical monochrome CRT electrode potentials.

Table 6.1 Monochrome CRT Electrode Potentials

Element	Potential to Cathode	Potential to Ground
Cathode	0	−15 kV
Control grid	−100 to +2 V	−15.1 to −14.998 kV
First anode	+400 V	−14.6 kV
Focus electrode	0 to +400 V	−14.6 to −15 kV
Second anode	+15 kV	0

6.1.5 Implosion Protection

Atmospheric pressure imposes severe pressures on the evacuated CRT bulb. The tube face is further exposed to the possibility of an accidental blow. To protect the viewer from possible harm, various implosion protection systems have been designed and incorporated into CRT devices:

- Laminated safety panel. A glass safety panel is formed and bonded to the tube faceplate with a transparent resin.

- Kimcode (*Kimball controlled devacuation*). A rim band of two pieces of stamped sheet metal is affixed around the skirt of the faceplate and secured by a tension band.

- Tension band. A metal tension band is installed around the skirt of the faceplate and secured by a clip (or welds).

- Shelbond. A resin-filled steel shell is affixed around the skirt of the faceplate.

- Prewelded tension band. Tension bands are expanded by heating to enable installation around the panel skirt periphery.

Mounting ears may be included in all the above systems except the laminated safety panel. With these implosion systems, breakage of the faceplate causes the tube to devacuate without a violent implosion.

The implosion safety front panel may be etched on its exterior surface, producing a ground-glass effect to avoid specular reflections from ambient light. The panel may also provide for neutral-gray attenuation.

6.2 Phosphor and Screen Characteristics

Many materials, naturally occurring and synthetic, organic and inorganic, have the ability to give off light. For video applications, the materials of interest are crystalline inorganic solids that are stable under cathode-ray tube fabricating and operating conditions. These materials are generally powders having average particle sizes in the range of 5 to 15 μm. Figure 6.2 shows some typical particle size distributions. Because of defects and irregularities in the crystal lattice structure, these materials have the ability to absorb incident energy and convert this energy into visible light. This process involves the transfer of energy from the electron beam to electrons in the phosphor crystal. The phosphor electrons are

Cathode Ray Tube Fundamentals   167

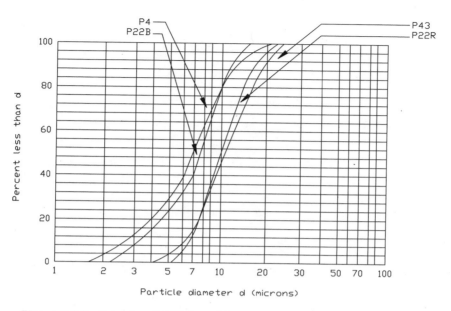

Figure 6.2 Particle size distribution of four representative phosphors.

thereby excited or raised to levels higher than the ground state. Light is emitted when the electrons return to more stable states. Figure 6.3 illustrates these changes in energy levels.

Phosphors are composed of a host crystal that comprises the bulk of the material and one or more *activators*, which may be present in amounts from parts per million to a few mole percent. Either the host or activator can determine the luminescent properties of a phosphor system. For example, in the zinc sulfide/cadmium sulfide:silver (ZnS:Ag/CdS:Ag) system, the emitted color ranges from blue at zero cadmium through green, to yellow, and into red as the cadmium content is increased.

In the commercial preparation of phosphors, the highly purified host and the required amount of activator are intimately mixed, normally with a flux, such as an alkali or alkaline earth halide or phosphate, which supplies a low-temperature melting phase. The flux controls the particle development and aids in the diffusion of the activator into the lattice. This mixture is then fired at high temperature, 1472 to 2192°F (800 to 1200°C), on a prescribed schedule, in order to develop the desired physical and luminescent properties. In some cases the firing is carried on under specific controlled atmospheres. After firing, the resultant cake is broken up, residual soluble materials are removed by washing, and any required coatings are applied.

6.2.1 Phosphor Screen Types

In a CRT display device, a phosphor screen is used to convert electron energy into radiant energy. The screen is composed of a thin layer of luminescent crystals—phosphors—that emit light when bombarded by electrons. This property is referred to as *cathodoluminescence*. It occurs when the energy of the electron beam is transferred to

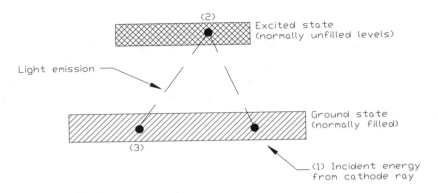

Figure 6.3 Energy transitions of electrons leading to luminescence.

electrons in the phosphor crystal, as explained in the previous section. The property of light emission during excitation is termed *fluorescence*. The property of emission immediately after excitation is removed is termed *phosphorescence*.

Standardization of phosphor types has been accomplished by registration of the various phosphors with the *Joint Electron Device Engineering Council* (JEDEC) of the Electronic Industries Association (EIA). Registered phosphors are designated by a number series known as *P numbers*, as shown in Table 6.2

Fabrication

Phosphor screens are usually fabricated by depositing small crystal grains on a glass faceplate to form a thin phosphor layer. The most common method for a monochrome device is *sedimentation*, also known as *phosphor settling*. This method allows the phosphor crystals to settle from a liquid suspension under the influence of gravity.

Electrophoretic screening is another method of depositing a phosphor screen. This process makes use of the effect of particle motion under the influence of an electric field. A phosphor powder is first fractionated to select the desired particle-size range. The phosphor is then placed in suspension in an electrolytic solution. Under suitable conditions of electrolyte, phosphor suspension, electrode geometry, and electrical uniformity of the substrate, the resulting phosphor deposit is uniform and compact. The electrophoretic screening process provides for higher resolution devices than those produced using phosphor settling.

The *monolayer screening* method consists of forming an ultrathin thermoplastic film on a glass faceplate. Phosphor particles are applied on the firm surface using a high-velocity air-brush while the thermoplastic is heated to near its melting point. The thickness of the thermoplastic determines the density of the phosphor layer. To form a monolayer screen, the film thickness must be approximately equal to the average grain size of the phosphor powder. This technique provides a means of controlling screen thickness with good reproducibility.

For certain applications where optimum image resolution is required, transparent phosphor screens prepared by *vacuum evaporation* may be employed. Activators may be

Table 6.2 Characteristics of Standardized Phosphor Types (*After* [2].)

EIA No.	Common Designation	Typical Use	Composition	Relative Efficiency	Typical CIE Coordinates				Decay[1]
					x	y	u'	v'	
P-4	WW	Black and white television	ZnS:Ag + $Zn_{(1-x)}Cd_xS$:Cu,Al	100	0.270	0.300	0.178	0.446	M/S
P-1	GJ	Projection green	$ZnSiO_4$:Mn	130	0.218	0.712	0.079	0.557	M
P-43	GY	Projection green	Gd_2O_2S:Tb	155	0.333	0.556	0.148	0.556	M
P-22R	X	Red direct view	Y_2O_3:Eu Y_2O_2S:Eu	65	0.640 0.625	0.340 0.340	0.441 0.429	0.528 0.525	M/S
P-22G	X	Green direct view	$Zn_{(1-x)}Cd_xS$:Cu,Al ZnS:Cu,Al	180	0.340 0.285	0.595 0.600	0.144 0.119	0.566 0.561	M
P-22B	X	Blue direct view	ZnS:Ag	25	0.150	0.065	0.172	0.168	M/S

[1] S = short, M = medium
Note that values given here are nominal and may change with measurement methods and source of phosphor.

coevaporated or diffused into the host material subsequent to the deposition. This is typically accomplished by heat treatment following the deposition. Thin-film phosphor screens are also fabricated by the *vapor-reaction* method, in which the phosphors are converted to a vapor state during processing and react with the gaseous atmosphere to provide film growth of the desired phosphor material.

Coating the phosphor-screen surface with a thin (100 to 300 nm) metallic film (usually aluminum) prevents charging of the phosphor and permits accurate control of the primary electron energy. The reflecting aluminum surface redirects the backward emission of light from the phosphor toward the observer and almost doubles the effective light output of the phosphor screen. Figure 6.4 illustrates the improvement. Another important function of *aluminizing* the screen is protection of the phosphor screen from ion-bombardment damage.

6.2.2 Luminescent Properties

Phosphors are commercially available with cathodoluminescent emission over the entire visible band, including ultraviolet and near-infrared. Table 6.3 lists the important characteristics of the most common phosphors used in video display applications. Absolute phosphor efficiency is measured as the ratio of total absolute energy emitted to the total excitation energy applied. When evaluating or comparing picture tubes, it is more meaningful to measure luminescence in footlamberts using a system whose responses match the eye. In addition to using a suitable detector, a number of other parameters must be controlled if meaningful measurements are to be obtained. Most important of these is the total energy at the screen, determined by:

- Anode voltage
- Cathode current
- Raster size

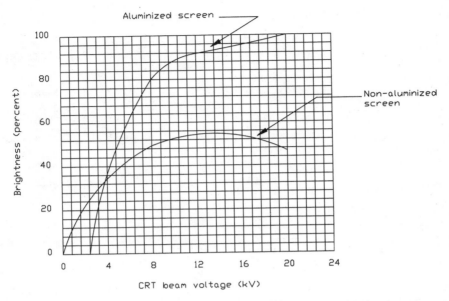

Figure 6.4 Light output of aluminized and non-aluminized phosphor screens. (*After* [1].)

Under normal CRT operating conditions, the luminescence of phosphors is essentially proportional to the beam current applied. However, when high beam currents are employed, some phosphors saturate. The same effect can be observed in a direct-view color video display in areas of highlight brightness, and with electron guns having small spot size. If linearity of the three primary phosphors is not closely matched, noticeable shift in white field color can result in highlight areas. Figure 6.5 shows some typical light output/beam current curves.

Thermal quenching is another mechanism that may result in a loss of phosphor efficiency. In most phosphors the energy transitions become less efficient as screen temperature increases. This phenomenon can be quite pronounced in projection systems where high power loading can be responsible for a large increase in screen temperature. Figure 6.6 shows some examples of thermal quenching. Quenching is a transient condition, and the screen returns to normal efficiency after being cooled.

Screen Burn and Aging

Aging is a nonreversible loss in phosphor efficiency caused by permanent damage to the crystal lattice. The susceptibility to burning varies among phosphor types, with higher-melting materials such as silicates generally more resistant to burning than materials with lower heats of formation, such as fluorides. Within given types, efficiency loss is proportional to the product of the beam current and the time applied. The term *coulomb aging* is often applied to this phenomenon. Screen burn is a discoloration or change in body color and can be visible on both excited and unexcited screens.

Table 6.3 Typical Characteristics of Common Phosphors

Type	Color* Fluorescent	Color* Phosphorescent	Persistence**	Intended Use
P1	YG	YG	M	Oscillography, radar
P2	YG	YG	M	Oscillography
P3	YO	YO	M	
P4	W	W	MS	Direct view television
P5	B	B	MS	Photographic
P6	W	W	S	
P7	B	Y	MS(B), L(Y)	Radar
P8	Replaced by P7			
P9	Not registered			
P10	Dark trace screen		VL	Radar
P11	B	B	MS	Photographic
P12	O	O	L	Radar
P13	RO	RO	M	Radar
P14	B	YO	MS(B), M(YO)	Radar
P15	UV	G	UV(VS), G(S)	Flying-spot scanner
P16	UV	UV	VS	Flying-spot scanner, photographic
P17	B	Y	S(B), L(Y)	Oscillography, radar
P18	W	W	M-MS	Projection television
P19	O	O	L	Radar
P20	YG	YG	M-MS	Storage tube
P21	RO	RO	M	Radar
P22	W(R,G,B)	W(R,G,B)	MS	Tricolor video (television)
P23	W	W	MS	Direct-view television
P24	G	G	S	Flying-spot scanner
P25	O	O	M	Radar
P26	O	O	VL	Radar
P27	RO	RO	M	Color TV monitor
P28	YG	YG	L	Radar
P29	P2 and P25 stripes			Radar, indicators
P30	Cancelled			
P31	G	G	MS	Oscillography, bright video
P32	PB	YG	L	Radar
P33	O	O	VL	Radar
P34	BG	YG	VL	Radar, oscillography
P35	G	B	MS	Oscillography
P36	YG	YG	VS	Flying-spot scanner
P37	B	B	VS	Flying-spot scanner, photographic
P38	O	O	VL	Radar
P39	YG	YG	L	Radar
P40	B	YG	MS(B), L(YG)	Low repetition rate (P12 and P16)
P41	UV	O	VS(UV), L(O)	Radar with light trigger

*Color: B = blue, P = purple, G = green, O = orange, Y = yellow, R = red, W = white, UV = ultraviolet
**Persistence to 10% decay level: VS = less than 1 ms, S = 1 ms to 10 ms, MS = 10 ms to 1 ms, M = 1 ms to 100 ms, L = 100 ms to 1 s

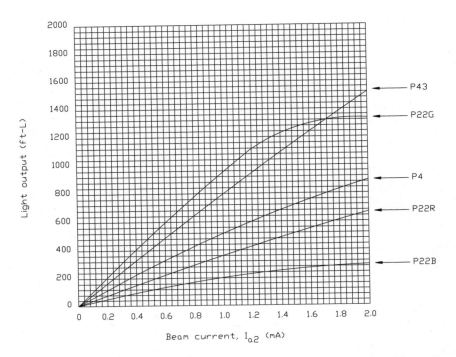

Figure 6.5 Linearity characteristics of five common phosphors. Conditions: E_{a2} = 25 kV. (1 ft·L = 3,426 cd/m²).

Spectral Emission

The visible effectiveness of a phosphor is measured by comparing its spectral-emission curve with a standard visibility function (eye-response curve). This efficiency is usually stated in lumens per radiated watt and is referred to as the *lumen equivalent*. It can be calculated using the eye response per wavelength interval, v, and the phosphor output per wavelength interval $P\lambda$ [3]:

$$\text{Lumen equivalent} = 680 \int_0^\infty vP\lambda \, d\lambda$$

The constant, 680, is the lumen content for 1 W of radiation at 555 nm, the wavelength of maximum eye response.

Chromaticity

The fundamental method of measuring color is to disperse radiant energy into its component parts using a prism or grating device such as a spectroradiometer. The resulting *spectral energy distribution* (SED) curve is a plot of relative energy as a function of wave-

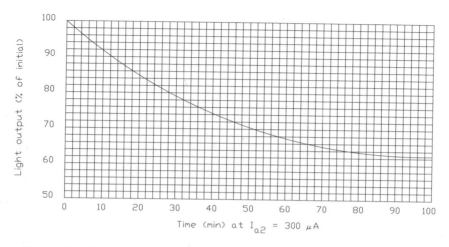

Figure 6.6 The loss in phosphor efficiency as the screen heats at high current operation (rare-earth green).

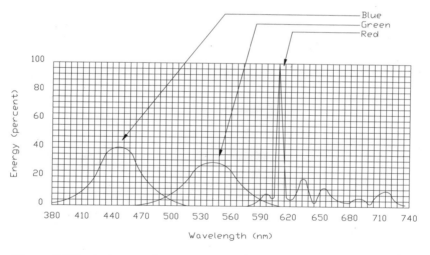

Figure 6.7 Typical spectral energy distribution of color primaries at equal current density.

length. Figure 6.7 shows a typical set of curves for a color CRT at equal current density for the primary phosphors. The figure also illustrates the two basic types of phosphor emission curves. The green and blue are *band emitters*, while the red is a *line emitter*.

The SED curve completely identifies the color of a light source, and colors can be matched or reproduced by exactly matching their SED curves. In practice, this is not an easy task, and materials having very different SED curves can be perceived as the same colors.

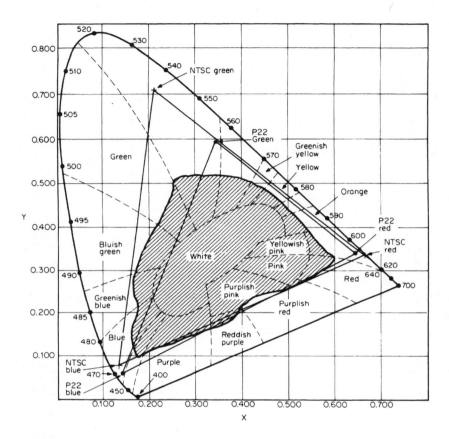

Figure 6.8 The Kelly chart of color designations for lights showing NTSC and current commercial primary phosphors and locus of dyes, paints, and pigments. (*After* [4].)

Under the CIE system, all colors can be designated in terms of two coordinates. (See Chapter 3.) Figure 6.8 shows such a representation. The curved perimeter of this chart designates the locus of pure color points. Within the chart, if any two points are picked as primary colors, all colors lying on a line connecting those points can be achieved by blending suitable amounts of those primaries. Furthermore, if any three primaries are chosen, all colors lying within the resulting triangle can be reproduced by suitable blends of the primaries. This blend may be the actual physical mixing of two phosphors—as in P4—or may be done electrically—as in a color picture tube.

Persistence

The luminescence of all phosphors can be divided into two parts:

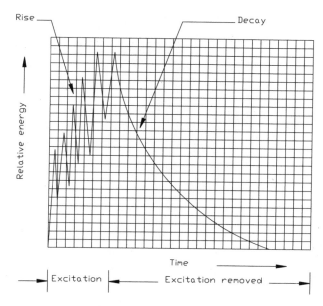

Figure 6.9 Typical phosphor rise time decay curve.

- *Fluorescence*, in which the emission continues only so long as the excitation is continued.
- *Phosphorescence*, in which the emission continues after the excitation is removed.

Persistence is a measure of how much phosphorescence a phosphor exhibits. It is generally defined as the time required, after excitation is removed, for the light output to decrease to a specified level. Figure 6.9 shows a typical buildup and decay curve. The shape of the persistence curve for most phosphors is dependent on the excitation conditions and the method used for the measurement. This method should relate to the final application of the display device. For special uses, persistence values for decay to 10 percent of the initial brightness may range from nanoseconds to several seconds. For television, typical values range from 20 μs to a few milliseconds. The main criterion is that the decay is fast enough so there is no objectionable smear of moving objects resulting from *afterglow*.

6.2.3 Screen Characteristics

Color picture tube *white field brightness* (WFB) has increased more than fivefold over the last three decades. While some of this increase is the result of improvements in phosphor efficiency, especially the introduction of the rare earth phosphors, many other factors have contributed to the improvement, including:

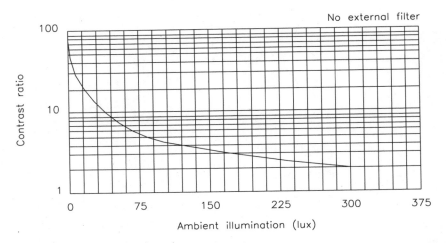

Figure 6.10 Contrast ratio as a function of ambient illumination for a given CRT brightness. (*After* [5].)

- Changes in chromaticity and color balance among the primary phosphors. The primaries currently used do not correspond to the original NTSC primaries, but the marginal loss in color rendition is more than offset by the significant increase in WFB.
- CRT screens are now metal backed or *aluminized* (as discussed previously in this chapter). This thin, 100 to 300 nm film serves a number of purposes. Acting as a mirror, it reflects light from the phosphor that would have otherwise been lost to the interior of the tube. The film also provides an electrically conductive path that eliminates screen charging and removes the need for phosphors to have good secondary emission characteristics. In addition, the aluminum film serves to protect the phosphor screen from ion bombardment, which would result in screen burn and loss in screen efficiency.
- Improvements in electron gun design and tube design, which have permitted tubes to be operated at higher anode-to-cathode voltages and currents.
- The *black matrix process* (see next section).

Contrast

Contrast ratio is normally defined as the ratio of the excited screen brightness to the level of brightness in the unexcited screen area. Unexcited screen brightness is composed of several factors. Because all common phosphors tend to be white-bodied or light in body color, they reflect ambient light efficiently. Reflected ambient light is a major component of the unexcited screen brightness. Figure 6.10 provides some measure of the decrease in contrast ratio with increasing ambient illumination. In addition, scattered electrons excite the "dark" areas, and the crystal nature of the phosphors tends to scatter light from the excited area into the nominally dark area.

Historically, contrast has been enhanced by placing neutral density filters between the phosphor screen and the viewer. Color picture tube faceplates are commonly available in transmissions ranging from about 50 to 90 percent. Light emitted from the phosphor screen passes through this faceplate once, while ambient light is attenuated twice, so that contrast is significantly improved. In some tube designs, further improvement is made by matching the absorptivity of the glass more closely to the emission of the phosphor.

While the use of neutral-density filters has improved contrast at the expense of tube brightness, the development of the *black matrix/black surround* process for color picture tubes has produced an increase in both light output and contrast at the same time. In this process, a layer of light-absorbing graphite is deposited so as to surround the phosphor dots or stripes. By reducing screen reflectivity, contrast is significantly improved. In the first version of this process, the so-called *beam-limited* or *positive guard band* process, a relatively small amount of graphite was applied to the screen, and the exposed phosphor area was larger than that defined by the aperture mask holes. An improved version, *window limited* or *negative guard band*, employs matrix windows—or phosphor areas—smaller than the electron beam. The result is that more graphite remains on the faceplate and contrast is further improved. This increase in screen contrast has permitted a significant reduction in glass tint, which provides a substantial increase in tube brightness.

Further increase in contrast has been achieved through the technique of encapsulating the phosphor particles within a shell of pigment. These pigments have normally been applied to the red and blue phosphors. The pigments do not significantly absorb the emitted light of the phosphor, while absorbing other frequencies of ambient light. Because it is difficult to find pigments that exactly match the phosphor's emission, there is usually some loss of brightness when improving the contrast.

In addition to the loss of picture quality resulting from reflections from the phosphor screen surface, picture viewability can be further degraded by reflections from the front surface of the tube face. Specular reflections of well-defined light sources are especially objectionable. They may be controlled in several ways. Converting the faceplate to a diffusing surface through grinding or etching reduces these specular reflections. Diffusing coatings may also be applied. Coatings that reduce specular reflections by means of optical interference are also available.

Modulation Transfer Function

The resolution and brightness output of a phosphor screen are two important parameters that influence the final image quality of the display. Resolution is assessed by various methods, one of which is the *modulation transfer function* (MTF). The MFT is defined as the ratio of the modulation at each spatial frequency present in the final image to the modulation of the same frequency in the object scaled to the final image size. The MTF indicates the ability of the phosphor screen to produce image detail. Figure 6.11 shows typical MTF performance for high-resolution screens prepared using the various methods previously discussed. The corresponding brightness output characteristics are shown in Figure 6.12.

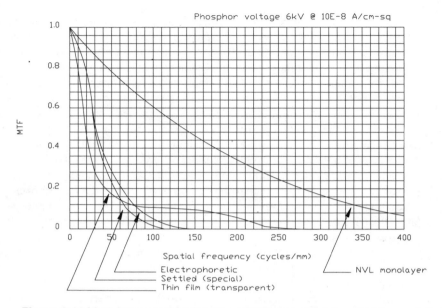

Figure 6.11 Modulation transfer function (MTF) characteristics of high-resolution phosphor screens prepared by various methods. (*After* [6].)

6.3 Electron Gun

The electron gun is basic to the structure and operation of any cathode-ray device, specifically display devices. In its simplest schematic form, an electron gun may be represented by the diagram in Figure 6.13, which shows a triode gun in cross section. Electrons are emitted by the cathode, which is heated by the filament to a temperature sufficiently high to release the electrons. Because this stream of electrons emerges from the cathode as a cloud rather than a beam, it is necessary to accelerate, focus, deflect, and otherwise control the electron emission so that it becomes a beam, and can be made to strike a phosphor at the proper location, and with the desired beam cross section.

6.3.1 Electron Motion

The laws of motion for an electron in a uniform electrostatic field are obtained from Newton's second law. The velocity of an emitted electron is given by the following:

$$v = \left\{ \frac{2eV}{m} \right\}^{1/2}$$

Where:
$e = 1.6 \times 10^{-19} C$
$m = 9.1 \times 10^{-28} g$

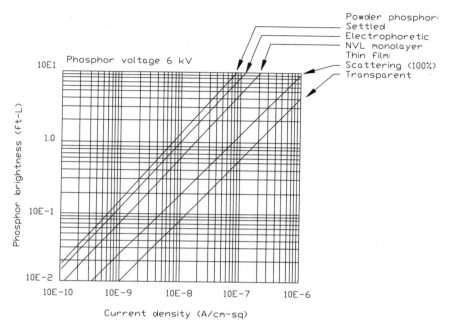

Figure 6.12 The brightness output of high-resolution phosphor screens prepared by various methods. Brightness is expressed as a function of current density at constant phosphor voltage. (*After* [6].)

$V = -Ex$, the potential through which the electron has fallen

When practical units are substituted for the values in the previous equation, the following results:

$$v = 5.93 \times 10^5 \, V^{\frac{1}{2}} \text{ m/s}$$

This expression represents the velocity of the electron. If the electron velocity is at the angle θ to the potential gradient in a uniform field, the motion of the electron is specified by the equation:

$$y = \frac{Ex^2}{4V_0 \sin^2 \theta} + \frac{x}{\tan \theta}$$

Where:
V_0 = the electron potential at initial velocity

This equation defines a parabola. The *electron trajectory* is illustrated in Figure 6.14, in which the following conditions apply:

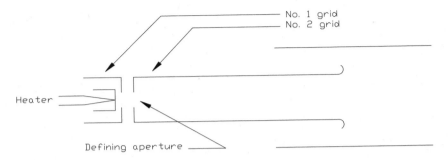

Figure 6.13 Triode electron gun structure.

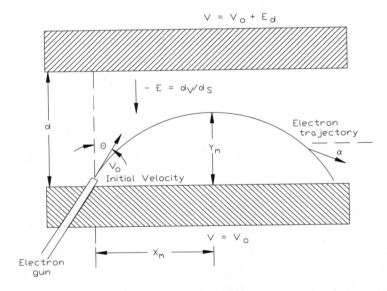

Figure 6.14 The electron trajectory from an electron gun using the parameters specified in the text. (*After* [7].)

- y_m = maximum height
- x_m = x displacement at the maximum height
- α = the slope of the curve

6.3.2 Tetrode Gun

The tetrode electron gun includes a fourth electrode, illustrated in Figure 6.15. The main advantage of the additional electrode is improved convergence of the emitted beam.

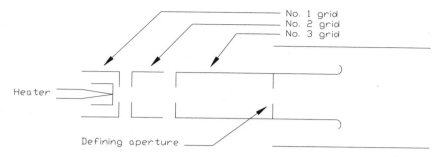

Figure 6.15 Basic structure of the tetrode electron gun.

6.3.3 Operating Principles

Nearly all currently available CRT electron guns have indirectly heated cathodes in the form of a small capped nickel sleeve or cylinder with an insulated coiled tungsten heater inserted from the back end. Most heaters operate at 6.3 Vac at a current of 300 to 600 mA. Low-power heaters are also available that operated at 1.5 V (typically 140 mA).

The cathode assembly is mounted on the axis of the modulating or *control grid* cylinder (or simply *grid*), which is a metal cup of low-permeability or stainless steel about 0.5-in in diameter and 0.375- to 0.5-in long. A small aperture is punched or drilled in the cap. The grid (G_1) voltage is usually negative with respect to the cathode (K).

To obtain electron current from the cathode through the grid aperture, there must be another electrode beyond the aperture at a positive potential great enough for its electrostatic field to penetrate the aperture to the cathode surface. Figure 6.16 illustrates a typical *accelerating electrode* (G_2) in relation to the cathode structure. The accelerating electrode may be implemented in any given device in a number of ways. In a simple accelerating lens in which successive electrodes have progressively higher voltages, the electrode may also be used for focusing the electron beam upon the phosphor, hence, it may be designated the *focusing* (or *first*) anode (A_1). This element is usually a cylinder, longer than its diameter and probably containing one or more disk apertures.

6.4 Electron Beam Focusing

General principles involved in focusing the electron beam are best understood by initially examining optical lenses and then establishing the parallelism between them and electrical focusing techniques.

6.4.1 Electrostatic Lens

Figure 6.17 shows a simplified diagram of an electrostatic lens. An electron emitted at zero velocity enters the V_1 region moving at a constant velocity (because the region has a constant potential). The velocity of the electron in that region is defined by the previously given equation:

182 Video Display Engineering

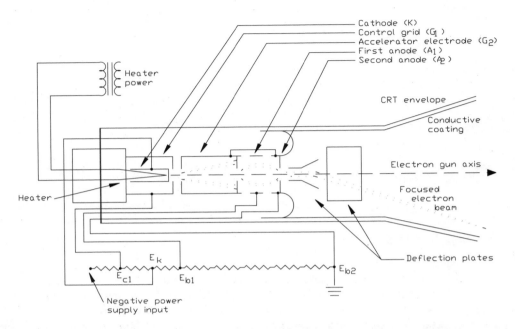

Figure 6.16 Generalized schematic of a CRT grid structure and accelerating electrode in a device using electrostatic deflection.

$$v = 5.92 \times 10^5 \, V^{1/2} \text{ m/s}$$

The velocity of the electron in the region is defined by the equation for the straight-line component, with V_1 replacing V. After passing through the surface into the V_2 region, the velocity changes to a new value determined by V_2. The only component of the velocity that is changed is the one normal to the surface, so that the following conditions are true:

$$v_t = v_1 \sin I_1$$

$$v_1 \sin I_1 = v_2 \sin I_2$$

Snell's law, also known as the law of refraction, has the form:

$$N_1 \sin I_1 = N_2 \sin I_2$$

Where:
N_1 = the index of refraction for the first medium
N_2 = the index of refraction for the second medium
I_1 = the angle of the incident ray with the surface normal
I_2 = the angle of the refracted ray with the surface normal

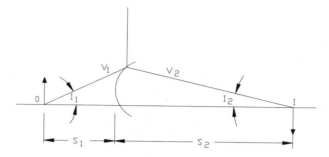

Figure 6.17 The basic principles of electron optics. (*After* [8].)

The parallelism between the optical and the electrostatic lens is apparent if appropriate substitutions are made:

$$V_1 \sin I_1 = V_2 \sin I_2$$

$$\frac{\sin I_1}{\sin I_2} = \frac{V_2}{V_1}$$

For Snell's law, the following applies:

$$\frac{\sin I_1}{\sin I_2} = \frac{N_2}{N_1}$$

Thus, the analogy between the optical lens and the electrostatic lens is apparent. The magnification m of the electrostatic lens is given by the following:

$$m = \frac{\left\{V_1/V_2\right\}^{1/2} S_2}{S_1}$$

(Symbols defined in Figure 6.17)

The condition of a thin, *unipotential lens*, where V_1 is equal to V_2, is illustrated in Figure 6.18. The following applies:

$$m = \frac{h_2}{h_1} = \frac{f_2}{X_1} = -\frac{X_2}{f_2}$$

The shape of the electron beam under the foregoing conditions is shown in Figure 6.19. If the potential at the screen is the same as the potential at the anode, the crossover is imaged at the screen with the magnification given by:

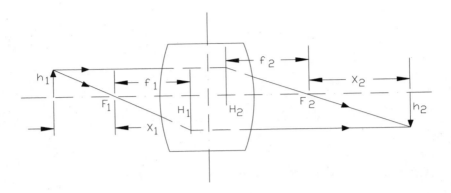

Figure 6.18 A unipotential lens. (*After* [8].)

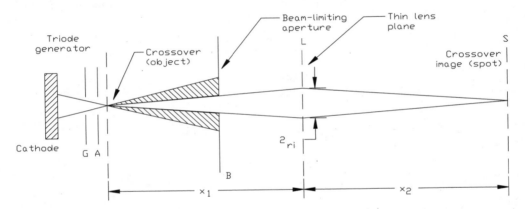

Figure 6.19 Electron beam shape. (*After* [9].)

$$m = \frac{x_2}{x_1}$$

The magnification can be controlled by changing this ratio, which, in turn, changes the size of the spot. This is one way to control the quality of the focus. Although the actual lens may not be "thin," and, in general, is more complicated than what is shown in the figure, the illustration is sufficient for understanding the operation of electrostatic focus.

The size of the spot can be controlled by changing the ratio of V_1 to V_2 or of x_2 to x_1 in the previous equations. Because x_1 and x_2 are established by the design of the CRT, the voltage ratio is the parameter available to the circuit designer to control the size or focus of the spot. It is by this means that focusing is achieved for CRTs using electrostatic control.

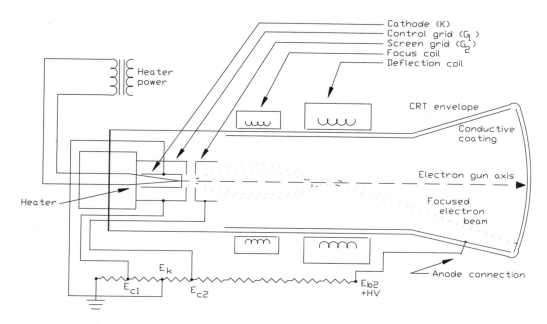

Figure 6.20 Generalized schematic of a CRT gun structure using electromagnetic focus and deflection. (*After* [10].)

6.4.2 Practical Applications

Figure 6.20 illustrates a common type of gun and focusing system that employs a screen grid (G_2), usually in the form of a short cup with an aperture facing the grid aperture. The voltage is usually maintained between 200 and 400 V positive. In an electrostatically focused electron gun, the screen grid is usually followed by the focusing anode. In a magnetically focused electron gun, the screen grid may be followed directly by the final anode.

In another type of electrostatically focused electron gun (illustrated in Figure 6.21) the grid is followed immediately by a long-apertured cylinder at the voltage of the principal anode (A_2). This element, called the *accelerator* or *preaccelerator*, is followed, in sequence, by either two short cylindrical electrodes or apertured disks. The last electrode and preaccelerator are connected within the tube. This set of three electrodes constitutes an *Einzel lens*. By proper design of the lens, the focal condition can be made to occur when the voltage on the central element is zero or a small positive voltage with reference to the cathode.

6.4.3 Electrostatic Lens Aberrations

There are five common types of aberration associated with electrostatic lenses for monochrome displays and an additional aberration associated with color displays:

- Astigmatism

186 Video Display Engineering

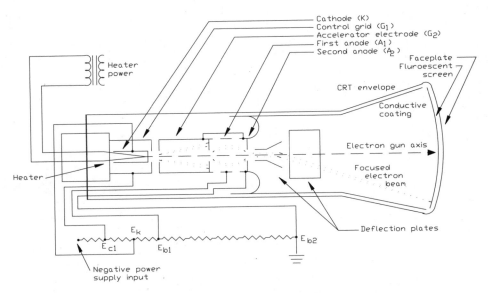

Figure 6.21 Generalized schematic of a CRT with electrostatic focus and deflection. An *Einzel* focusing lens is depicted. (*After* [10].)

- Coma
- Curvature of field
- Distortion of field
- Spherical aberration
- Chromatic aberration (for color systems)

Chromatic aberration, illustrated in Figure 6.22, is analogous to the effect in geometrical optics resulting in light of different wavelengths having different focal lengths. In an electrostatic lens, electrons with different velocities will have different focal points. However, because electron velocity is different only insofar as the electrons leave the cathode with different emission velocities, the effect is generally small at the accelerating potentials that are used, and the error is usually not significant.

Coma applies to images and objects not on the axis of the lens system. Figure 6.23 illustrates circles imaged in a distorted form. Coma may be reduced by using less of the lens center, but this reduces the amount of beam current and, therefore, available luminance; it may not be a desirable approach.

Astigmatism results from objects positioned off the axis lines toward the axis having different focal lengths than lines that are perpendicular to them. This effect is well known in geometrical optics and is shown in Figure 6.24. From this figure it can be seen that compromises must be made when focusing the entire image. Changing the focusing voltage changes the portion of the image that is sharply focused, while the rest of the image may be blurred.

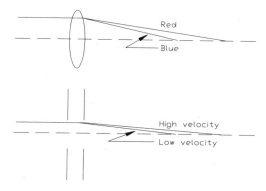

Figure 6.22 Illustration of chromatic aberration. (*After* [7].)

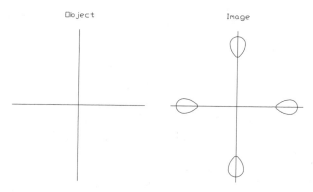

Figure 6.23 Illustration of coma. (*After* [7].)

Curvature of field is usually associated with astigmatism but is a more noticeable effect, resulting from the image lying on a curved surface for an object that is in a plane at right angles to the axis. This results in concentric circles that can be adjusted in the image plane so only one radial distance is sharp. Thus, if the center is focused, the outside will be unfocused, or the opposite, if the outer circle is focused.

Distortion of field results from variations in linear magnification with the radial distance. These are the well-known *pincushion* and *barrel* distortions; the former results from an increase in magnification and the latter from a decrease. The distortions are illustrated in Figure 6.25.

Spherical aberration is a distortion where parallel rays entering the lens system have different focal lengths depending on the radial distance of the ray from the center of the lens. This effect is shown in Figure 6.26. It is perhaps the most serious of all the aberrations. It can be seen from the figure that the focal length becomes smaller as the distance increases. This is known as *positive spherical aberration* and is always found when electron lenses are used. The focal length increases slowly at first, and then more rapidly as the radial distance in-

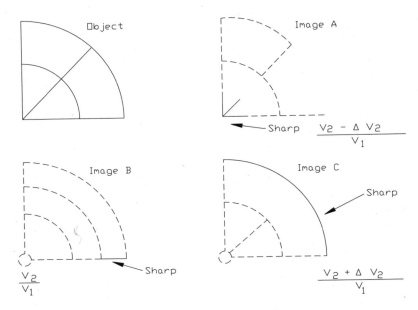

Figure 6.24 Illustration of astigmatism. (*After* [7].)

Figure 6.25 Pincushion and barrel distortion. (*After* [7].)

creases. This type of aberration is always positive in electron lenses, and cannot be eliminated by adding a lens with equal *negative spherical aberration*, as is the case with optical systems. However, it is possible to reduce the effect somewhat by using dual-cylinder lenses with a high-potential inner cylinder and a lower-potential outer cylinder.

6.4.4 Magnetic Focusing

It is well known that electron beams can be focused with magnetic fields as well as with electrostatic fields, but the analogy with optical systems is not as apparent. When elec-

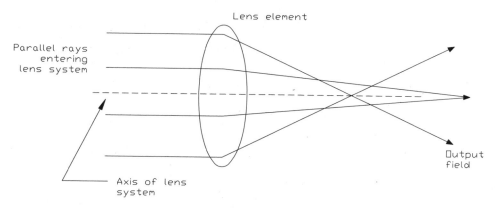

Figure 6.26 Spherical aberration. (*After* [11].)

trons leave a point on a source with the principal component of the velocity parallel to the axis of a long magnetic field, they travel in helical paths and come to a focus at a further point along the axis. The helical paths have essentially the same pitch. Supporting equations show that the pitch is relatively insensitive to θ for small angles, and the electrons will return to their original relative positions at some distance P on the magnetic path that is parallel to the electron beam. Thus, spreading of the beam is avoided, but there is no reduction in the initial beam diameter. Focusing is achieved by changing the current in the focusing coil until the best spot size is achieved. The manner in which the focus coil is placed around the neck of the CRT is shown in Figure 6.27. The focal length of such a coil is given by the following equation [12]:

$$f = \frac{4.86 V_d}{N^2 I^2}$$

Where:
f = focal length
d = diameter of wire loop
NI = current in ampere-turns
V = potential of region

This equation can be used for a short coil with a mean diameter of d that has N turns and a current equal to I.

The image rotation θ is expressed by:

$$\theta = \frac{0.19\, NI}{V^{1/2}}$$

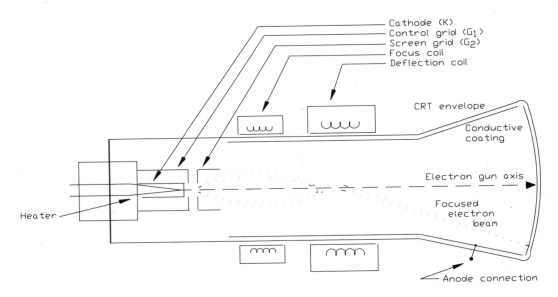

Figure 6.27 Magnetic-focusing elements on the neck of a CRT.

6.4.5 Practical Applications

Magnetic focusing is rarely used in common video displays; the majority of CRTs are designed with electrostatic focus elements. Magnetic focusing, however, can provide superior resolution compared with electrostatic focusing. The gun structure is simpler than what is needed for electrostatic focusing. Only the cathode, control electrode, and accelerating electrode are required, with the focus coil usually located externally on the neck of the CRT. A constant current source must be provided for the magnetic focus coil, which can be varied to control the focal point.

A common method of magnetic focusing employs a short magnetic lens that operates by means of the radial inhomogeneity of the magnetic field, and can have both the object and image points distant from the lens. The typical short magnetic lens or focus coil consists of a large number of turns of fine wire with a total resistance of several hundred ohms, wound on a bobbin of insulating paper or plastic. The bobbin and coil are almost totally enclosed in a soft-iron shell, except for an annular gap of about 0.375-in at one end of the core tubing.

6.4.6 Magnetic Lens Aberrations

A magnetic lens is subject to the same aberrations as an electrostatic lens. A magnetic lens may also suffer from an additional distortion that is associated with the rotation of the image. This distortion is called *spiral distortion* and is illustrated in Figure 6.28. Spiral distortion results from different parts of the image being rotated different amounts as a function of their radial position. The effect is reduced by using very small apertures, or is essentially eliminated by having a pair of lenses that rotate in different directions. There is

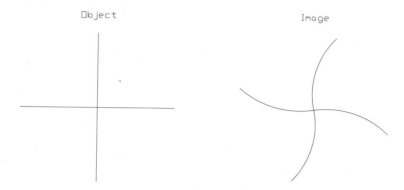

Figure 6.28 Spiral distortion in magnetic-lens images. (*After* [7].)

also the possibility of distortion resulting from stray fields or ripple in the current driving the focus coil. Current ripple causes a point to become a blurred spot, whereas stray fields cause a point to elongate to a line. Both of these effects can be minimized by careful design of the current source and the focus coil, respectively.

Beam Crossover

The *beam crossover* is used as the object whose image appears on the screen of the CRT. Therefore, the location and size of the crossover are important in determining the minimum spot size attainable by means of the focusing techniques previously described in this section. While the exact values are difficult to determine, a good approximation can be achieved by assuming a spherical field in the vicinity of the cathode. This idealized arrangement is shown in Figure 6.29. The radius of the crossover is given by [7]:

$$r_0 = \frac{2 r_c}{\sin 2\theta \left\{ V_2 / V_1 \right\}^{1/2}}$$

Where:
r_0 = the crossover radius
r_c = the crossover potential
V_2 = the crossover potential
V_e = voltage equivalent of emission velocity
θ = cathode half-angle viewed from the crossover

The crossover radius changes for different velocities of emission. Therefore, because the electrons will be emitted at all possible velocities, an average value of V_e must be used. In addition, the equation is valid only for small values of θ (less than 20°). It should be noted that these effects are controlled by the tube design and are not available for user manipulation after a given CRT has been selected.

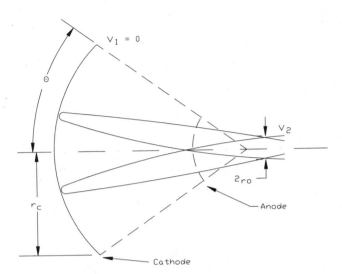

Figure 6.29 Idealized cathode with a spherical field. (*After* [7].)

6.5 Types of CRT Devices

The function of a CRT is to convert an electrical signal into a visual display. Within that general description is a wide range of applications, including:

- Video display
- Computer terminal display
- Instrumentation (oscilloscope)
- Radar
- Image recording
- Specialized military applications
- Medical imaging

This section will concentrate on applications for monochrome devices. Chapter 8 discusses color CRTs.

6.5.1 Monochrome CRT

Monochrome picture tubes were the workhorse of the television industry from the 1940s until the emergence of color as the dominant technology in the mid 1960s. Early developmental efforts focused on larger screen size and cost-effective manufacturing techniques. More recent efforts include smaller screen sizes, higher deflection angles, and smaller neck diameters. Much of this work is driven by the computer industry and specialized applications, such as computer-based machine controllers and instrumentation.

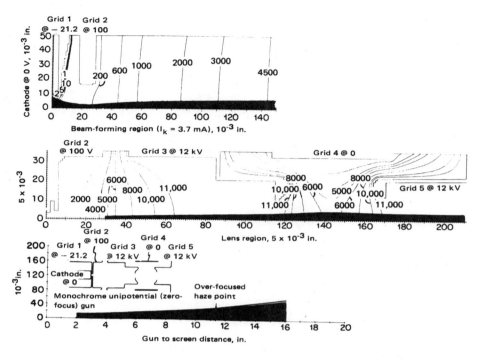

Figure 6.30 Monochrome electron gun configuration and the corresponding field and beam plots. (*From* [4]. *Used with permission*.)

6.5.2 Monochrome Electron Gun

With the need for low cost at an acceptable performance level, monochrome electron guns typically use a low focus voltage unipotential design similar to that shown in Figure 6.30. In this design a focus voltage at, or near, ground potential is a cost-saving feature. The gun model shown is divided into three sections:

- Beam-forming region
- Lens region
- Gun to screen region

From this model the analogy to light optics can be seen. The waist of the beam, in the beam-forming region, is imaged by the main lens to the screen of the tube. The contribution to final spot size of this imaged crossover may be calculated from the well-known magnification equation:

$$S_i = S_c \times \frac{V_o}{V_i} \times \frac{1}{2} \times \frac{D_i}{D_o}$$

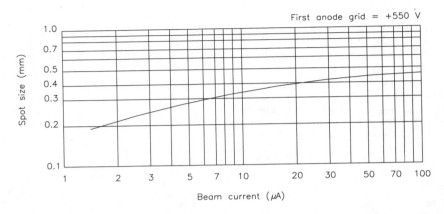

Figure 6.31 Variation in CRT spot size as a function of beam current. (*After* [5].)

Where:
S_i = image size
S_c = crossover size
V_o = potential at the crossover
V_i = potential at the screen
D_i = image distance (measured from the main lens)
D_o = object distance (measured from the main lens)

The size of the crossover results from the following:

- Thermal velocity of electrons leaving the cathode
- Imperfect focusing of the rays in the beam-forming region
- Electron self-repulsion in the beam-forming region

The total spot size at the screen is determined by:

- The magnified crossover previously discussed
- Spherical aberration of the main lens
- Electron self-repulsion from the crossover to the screen

The low-focus unipotential gun modeled has the advantage of not only a low-cost focus supply, but a small beam diameter in the lens and yoke region, resulting from rapid acceleration in the grid-2/grid-3 region. With this small beam bundle, the tube suffers less from deflection defocusing than designs having larger beam bundles.

6.5.3 Resolution Spot Size

CRT spot size is a function of many factors. The following generalizations, however, can be made:

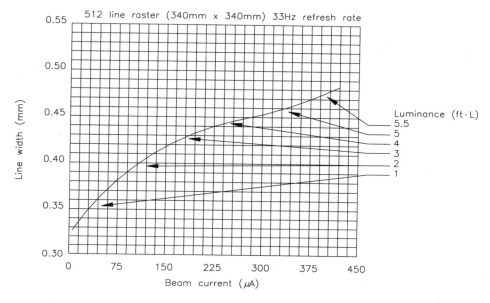

Figure 6.32 The interdependence of beam current, line width, and brightness in a CRT. (*After* [5].)

- Spot size varies inversely with neck diameter (lens diameter).
- With increasing anode voltage, beam current can be increased in direct proportion to the anode voltage, for a given spot size.
- Spot size varies directly with gun-to-screen distance—a function of the image distance deflection angle.

Display resolution is intimately tied with spot size and raster line width. As illustrated in Figure 6.31, spot size increases with increasing beam current. The actual minimum spot size of the CRT, however, is determined largely by the achievable manufacturing tolerances of the gun assembly. Highly accurate electron optics, including the focus and deflection systems, are also required for high-resolution display.

Previous sections have discussed electron lens correction techniques. While such correction is commonly used—and in some cases is unavoidable—it is accepted practice that the less correction applied to the device, the better. Lens correction should be used to offset the unavoidable distortions resulting from electron optics, not manufacturing tolerances in the electron gun and deflection systems.

Figure 6.32 shows the approximate relationship between resolution and luminous intensity as a function of beam current. Note how line width increases as the luminance output increases. In some applications, this characteristic forces a tradeoff between the ultimate resolution of the device and the maximum usable screen brightness.

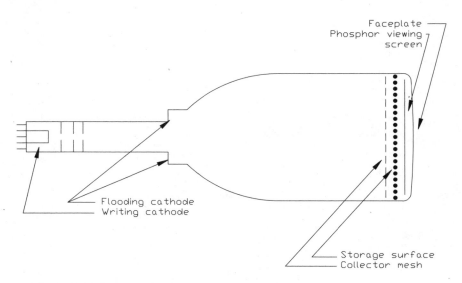

Figure 6.33 Construction of a typical storage CRT.

6.5.4 Storage CRT

The storage CRT produces a visual display of controllable duration. The tube typically has two electron guns, a phosphor viewing screen, and two or more fine-mesh screens, as illustrated in Figure 6.33. One of the electron guns is referred to as the *writing gun* and the other as the *flood gun*. The screen nearest to the guns is the *collector mesh*. The second mesh is the *storage mesh* and is coated with a thin dielectric material to form a surface on which the electrons store information.

The writing gun emits a small electron beam that is modulated by the information to be stored by the tube. This information may be applied to the control grid for intensity modulation, or to the deflection electrodes for spatial modulation. The storage surface is scanned by the high-resolution beam, which strikes the screen surface. A positive-charge image corresponding to the input signal is created on the storage surface (storage mesh) by secondary-emission effects. The image remains on the storage surface until it decays from the neutralizing action of gas ions, or is erased by a command from the operator.

The flood gun provides a flow of low-energy electrons, caused by the collimation system to arrive orthogonally and uniformly over the storage surface. These electrons initially charge the storage surface to the flood gun cathode potential, after which all flood gun electrons are accelerated into the high voltage field generated by the viewing screen. These electrons strike the phosphor on the viewing screen. The resulting luminance level is set by beam power and phosphor efficiency.

The screen is erased by applying a positive potential to the backing electrode that is equal to the cutoff potential of the storage surface. This positive pulse causes the storage surface to rise in potential, but the surface is charged down by the flood gun electrons. When the posi-

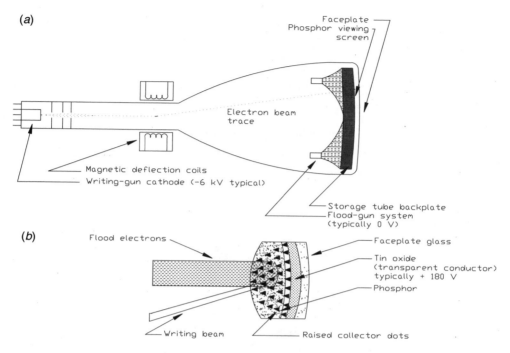

Figure 6.34 A typical bistable storage CRT: (*a*) basic construction, (*b*) detail of screen storage system.

tive pulse is removed, the backing electrode and storage surface drop equally; the cutoff condition then exists.

Increased storage sensitivity can be obtained through the addition of a third mesh coated with a dielectric material similar to that on the storage mesh and placed between the two meshes of the standard tube. This high-speed target is optimized to have high sensitivity, but consequently short retention characteristics.

Depending upon the device, after an image is written to the screen it may be retained for at least several minutes, and, in some cases, several hours at usable luminance levels.

The storage CRT, which saw considerable use in instrumentation applications for many years, has now almost entirely been replaced by digital storage techniques that rely on computer-based systems to quantize, store, and manipulate input waveforms and/or images.

Bistable Storage Tube

The *bistable storage tube* is a variation on the basic concept, which allows charge storage in a CRT without the addition of a collector and storage mesh. The primary advantages of this approach are (1) construction of the device is simpler and (2) display resolution is not limited by the pitch of the mesh. Figure 6.34 shows a typical bistable storage CRT. Two guns are used and the phosphor screen is structured to have collector islands surrounded by a phosphor layer deposited on a transparent film. The phosphor acts as both a dielectric storage surface and as the light emitter. The phosphor can be maintained at two stable potentials through the action of secondary emission from the phosphor layer. The bistable storage tube is similar to the mesh-type in operation, with the phosphor acting as part of the storage target as well as the light-emitting surface.

Viewing in the stored mode is accomplished by increasing the number of flood-beam electrons reaching the target. These electrons land with more energy on those target areas with stored information, so that a visible output pattern is created.

As with the storage tube, the bistable storage tube is presented here as a matter of historical interest.

6.5.5 Computer Display Terminal Devices

CRTs for monochrome computer display applications are similar to tubes used in high-resolution video monitors. However, because the display is principally alphanumeric and vector-graphical, the linearity of the beam-modulation characteristics is less important. Requirements include well-focused round spots with minimum spot growth or deflection aberrations from the center to the useful edges of the display area. High legibility is of primary importance, implying high contrast. White-emitting phosphors are popular, but not necessary, so that highly efficient, high-visual response phosphors emitting in the yellow or green spectral regions are applicable.

Some years ago, the storage CRT found considerable use in graphics terminals. Attributes include an absence of flicker and high resolution. The *direct-view storage tube* (DVST) is one such device. Because of its inherent storage capabilities, the screen does not need to be refreshed; the image is stored as a distribution of charges on the inside surface of the screen. Because refresh is not required, the DVST can display complex images without the high scan rate and wide bandwidth required by a conventional CRT. The major disadvantage of the DVST is that modifying any part of an image requires redrawing the entire screen to establish a new charge distribution in the device. This redraw process can be slow (several seconds), if a complex image is involved.

The DVST has now been replaced by advanced raster display systems.

6.5.6 Oscilloscope CRT Devices

The oscilloscope is the most common instrumentation application for a CRT. Such applications require a display that includes the following attributes:

- Bright display

- Sharp image

- Responsive to rapid changes in deflection potential

- Single-line trace with a minimum of deflection defocusing or astigmatism

The requirement for rapid deflection to display arbitrary waveforms dictates the use of electrostatic deflection, at least for the vertical direction. For general use, both horizontal and vertical axes employ electrostatic deflection. Because the included deflection angle must be small, usually less than 45° (to preserve good spot size and shape), oscilloscope CRTs are relatively long compared with the face diameter. The phosphors generally used for oscilloscope CRTs include P1 (green, medium persistence) or P2 (yellow-green, medium persistence).

Storage CRTs were widely used in instrumentation applications—mainly storage oscilloscopes—until digital technology made the storage of complex input signals practical and affordable. Most instrumentation applications for the storage tube, therefore, have shifted to computer-based storage, manipulation, and display systems.

6.5.7 Radar CRT Devices

Except for the *A-scope* radar display, which is essentially the same as an oscilloscope display, most radar displays consist of a two-dimensional coordinate image with beam-intensity modulation. Because the coordinate scans are mathematically regular and at preselected rates, electromagnetic deflection is generally used, which permits greater deflection angles and consequently shorter tubes to be used for a given face diameter.

In modern radar displays, it is necessary to include alphanumeric characters, symbols, and vectors in the image along with the radar information. Shaped-beam tubes, or a multiple-beam tube, in which one beam is devoted to the tracing of characters or symbols and the other to the plan-position-indicator (PPI) display, can be used for this purpose.

Long-persistence phosphors are generally used in CRTs for radar displays, because it is desirable to be able to see the situation in the entire covered area at any given time. Here again, digital technology is rapidly changing the display device requirements for advanced radar systems.

6.5.8 Recording CRT Devices

Cathode ray tubes for recording or transcribing information on photographic or otherwise sensitized film typically use very-high-resolution (VHR) or ultrahigh resolution (UHR) types. The majority of these types have nominal faceplate diameters of 4- or 5-in. The spot diameter of a VHR and UHR tubes range from approximately 0.0015-in down to 0.00033-in or smaller.

The displayed information is transferred to the recording medium either by an external focusing optical system or by direct contact with a fiber-optic faceplate, requiring no focusing.

6.6 CRT Measurement Techniques

A number of different techniques have evolved for measuring the static performance of CRT display devices and systems [13]. Most express the measured device performance in a unique figure of merit or metric. While each approach provides useful information, the lack of standardization in measurement techniques makes it difficult or impossible to directly compare the performance of a given class of CRT.

Regardless of the method used to measure performance, the operating parameters must be set for the anticipated operating environment. Key parameters include:

- Input signal level

- System/display line rate
- Luminance (brightness)
- Contrast
- Image size
- Aspect ratio

If the display is used in more than one environmental condition—such as under day and night conditions—a set of measurements is appropriate for each application.

6.6.1 Subjective CRT Measurements

Three common subjective measurement techniques are used to assess the performance of a CRT [13]:

- *Shrinking raster*
- *Line width*
- *TV limiting resolution*

Predictably, subjective measurements tend to exhibit more variability than objective measurements. While not generally used for acceptance testing or quality control, subjective CRT measurements provide a fast and relatively simple means of performance assessment. Results are usually consistent when performed by the same observer. However, results for different observers are often not consistent, because different observers use different visual criteria to make their judgments.

Shrinking raster and line width techniques are used to estimate the vertical dimension of the CRT beam spot size (*footprint*). There are several underlying assumptions with this approach:

- The spot is assumed to be symmetrical and Gaussian in the vertical and horizontal planes.
- The display MTF calculated from the spot size measurement results in the best performance envelope that can be expected from the CRT.
- The modulating electronics are designed with sufficient bandwidth so that spot size is the limiting performance parameter.
- The modulation contrast at low spatial frequencies approaches 100 percent.

Depending on the application, not all of these assumptions are valid:

- **Assumption 1**. Verona [13] has reported that the symmetry assumption is generally not true. The vertical spot profile is only an approximation to the horizontal spot profile; most spot profiles exhibit some degree of astigmatism. However, significant deviations from the symmetry and Gaussian assumptions result in only minor deviations from the projected performance when the assumptions are correct.

- **Assumption 2**. The optimum performance envelope assumption infers that other types of measurements will result in the same, or lower, modulation contrast at each spatial frequency. The MTF calculations based on a beam footprint in the vertical axis indicate the optimum performance that can be obtained from the display because finer detail (higher-spatial frequency information) cannot be written onto the screen smaller than the spot size.

- **Assumption 3**. The modulation circuit bandwidth must be sufficient to pass the full incoming video signal. Typically, the video circuit bandwidth is not a problem with current technology circuits, which are usually designed to provide significantly more bandwidth than the CRT is capable of displaying. However, in cases where this assumption is not true, the calculated MTF based purely on the vertical beam profile will be incorrect. The calculated performance will be better than the actual performance of the display.

- **Assumption 4**. The calculated MTF is normalized to 100 percent modulation contrast at zero spatial frequency and ignores the light scatter and other factors that degrade the actual measured MTF. Independent modulation contrast measurements at a low-spatial frequency can be used to adjust the MTF curve to correct for the normalization effects.

Shrinking Raster Method

The shrinking raster measurement procedure is relatively simple. Steps include the following [13]:

- The brightness and contrast controls are set for the desired peak luminance with an active raster background luminance (one percent of peak luminance) using a stair-step video signal.

- While displaying a flat field video signal input corresponding to the peak luminance, the vertical gain/size is reduced until the raster lines are barely distinguishable.

- The raster height is measured and divided by the number of active scan lines to estimate the average height of each scan line. The number of active scan lines is typically 92 percent of the line rate. (For example, a 525 line display has 480 active lines, an 875 line display has 817, and a 1025 line display has 957 active lines.)

The calculated average line height is typically used as a stand-alone metric of display performance.

The most significant shortcoming of the shrinking raster method is the variability introduced through the determination of when the scan lines are *barely distinct* to the observer. Blinking and eye movements often enhance the distinctness of the scan lines; lines that were indistinct become distinct again. Furthermore, while the shrinking raster procedure can produce acceptable results on large format CRT displays, it is less accurate for miniature devices (1-in diameter and less). The nominal line spacing for a 1-in CRT operating with full raster (at 875 line-rate video) is approximately 15.6 to 15.8 µm. Because the half-intensity spot width is about 20 to 22 µm, there is a significant amount of spot overlap between the scan lines before the raster height is reduced.

Line Width Method

The line-width measurement technique requires a microscope with a calibrated graticule [13]. The focused raster is set to a 4:3 aspect ratio and the brightness and contrast controls are set for the desired peak luminance with an active raster background luminance (one percent of peak luminance) using a stair-step video signal. A single horizontal line at the anticipated peak operating luminance is presented in the center of the display. The spot is measured by comparing its luminous profile with the graticule markings. As with the shrinking raster technique, determination of the line edge is subjective.

TV Limiting Resolution Method

This technique involves the display of two-dimensional, high-contrast bar patterns or lines of various size, spacing, and angular orientation [13]. The observer subjectively determines the limiting resolution of the image by the smallest set of bars that can be resolved. There are several potential errors with this technique, including:

- A phenomenon called *spurious resolution* can occur that leads the observer to overestimate the limiting resolution. Spurious resolution occurs beyond the actual resolution limits of the display. It appears as fine structures that can be perceived as line spacings closer than the spacing at which the bar pattern first completely blurs. This situation arises when the frequency response characteristics fall to zero, then go negative, and perhaps oscillate as they die out. At the bottom of the negative trough, contrast is restored, but in reverse phase (white becomes black and black becomes white).

- The use of test charts imaged with a video camera can lead to incorrect results because of the addition of camera resolution considerations to the measurement. Electronically-generated test patterns are more reliable image sources.

The proper setting of brightness and contrast is required for this measurement. Brightness and contrast controls are adjusted for the desired peak luminance with an active raster background luminance (one percent of peak luminance) using a stair-step video signal. Too much contrast will result in an inflated limiting resolution measurement; too little contrast will result in a degraded limiting resolution measurement.

Electronic resolution pattern generators typically provide a variety of resolution signals from 100 to 1000 *TV lines/picture height* (TVL/PH) or more in a given multiple (such as 100). Figure 6.35 illustrates an electronically-generated resolution test pattern for high-definition video applications.

Application Considerations

The subjective techniques discussed in this section, with the exception of TV limiting resolution, measure the resolution of the *display*. The TV pattern test measures *image* resolution, which is quite different [13].

For example, consider a video display in which the scan lines can just be perceived—about 480 scan lines per picture height. This indicates a *display* resolution of at least 960 TV lines—counting light *and* dark lines, per the convention. If a pattern from an electronic generator is displayed, observation will show the image beginning to deteriorate

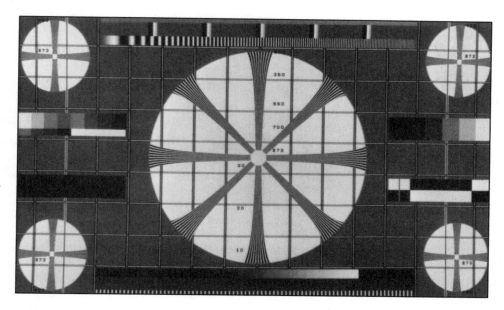

Figure 6.35 Wide aspect ratio resolution chart produced by an electronic signal generator. (*Courtesy of Tektronix.*)

at about 340 TV lines. This characteristic is the result of beats between the image pattern and the raster, with the beat frequency decreasing as the pattern spatial frequency approaches the raster spatial frequency. This ratio of 340/480 = 0.7 (approximately) is known as the *Kell factor*. Although debated at length, the factor does not change appreciably in subjective observations.

6.6.2 Objective CRT Measurements

Four common types of objective measurements can be performed to assess the capabilities of a CRT [13]:

- *Half power width*
- *Impulse Fourier transform*
- *Knife edge Fourier transform*
- *Discrete frequency*

While more difficult to perform than the subjective measurements discussed so far, objective CRT measurement techniques offer greater accuracy and better repeatability. Some of the procedures require specialized hardware and/or software.

Half Power Width Method

The half power width technique is appropriate for large displays (9-in and larger), but it is not reliable when used to measure line width on a miniature CRT [13]. A single horizontal line is activated with the brightness and contrast controls set to a typical operating level (as discussed in the previous section). The line luminance is equivalent to the highlight luminance (maximum signal level). The central portion of the line is imaged with a microscope in the plane of a variable width slit. The open slit allows all the light from the line to pass through to a photodetector. The output of the photodetector is displayed on an oscilloscope. As the slit is gradually closed, the peak amplitude of the photodetector signal decreases. When the signal drops to 50 percent of its initial value, the slit width is recorded. The width measurement divided by the microscope magnification represents the *half power width* of the horizontal scan line. The half power width of a miniature CRT may be measured using a scanning microphotometer and software to perform numerical integration on the luminance profile.

The half power width is defined as the distance between symmetrical integration limits, centered about the maximum intensity point, which encompasses half of the total power under the intensity curve. The half power width is not the same as the half intensity width measured between the half intensity points. The half intensity width is theoretically 1.75 times greater than the half power width for a Gaussian spot luminance distribution.

It should be noted that the half power line width technique relies on line width to predict the performance of the CRT. Many of the precautions outlined in the previous section also apply here. The primary difference, however, is that line width is measured under this technique objectively, rather than subjectively.

Fourier Transform Methods

The impulse Fourier transform technique involves measuring the luminance profile of the spot and then taking the Fourier transform of the distribution to obtain the MTF [13]. The MTF, by definition, is the Fourier transform of the line spread function. Commercially available software may be used to perform these measurements using either an impulse or knife edge as the input waveform. Using the vertical spot profile as an approximation to the horizontal spot profile is not always appropriate, and the same reservations expressed in the previous section apply here.

The measurement is made by generating a single horizontal line on the display with the brightness and contrast set as discussed previously. A microphotometer with an effective slit aperture width approximately one tenth the width of the scan line is moved across the scan line (the long slit axis parallel to the scan line). The data taken is stored in an array, which represents the luminance profile of the CRT spot, distance versus luminance. The microphotometer is calibrated for luminance measures and for distance measures in the object plane. Each micron step of the microphotometer represents a known increment in the object plane. The software then calculates the MTF of the CRT based on its line spread from the calibrated luminance and distance measurements. Finite slit width corrections may also be made to the MTF curve by dividing it by a measurement system MTF curve, obtained from the luminance profile of an ideal knife edge aperture or a standard source.

The knife edge Fourier transform measurement may be conducted using a low-spatial frequency vertical bar pattern (5 to 10 cycles) across the display with the brightness and contrast controls set as discussed previously. The frequency response of the square wave pattern generator and video pattern generator should be greater than the frequency response of the display system (100 MHz is typical). The microphotometer scans from the center of a bright bar to the center of a dark bar (left to right), measuring the width of the boundary and comparing it to a knife edge. The microphotometer slit is oriented vertically, with its long axis parallel to the bars. The scan is usually made from a light bar to a dark bar in the direction of spot movement. This procedure is preferred because waveforms from scans in the opposite direction may contain anomalies. When the beam is turned on in a square wave pattern, it tends to overshoot and oscillate. This behavior produces artifacts in the luminance profile of the bar edge as the beam moves from an off to an on state. In the on-to-off direction, however, the effects are minimal and the measured waveform does not exhibit the same anomalies that can corrupt the MTF calculations.

The bar edge (knife edge) measurement, unlike other techniques discussed so far, uses the horizontal spot profile to predict display performance. All of the other techniques use the vertical profile as an approximation of the more critical horizontal spot profile. The bar edge measurement will yield a more accurate assessment of display performance because the displayed image is being generated with a spot scanned in the horizontal direction.

Discrete Frequency Method

The discrete sine wave frequency response measurement technique provides the most accurate representation of display performance [13]. With this approach there are no assumptions implied about the shape of the spot, the electronics bandwidth, or low-frequency light scatter. Discrete spatial frequency sine wave patterns are used to obtain a discrete spatial frequency MTF curve that represents the signal-in to luminance-out performance of the display.

The measurement begins by setting the brightness and contrast as previously discussed, with black level luminance at one percent of the highlight luminance. A sine wave signal is produced by a function generator and fed to a pedestal generator, where it is converted to an RS-170A or RS-343 signal and then applied to the CRT. The modulation and resulting spatial frequency pattern is measured with a scanning microphotometer. The highlight and black level measurements are used to calculate the modulation constant for each spatial frequency from 5 *cycles/display width* to the point that the modulation constant falls to less than one percent. The modulation constant values are then plotted as a function of spatial frequency, generating a discrete spatial frequency MTF curve.

While the discrete sine wave technique is hardware intensive and time-consuming, it produces the most definitive display assessment. The other objective techniques, while less complex and faster to perform, lack the fidelity of the discrete frequency approach.

6.7 Viewing Environment Considerations

The environment in which a video display device is viewed is an important criterion for critical viewing situations. Applications in which color purity and adherence to set standards are important require a standardized (or at least consistent) viewing environment.

For example, textile colors viewed on a display with a white surround will appear different than the same colors viewed with a black surround. By the same token, different types of ambient lighting will make identical colors appear different on a given display.

The Society of Motion Picture and Television Engineers (SMPTE) has addressed this issue with RP 166-1995, which specifies the environmental and surround conditions that are required in television or video program review areas for the "consistent and critical evaluation" of conventional television signals [14]. Additionally, the Recommended Practice is designed to provide for repeatable color grading or correction. A number of important parameters are specified in RP 166-1995, including the following:

- The distance of the observer from the monitor screen should be 4 to 6 picture heights for standard-definition television (SDTV) displays.

- The observer should view the monitor screen at a preferred angle in both the horizontal and vertical planes of 0° ±5° and, in any event, no greater than ±15° from the perpendicular to the midpoint of the screen.

- The viewing area decor should have a generally neutral matte impression, without dominant colors.

- Surface reflectances should be nonspecular and should not exceed 10 percent of the peak luminance value of the monitor white.

The Recommended Practice suggests placing the monitor in a freestanding environment 2.5 to 5 screen heights in front of the wall providing the visual surround. Another acceptable approach is to mount the monitor in a wall with its face approximately flush with the surface of the wall. It is further recommended that all light sources in use during picture assessment or adjustment have a color quality closely matching the monitor screen at reference white (i.e., D_{65}).

It is often necessary to have black-and-white monitors surrounding one or more color monitors in a studio control room. According to RP 166-1995, the black-and-white monitors should be the same color temperature as the properly adjusted color monitor(s), 6500K. Black-and-white monitors are normally equipped with P4 phosphors, at about 9300K. This cooler color temperature prevents the background surrounding the color monitors from remaining neutral. Most black-and-white video monitors can be ordered with 6500K phosphors.

6.8 Picture Monitor Alignment

The proper adjustment and alignment of studio picture monitors is basic to video quality control. Uniform alignment throughout the production chain also ensures consistency in color adjustment, which facilitates the matching of different scenes within a program that may be processed at different times and in different facilities. The SMPTE has addressed this requirement for conventional video through RP 167-1995. The Recommended Practice offers a step-by-step process by which color monitors can be set. Key elements of RP 167-1995 include the following [15]:

- **Initial conditions**. Setup includes allowing the monitor to warm up and stabilize for 20 to 30 minutes. The room ambient lighting should be the same as it is when the monitor is in normal service, and several minutes must be allowed for visual adaptation to the operating environment.

- **Initial screen adjustments**. The monitor is switched to the setup position, in which the red, green, and blue screen controls are adjusted individually so that the signals are barely visible.

- **Purity**. Purity, the ability of the gun to excite only its designated phosphor, is checked by applying a low-level flat-field signal and activating only one of the three guns at a time. The display should have no noticeable discolorations across the face.

- **Scan size**. The color picture monitor application establishes whether the *overscan* or *underscan* presentation of the display will be selected. An underscanned display is one in which the active video (picture) area, including the corners of the raster, is visible within the screen mask. Normal scan brings the edges of the picture tangent to the mask position. Overscan should be no more than 5 percent.

- **Geometry and aspect ratio**. Display geometry and aspect ratio are adjusted with the crosshatch signal by scanning the display device with the green beam only. Correct geometry and linearity are obtained by adjusting the pincushion and scan-linearity controls so that the picture appears without evident distortions from the normal viewing distance.

- **Focus**. An ideal focus target is available from some test signal generators; if it is unavailable, multiburst, crosshatch, or white noise can be used as tools to optimize the focus of the displayed picture.

- **Convergence**. Convergence is adjusted with a crosshatch signal; it should be optimized for either normal scan or underscan, depending upon the application.

- **Aperture correction**. If aperture correction is used, the amount of correction can be estimated visually by ensuring that the $2T \sin^2$ pulse has the same brightness as the luminance bar or the multiburst signal when the 3 and 4.2 MHz bursts have the same sharpness and contrast.

- **Chrominance amplitude and phase**. The chrominance amplitude and phase are adjusted using the SMPTE color bar test signal and viewing only the blue channel. Switching off the comb filter, if it is present, provides a clear blue channel display. Periodically, the red and green channels should be checked individually in a similar manner to verify that the decoders are working properly. A detailed description of this procedure is given in [14].

- **Brightness, color temperature, and gray scale tracking**. The 100-IRE window signal is used to supply the reference white. Because of typical luminance shading limitations, a centrally placed PLUGE [16] signal is recommended for setting the monitor brightness control. The black set signal provided in the SMPTE color bars also can be used for this purpose.

- **Monitor matching**. When color matching two or more color monitors, the same alignment steps should be performed on each monitor in turn. Remember, however, that monitors cannot be matched without the same phosphor sets, similar display uniformity characteristics, and similar sharpness. The most noticeable deviations on color monitors are the lack of uniform color presentations and brightness shading. Color matching of monitors for these parameters can be most easily assessed by observing flat-field uniformity of the picture at low, medium, and high amplitudes.

For complete monitor-alignment procedures, see [15].

As more experience is gained with digital television (DTV)-based systems, operating parameters such as those detailed in this section will no doubt be updated to take into consideration the unique attributes and requirements of HDTV.

6.9 References

1. Fink, Donald, (ed.): *Television Engineering Handbook*, McGraw-Hill, New York, N.Y., 1957.
2. Skolnik, M. Il, (ed.):*Radar Handbook*, McGraw-Hill, New York, N.Y., pp. 6–8, 1970.
3. Moon, P.: *The Scientific Basis of Illuminating Engineering*, Dover, New York, N.Y., 1961.
4. Whitaker, Jerry C., and K. Blair Benson: *Standard Handbook of Video and Television Engineering*, 3rd ed., McGraw-Hill, New York, N.Y., 2000.
5. Richards, C. J.: *Electronic Display and Data Systems: Constructional Practice*, McGraw-Hill, New York, N.Y., 1973.
6. Diakides, N. A.: "Phosphors," *Proc. Soc. Photo-Opt. Instrum. Eng.*, vol 42, August 1973.
7. Spangenberg, K. R.: *Vacuum Tubes*, McGraw-Hill, New York, N.Y., 1948.
8. Sherr, S.: *Electronic Displays*, Wiley, New York, N.Y., 1979.
9. Moss, Hilary: *Narrow Angle Electron Guns and Cathode Ray Tubes*, Academic, New York, N.Y., 1968.
10. *Cathode Ray Tube Displays*, MIT Radiation Laboratory Series, vol. 22, McGraw-Hill, New York, N.Y., 1953.
11. Zworykin, V. K., and G. Morton: *Television*, 2d ed., Wiley, New York, N.Y., 1954.
12. Pender, H., and K. McIlwain (eds.): *Electrical Engineers Handbook*, Wiley, New York, N.Y., 1950.
13. Verona, Robert W.: "Comparison of CRT Display Measurement Techniques, in *Helmet-Mounted Displays III*, Thomas M. Lippert (ed.), Proc. SPIE 1695,SPIE, Bellingham, Wash., pp. 117–127, 1992.
14. "Critical Viewing Conditions for Evaluation of Color Television Pictures," SMPTE Recommended Practice RP 166-1995, SMPTE, White Plains, N.Y., 1995.
15. "Alignment of NTSC Color Picture Monitors," SMPTE Recommended Practice RP 167-1995, SMPTE, White Plains, N.Y., 1995.

16. Quinn, S. F., and C. A. Siocos: "PLUGE Method of Adjusting Picture Monitors in Television Studios—A Technical Note," *SMPTE Journal*, SMPTE, White Plains, N.Y., vol. 76, pg. 925, September 1967.

6.10 Bibliography

Anstey, G., and M. J. Dore: "Automatic Measurement of Cathode Ray Tube MTFs," Royal Signals and Radar Establishment, 1980.

Barten, P. J. G.: "Spot Size and Current Density Distribution of CRTs," *Proceedings of the Society for Information Display*, Society for Information Display, San Jose, Calif., vol. 25, no. 3, pp. 155–159, 1984.

Bedell, R. J.: "Modulation Transfer Function of Very High Resolution Miniature Cathode Ray Tubes," *IEEE Transactions on Electron Devices*, vol. ED-22, no. 9, pp. 793–796, September 1975.

Birks, J. B.: "Electrophoretic Deposition of Insulating Materials," *Progr. Dielectrics*, vol. 1, 1959.

Boers, J.: "Computer Simulation of Space Charge Flows," Rome Air Development Command RADC-TR-68-175, University of Michigan, 1968.

Casteloano, Joseph A.: *Handbook of Display Technology*, Academic, New York, N.Y., 1992.

Cerulli, N. F.: "Method of Electrophoretic Deposition of Luminescent Materials and Product Resulting Therefrom," U.S. Patent 2,851,408, September 9, 1958.

Crost, Munsey E.: "Display Devices and the Human Observer," *Proc. Interlab. Sem. Component Technol.* Pt. 1, *R&D Tech. Rep.* ECOM-2865. U.S. Army Electronics Command, Fort Monmouth, N.J., August, 1967.

Curie, D., and G. F. J. Garlick: *Luminescence in Crystals*, Wiley, New York, N.Y., 1963.

Dickenson, W.: "Monochrome Picture Tubes—Status Report," *IEEE Trans. Broadcast TV Receivers*," vol. BTR-13, no. 3, pp. 46–48, 1967.

Donofrio, R.: "Color in Color TV—A Phosphor Approach," *Color Engineering*, pp. 11–14, February 1971.

Donofrio, R.: "Image Sharpness of a Color Picture Tube by Modulation Transfer Techniques," *IEEE Tran. Broadcast Television Receivers*, vol. BTR-18, no. 1, p. 16, February 1972.

Donofrio, R.: "Low Current Density Aging," *Proc. Electrochemical Society*, May 12, 1981.

EG&G Gamma Scientific Inc.: Digital Spatial Scanning System Software, SPATL-C11/MTF Ver. 5.29, San Diego, CA 92123.

Fink, Donald, and Donald Christiansen (eds.): *Electronics Engineers Handbook*, 3rd ed., McGraw-Hill, New York, N.Y., 1989.

Fiore, J., and S. Kaplan: "A Second Generation Color Tube Providing More than Twice the Brightness and Improved Contrast," Spring Conf. Broadcast and Television Receivers, IEEE, June 1969.

Foley, James D., et al.: *Computer Graphics: Principles and Practice*, 2nd ed., Addison-Wesley, Reading, Mass., 1991.

Gallaro, A. V., and R. A. Hedler: "Process for Forming a Color CRT Screen Structure Having Optical Filter Therein," U.S. Patent 3,884,694, 1973.

Gerritson, J.: "Soft Flash Picture Tubes," *IEEE Trans. Consumer Electronics*, vol. CE-24, no. 4, pp. 560–561, November 1978.

Goldberg, P.: *Luminescence of Inorganic Solids*, Academic, New York, N.Y., 1966.

Hunter, R.: *The Measurement of Appearance*, Wiley, New York, N.Y., 1975.

IEEE Standard Dictionary of Electrical and Electronics Terms, 2nd ed., Wiley, New York, N.Y., 1977.

Infante, C.: "On the Resolution of Raster-Scanned CRT Displays," *Proceedings of the Society for Information Display*, Society for Information Display, San Jose, Calif., vol. 26, no. 1, pp. 23–36, 1985.

Jenkins. A. J.: "Modulation Transfer Function (MTF) Measurements on Phosphor Screens," *Assessment of Imaging Systems: Visible and Infrared* (Sira), SPIE, Bellingham, Wash., vol. 274, pp. 154–158, 1981.

Jordan, Edward C., ed.: *Reference Data for Engineers: Radio, Electronics, Computer, and Communications*, 7th ed., Howard W. Sams, Indianapolis, IN, 1985.

Judd, D., and G. Wyszecki: *Color in Business, Science and Industry*, Wiley, New York, N.Y., 1975.

Kallman, H. P., and G. M. Spurch: *Luminescence of Organic and Inorganic Materials*, Wiley, New York, N.Y., 1962.

Kobari, Y., et al.: "A Novel Arc Suppression Technique for CRTs," *IEEE Trans. Consumer Electronics*, vol. CE-26, no. 3, pp. 446–450, August 1980.

Koller, L. R.: "Thin Film Phosphors," Electrochemical Society Meeting, Washington D.C., May 13, 1957.

Kucherrov, G. V., et. al.: "Application of the Modulation Transfer Function Method to the Analysis of Cathode-Ray Tubes," *Radio Engineering and Electronics Physics*, vol. 19, pp. 150–152, February 1974.

Langmuir, D.: "Limitations of Cathode Ray Tubes," *Proc. IRE*, vol. 25, pp. 977–991, 1937.

Lehmann, W.: "Method of Forming a Uniform Layer of Luminescent Material on a Surface," U.S. Patent 2,798,821, July 1957.

Leverenz, H. W.: *An introduction to Luminescence of Solids*, Dover, New York, N.Y., 1968.

Luxenberg, H. R., and R. L. Kuehn (eds.): *Display Systems Engineering*, McGraw-Hill, New York, N.Y. 1968.

Mannos, J. L., and R. W. Tracy: "Cathode-Ray Tube (CRT) Softcopy Image Display Evaluation," *Advances in Display Technology*, SPIE, Bellingham, Wash., vol. 199, pp. 146–150, 1979.

Middlebrook, B., and M. Day: "Measure CRT Spot Size to Pack More Information into High-Speed Graphic Displays: You Can do it with the Vernier Line Method," *Electronic Design*, vol. 15, pp. 58–60, July 19, 1975.

Morgan, M. J., and R. J. Watt: "The Modulation Transfer Function of a Display Oscilloscope: Measurements and Comments," *Vision Research*, vol. 22, pp. 1083–1085, Great Britain, 1981.

Nix, L.: "Spot Growth Reduction in Bright, Wide Deflection Angle CRTs," *SID Proc.*, Society for Information Display, San Jose, Calif., vol. 21, no. 4, pg. 315, 1980.

"Optical Characteristics of Cathode Ray Tube Screens," JEDEC Publication 16B, Electron Tube Council, Washington, D.C., 1971.

"Optical Characteristics of Cathode Ray Tube Screens," JEDEC Publication 16C, Electron Tube Council, Washington, D. C., 1971.

"Optical Characteristics of Cathode Ray Tube Screens," TEPAC Publication 116, Electronic Industries Association, Washington, D. C., 1980.

Pakswer, S., and P. J. Intiso: *Journal of the Electrochemical Society*, vol. 99, 1952.

Pfahnl, A.: "Aging of Electronic Phosphors in Cathode Ray Tubes," *Advances in Electron Tube Techniques*, Pergamon, New York, N.Y., pp. 204–208, 1961.

Poole, H. H.: *Fundamentals of Display Systems*, Spartan, Washington, D.C., 1966.

Rash, C. E., and R. W. Verona: "Temporal Aspects of Electro-Optical Imaging Systems," *Imaging Sensors and Displays*, SPIE, Bellingham, Wash., vol. 765, pp. 22–25, 1987.

Richmond, J. C.: "Image Quality of Photoelectric Imaging Systems and its Evaluation," *Proceedings of Symposium on Photo-Electronic Image Devices*, 6th ed., National Bureau of Standards, Washington, D. C., pp. 519–538, 1976.

Rychlewski, T., and R. Vogel: "Phosphor Persistence in Color Television Screens," *Electrochemical Technology*, vol. 4, no. 1-2, pp. 9–12, January–February 1966.

Sadowski, M.: *RCA Review*, vol 95, 1957.

Say, D.: "The High Voltage Bi-Potential Approach to Enhanced Color Tube Performance," *IEEE Trans. Consumer Electronics*, vol. CE-24, no. 1, 1978.

Schwartz, J., and M. Fogelson: "Recent Developments in Arc Suppression for Picture Tubes," *IEEE Trans. Consumer Electronics*, vol. CE-25, no. 1, pp. 82–90, February 1979.

Sherr, S.: *Fundamentals of Display Systems Design*, Wiley, New York, N.Y., 1970.

True, R.: "Space Charge Limited Beam Forming Systems Analyzed by the Method of Self-Consistent Fields with Solution of Poisson's Equation on a Deformable Relaxation Mesh," Ph.D. thesis, University of Connecticut, Storrs, 1968.

Underwriters Laboratory Report UL492.8, January 25, 1974.

Verona, R. W., H. L. Task, V. C. Arnold, and J. H. Brindle: "A Direct Measure of CRT Image Quality," U. S. Army Aeromedical Research Laboratory, USAARL Report No. 79-14, 1979.

Vogel, R.: "Contrast Measurements in Color T.V. Tubes," IEEE Conf. Chicago, 1970.

Wintringham, W.: "Color Television and Colorimetry," *Proc. IRE*, vol. 39, pp. 1135–1172, 1951.

Wyszecki, G., and W. Stiles: *Color Science*, Wiley, New York, N.Y., 1967.

Chapter 7

CRT Deflection Systems

7.1 Introduction

There are two basic methods of deflection of an electron beam in a cathode ray tube:[1]

- A transverse electrostatic field
- A transverse electromagnetic field

The choice of electrostatic or electromagnetic deflection is influenced by a number of factors including:

- The required electron beam deflection speed. At deflection intervals of less than 10 µs, electrostatic deflection is usually considered superior to electromagnetic deflection. At deflection intervals of less than 5 µs, electrostatic deflection is used almost exclusively.

- Electron beam spot size. High-resolution displays typically utilize electromagnetic deflection systems. For applications requiring resolution of more than 600 television lines, and for luminance above 150 candela per square meter (*nit*), electromagnetic deflection is preferred because of the higher accelerating potential that may be used.[2] This higher potential permits smaller practical spot size and higher luminance output from the CRT.

- Tube geometry. Devices using electromagnetic deflection are typically shorter than their electrostatic deflection counterparts. A wide-angle magnetic deflection tube can be 30 percent shorter than an equivalent electrostatic deflection device.

Beyond the deflection region, the electron beam travels in a field-free space until it impinges on the phosphor screen. This condition implies that the anode (A_2) voltage is the highest potential applied to the device. However, in a large percentage of electrostatic deflection CRTs and some specialized electromagnetic deflection devices, an additional acceleration

---

[1] Portions of this chapter were adapted from: Jerry C. Whitaker and K. B. Benson (eds): *Standard Handbook of Video and Television Engineering*, 3rd ed., McGraw-Hill, New York, 2000. Used with permission.
[2] 1 nit = 1 candela per square meter (cd/m^2)

Table 7.1 Comparison of Common Electromagnetic and Electrostatic Deflection CRTs (*After* [1].)

Parameter	Magnetic Deflection	Electrostatic Deflection
Deflection settling time	10 μs	< 1 μs to one spot diameter
Small-signal bandwidth	2 MHz	5 MHz
Video bandwidth	15–30 MHz	25 MHz
Linear writing speed	1 μs/cm	25 cm/μs
Resolution	1000 TV lines or more	17 lines/cm
Luminance	300 nits	150–500 nits
Spot size	0.25 mm (53 cm CRT)	0.25–0.38 mm
Accelerator voltage	10 kV	28 kV
Phosphor	Various	P-31 and others
Power consumption	250 W	130–140 W
Off-axis deflection	55°	20°
Physical length	Up to 30% shorter than electrostatic deflection	
Typical applications	Video display	Waveform display

voltage is applied to the electron beam beyond the deflection region. This configuration is referred to as *post-deflection acceleration* (PDA). Implementation of PDA varies from one tube type to the next. The purpose of such a system is to efficiently convey the electron beam from the deflection region to the phosphor screen.

Table 7.1 compares the principle operating parameters of electromagnetic and electrostatic deflection CRTs. The first difference is the longer vacuum envelope required for the electrostatic type. This imposes packaging limitations on the assembly that contains the tube. Related to the increased length are the narrower deflection angles available in electrostatic types and the higher focus voltage required, as well as the need for a post-accelerator voltage. Of greatest significance, however, are the higher luminance and smaller spot size typically attainable with electromagnetic deflection. The one characteristic not shown in the table is the faster deflection speed possible with the electrostatic deflection tube, which can be as low as 1 μs, compared with the 10 μs that is possible with a typical electromagnetic deflection tube. This parameter is not included because it is influenced by the choice of deflection amplifier and may be higher or lower, depending on the type of amplifier used. However, with the use of typical amplifier designs, the 10/1 advantage is not unusual.

7.2 Electrostatic Deflection

An electrostatic deflection system generally consists of metallic deflection plates used in pairs within the neck of the CRT. Figure 7.1 shows the basic construction of an electrostatic deflection device. The simplest design incorporates flat rectangular parallel plates facing each other, with the electron beam directed along the central plane between them. The deflection plates are located in the field-free space within the second-anode region. The plates are essentially at second-anode voltage when no deflection signal is applied.

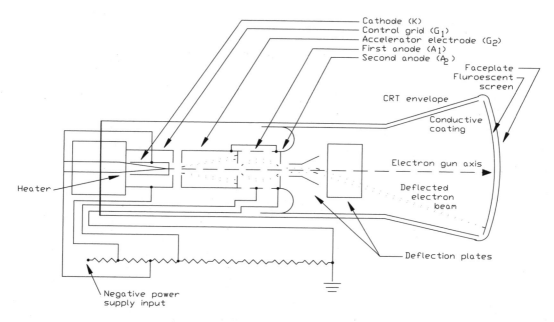

Figure 7.1 Electrostatic deflection CRT.

Deflection of the electron beam is accomplished by establishing an electrostatic field between the plates.

Most electrostatic deflection devices do not exhibit excessive deflection defocusing until the beam deflection angle off-axis exceeds 20°. This limitation prevents the use of high deflection angles common in magnetic deflection devices. Deflection angles of 55° off-axis are common in magnetic deflection CRTs.

Most electrostatic deflection CRTs are used to display electronic waveforms as a function of time. To accomplish this task, it is necessary to generate a sweep signal representing the passage of time, and to superimpose on the signal an orthogonal deflection representing signal amplitude. This is typically accomplished through the use of two pairs of deflection plates. The second pair of deflection plates must have a sufficiently wide entrance window to accept the maximum deflection of the beam produced by the first pair of plates. Design tradeoffs include overall deflection sensitivity and deflection plate capacitance. The plates (as shown in Figure 7.1) diverge to accommodate their own deflection of the electron beam. For a given deflection voltage, the magnitude of deflection is inversely proportional to the anode voltage.

7.2.1 Principles of Operation

Figure 7.2 shows the basic arrangement of an electrostatic deflection CRT. Electron trajectory is illustrated in Figure 7.3. These figures provide the basis for deriving equations that describe the principles of operation for electrostatic deflection. Assuming that an electron

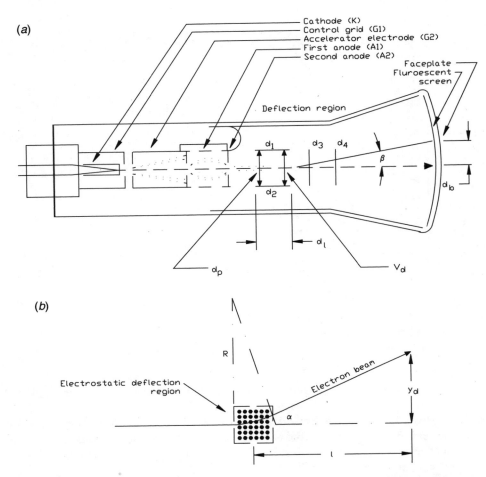

Figure 7.2 The elements of an electrostatic deflection CRT: (a) overall tube geometry, (b) detail of deflection region.

enters the deflection field between the deflection plates at right angles (θ equal to 90°) then:

$$\tan\theta = \frac{V_d\, d_l}{2 d_p V_b}$$

Where:
V_d = the voltage between the deflection plates
d_l = length of the plates
d_p = distance between the plates

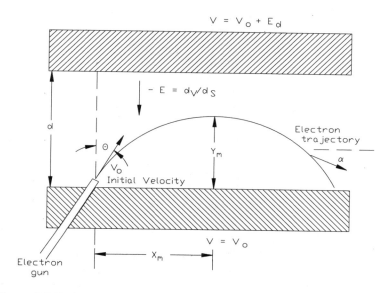

Figure 7.3 Trajectory of an electron beam. (*After* [2].)

V_b = beam voltage

Because the center of deflection is at the center of the field, tan θ is approximately equal to y_d/l, where y_d = the deflection distance and l = the length of the deflection field, shown in Figure 7.2(*b*). It follows, then:

$$y_d = \frac{l \, d_l \, V_d}{2 d_p V_b}$$

This equation holds true for parallel plates and neglects fringe effects. For nonparallel plates, the gradient is [2]:

$$\frac{dv}{dy} = \frac{V_d}{a_1 + (a_2 - a_1) X/d_l}$$

Where:
a_1 = plate separation at the entrance end
a_2 = plate separation at the departure end
X = distance for the beam in the field of the plates

The deflection equation then becomes [2]:

$$y = \frac{d_l V_d \ln\{a_2/a_1\}}{2V_b a_1 \{a_2/a_1\} - 1}$$

When a_1 is equal to a_2 (the plates are flat), the foregoing equation reduces to [2]:

$$yd = \frac{l\, d_l V_d}{2 d_p V_b}$$

(As shown previously)

7.2.2 Acceleration Voltage Effects

In an electrostatic deflection CRT, the deflection distance is directly proportional to the deflection voltage (V_d) and inversely proportional to the acceleration or beam voltage (V_b). This situation leads to the requirement that the deflection voltage must be increased a proportional amount when the beam voltage is increased. In an electromagnetic deflection CRT, deflection is proportional to the square root of the beam voltage. This leads to two considerations when electrostatic and magnetic deflection are compared. First, it can be shown that the beam spread as a function of beam voltage (in kilovolts) and current (in milliamperes) is given by [3]:

$$K^{1/2} = \frac{32.2\, r_0\, V^{3/4}}{I^{1/2}}$$

Where:
V = beam voltage in kV
I = beam current in mA

This expression is represented in the nomograph shown in Figure 7.4, where r_0 = the crossover spot size and z = the position of the beam. Based on the data presented, it is clear that the beam size is affected by the beam voltage, and the higher the beam voltage, the smaller the spot. However, increasing the beam voltage increases the required deflection voltage by a proportional amount, so that magnetic deflection can use larger beam voltages and, therefore, attain smaller spot sizes.

A second effect of beam voltage is that of light output from the phosphor, or luminance. Phosphor luminance is given by the empirical expression [4]:

$$L = A f(p) V^n$$

Where
L = phosphor luminance

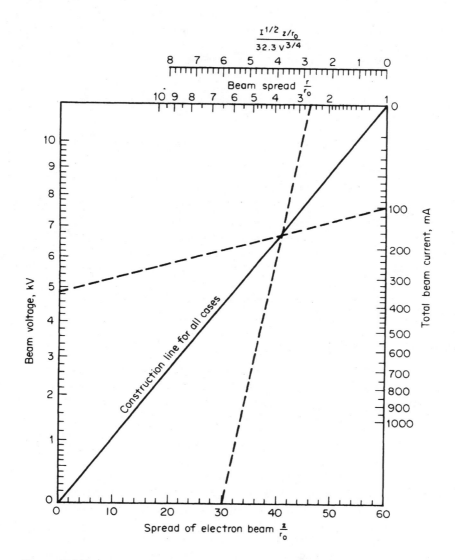

Figure 7.4 Beam spread nomograph. (*From* [1]. *Used with permission.*)

$f(p)$ = function of current
V = accelerating voltage
A = constant
n = 1.5 to 2

The phosphor output or luminance may also be expressed by:

$$L = \frac{k_b I_b V_b^n}{A}$$

Where:
k_b = a proportionality factor termed *luminous efficiency*
I_b = beam current
V_b = beam voltage
A = the area of the phosphor surface

The proportionality factor, which typically ranges from 5 to 62 for the various phosphors, is given in terms of lumens per watt.

The preceding equation clearly demonstrates the effect of beam voltage on light output and illustrates the desirability of maintaining the beam voltage at the highest value consistent with the other requirements, such as deflection sensitivity and focus voltage. The equation is approximately correct over the linear region, but does not hold true when phosphor saturation occurs. It is not possible, therefore, to increase phosphor luminance merely by raising the beam voltage.

7.2.3 Post-Deflection Accelerator

A post-deflection accelerator electrode is used in a number of electrostatic deflection CRTs. This element generally takes the form of a wide graphite band around the inside of the envelope funnel just behind the faceplate. It is usually connected to the aluminum film if the phosphor is aluminized. This element is also known as the *post-accelerator* or *third anode* (A_3).

The *spiral accelerator* is another type of post-deflection element. It can be used with a considerably higher A_3/A_2 voltage ratio without detrimental effects caused by the localized electron lens. The spiral accelerator consists of a high-resistance narrow circumferential spiral stripe of graphite of low screw pitch painted over a substantial length of the inside of the tube envelope funnel. The ends of the spiral are electrically connected to the A_2 and A_3 terminals. In operation, the spiral accelerator requires a small continuous direct current to establish a nearly uniform potential variation from the A_2 to the A_3 voltage. In effect, the element constitutes a "thick" electron lens, and because there are no abrupt changes in potential, the trace-distortion effects are much smaller than in a thin lens.

7.3 Electromagnetic Deflection

In contrast with electrostatic deflection systems, the components in an electromagnetic deflection system are almost universally located outside the tube envelope, rather than inside the vacuum. Because the neck of the CRT beyond the electron gun is free of obstructions, a larger-diameter electron beam can be used in the magnetic deflection CRT than in the electrostatic deflection device. This permits greater beam current to the phosphor screen and, consequently, a brighter picture. Deflection angles of 110° (55° off-axis) are commonly used in video display tubes without excessive spot defocusing. Large deflection angles permit the CRT to be constructed with a shorter bulb section for a given screen size.

The electromagnetic deflection yoke is most suitable for repetitive types of scanning, such as raster scans in which parallel scan lines are swept out in a rectangular area. The yoke is also well suited for *plan-position-indicator* (PPI) scans in which a radial scan line is directed outward from the center of the screen to cover a circular area.

Electromagnetic deflection has also been used for random address displays. The principal design challenge with random deflection is the inductance of the deflection coils. For any specific field strength an ampere-turns product must be established. Therefore, low inductance implies high current, which may be difficult to supply, especially for large bandwidth signals. Normally, for each axis the yoke includes two coils, each bent into a saddle shape and extending halfway around the CRT neck.

PPI deflection may be accomplished with a single axis yoke that is rotated physically by an external motor, or self-synchronous repeater driven by the radar antenna. With this arrangement, a constant-amplitude triggered linear-sawtooth wave is applied to the yoke. Another approach to PPI deflection employs a stationary yoke with two orthogonal deflection axes. One axis receives a current waveform of the linear sawtooth with its amplitude coefficient varying according to the algebraic sine of the antenna rotation angle; the other axis receives a similar waveform, varying according to the algebraic cosine of the rotation angle.

7.3.1 Principles of Operation

The basic magnetic deflection equation is derived from the expression of the behavior of an electron stream in the presence of a magnetic field [5]. For a uniform magnetic field, the radius r of the electron path when the electron enters the field at right angles is given by:

$$r = \frac{3.38 \times 10^{-6} V^{1/2}}{B_m} \text{ m}$$

Where:
B_m = magnetic flux density
V = potential at initial velocity

Figure 7.5 illustrates the basic parameters of electron beam deflection by a magnetic field. Assuming that the magnetic field is uniform within the area delineated by the dots, the electron beam will follow the arc of a circle whose radius R is given by:

$$R = \frac{3.38 \times 10^{-6} V^{1/2} \sin\theta}{B_m}$$

Where:
θ = the angle at which the electron enters the field

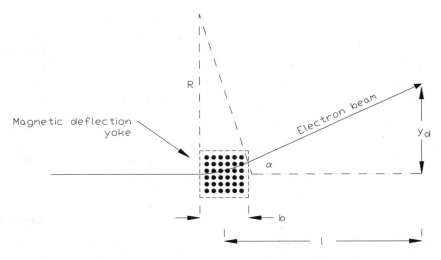

Figure 7.5 Principle quantities of magnetic deflection. (*After* [2].)

The angle at which the beam leaves the magnetic field is then given by:

$$\sin\alpha = 2.97 \times 10^5 \frac{lB_m}{V^{1/2}}$$

The deflection is related to this angle by the following:

$$yd = l\tan\alpha$$

$\tan\alpha$ is approximately equal to yd/l because the center of deflection is at the center of the field. Substituting terms provides:

$$yd = 2.97 \times 10^5 \frac{l^2 B_m}{V^{1/2}}$$

This equation assumes that $\sin\alpha$, $\tan\alpha$, α and α are all equivalent, which holds true for the small values of α. For cases where this equivalence does not hold, the deflection is not directly proportional to the current because $\sin\alpha$ rather than $\tan\alpha$ is proportional to the current. This leads to the need for corrective circuitry to compensate for the error.

7.3.2 Flat-Face Distortion

Flat-face distortion results from the difference between the radius of curvature of the deflected beam and the actual radius of the display surface [5]. This distortion is illustrated in

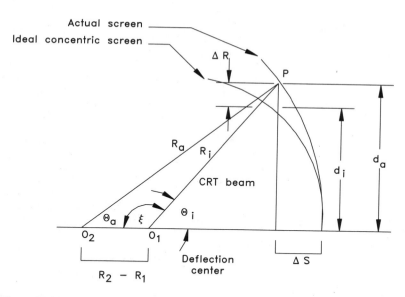

Figure 7.6 Linearity distortion resulting from CRT screen curvature. (*After* [6].)

Figure 7.6 for the general case. The ratio of true deflection to deflection on an ideal surface that has its deflection center at the center of curvature, is given by:

$$\frac{d_a}{d_i} = \frac{R_a \sin\theta_a}{R_i \sin\theta_i}$$

The deflection angle may be expressed in terms of current through the yoke by substituting for B_m in the equation that describes the angle at which an electron beam leaves a magnetic field (which follows) and using a constant to replace all other terms.

$$\sin\alpha = 2.97 \times 10^5 \frac{l B_m}{V^{1/2}}$$

After making the substitutions described, including substituting θ for α, the equation becomes:

$$\sin\theta = KI$$

It is clear that only in the ideal case will the deflection be directly proportional to the yoke

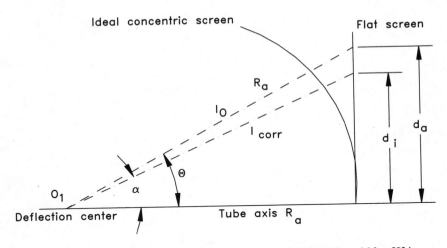

Figure 7.7 The mechanics of flat-face CRT linearity distortion. (*After* [3].)

current. If the general case is then applied to the specific case of the flat-faced screen, as illustrated in Figure 7.7, the equation becomes:

$$\frac{d_a}{d_i} = \frac{R_a \tan\theta}{R_a \sin\theta}$$

The equation can be simplified to the following:

$$\frac{d_a}{d_i} = \frac{1}{\cos\theta}$$

If the equation is rewritten as:

$$\frac{d_a}{d_i} = \frac{1}{\left(1-\sin^2\theta\right)^{1/2}}$$

it may be further reduced by using the equation:

$$\sin\theta = KI$$

(described previously). By then expanding the denominator, the following results (if all but the first two terms of the expansion are neglected):

$$d_a = d_i \left\{ 1 + \frac{K^2 I^2}{2} \right\}$$

The resulting indicated deflection error can be compensated for dynamically by introducing special circuitry, which is described later in this chapter.

7.3.3 Deflection Defocusing

Another effect resulting from deflection of the electron beam is the defocusing that occurs when the beam is moved from the center to some other location on the CRT screen [5]. The changes in focus result from the edges of the beam entering the magnetic field at different angles because of convergence. The edges of the beam will be deflected by the same radius of curvature and leave the field with a new convergence angle. This effect can be reduced through the use of dynamic focusing circuits.

7.3.4 Deflection Yoke

In order to understand the operation and function of the deflection yoke, it is necessary to examine its underlying structure. The definition of field strength is given by:

$$H D_a = ni$$

Where:
H = field strength (H)
D_a = yoke diameter (m)
n = number of turns
i = current through the yoke (A)

The length of the field is expressed by [7]:

$$l = r \sin \theta$$

Where:
l = length of the uniform magnetic field
r = radius of the electron path
θ = the deflection angle

The dimensions of the yoke are set by the maximum deflection angle and the outside diameter of the CRT neck. These establish the maximum length and minimum inside diameter of the yoke. Given these parameters and the accelerating voltage of the CRT, supporting math may be reduced to provide a simple relationship between the yoke inductance and current:

$$k = \frac{LI^2}{2}$$

Where:
k = the yoke energy constant for a given deflection angle and accelerating voltage (µs)
L = yoke inductance (µH)
I = yoke current to deflect the beam through the half-angle selected by the accelerating voltage (A)

The maximum voltage induced across the yoke is found from the following:

$$e = L\frac{di}{dt}$$

Where:
e = the induced potential (V)
L = yoke inductance (H)
di/dt = rate of change of current (A/s)

Another factor that determines the upper operating limit for the yoke is the retrace time, which must be less than the approximately 10 µs allowed for NTSC video. The natural retrace time for a yoke can be found from the following:

$$T_r = \pi \left(L_y C_y\right)^{1/2} \leq 10\,\mu s$$

Where:
L_y = yoke inductance (H)
C_y = yoke capacitance (F)
T_r = retrace time (s)

Using a shunt capacitance value of 330 pF leads to a maximum yoke inductance of 30 mH. This applies to the horizontal-deflection yoke. The vertical-deflection yoke can be larger because of the longer retrace time of about 2 ms for NTSC, leading to yoke inductances as high as 1 H.

For a specific yoke with a constant anode voltage, the current required for deflection varies as the sine of the deflection angle:

$$\frac{I_2}{I_1} = \frac{\sin\theta_2}{\sin\theta_1}$$

It follows that:

$$I_2 = I_1 \frac{\sin\theta_2}{\sin\theta_1}$$

Where:
I_1 = the given current
I_2 = new current
θ_1 = the given deflection angle
θ_2 = the new deflection angle

If the deflection yoke is given, and the deflection angle is constant, then the deflection current will vary as the square root of the anode voltage:

$$\frac{I_2}{I_1} = \left\{\frac{V_2}{V_1}\right\}^{1/2}$$

Where:
V_1 = the given anode voltage
V_2 = new anode voltage

The step response of the yoke can be found from [3]:

$$I = \frac{E}{R\left\{1 - e^{(-Rt/L)}\right\}}$$

Where:
E = the voltage across the yoke
R = yoke resistance
t = settling time
I = current through the yoke
L = yoke inductance

If R/L is considerably smaller than 1, this equation reduces to:

$$I = \frac{E\,R\,t}{R\,L}$$

It can be further simplified to:

$$t = \frac{I\,L}{E}$$

This commonly used equation for *settling time* is accurate to about 1 percent for settling time to 99 percent of the final value. However, it may be in error by as much as 25 percent for settling time to 99.9 percent of the final value, or if R/L is not sufficiently small. However, it is usually adequate for most calculations and is in general use.

All the proportionalities of yoke application may be combined in a single equation as follows:

$$I_2 = I_1 \left\{\frac{L_1}{L_2}\right\}^{1/2} \left\{\frac{E_2}{E_1}\right\}^{1/2} \frac{\sin\theta_2}{\sin\theta_1}$$

Yoke Selection Parameters

The selection of a yoke for a given application involves the consideration of a number of parameters, some of which require design tradeoffs. The typical selection procedure includes the following steps:

- Select a CRT and determine the maximum deflection angle
- Determine yoke inside diameter (ID) dimensions from the CRT neck size
- Find the yoke energy constant from the yoke manufacturer, based on the ID and deflection angle
- Establish the anode voltage
- Determine the half-angle deflection for the two axes
- Calculate the energy constants
- Set the minimum time for the half-angle deflection
- Find the maximum allowable induced voltage
- Calculate the maximum allowable yoke inductance

For any given application, a number of choices usually exist, although some may be more practical to realize in hardware than others.

7.4 Distortion Correction Circuits

The real-world restrictions of manufacturing tolerances and practical deflection systems result in devices that deviate from the ideal case. This deviation usually requires external correction circuitry to permit the device to operate within tolerances demanded by the end-user. Depending on the type of CRT and the intended application, one or more correction signals may be applied to the deflection elements.

7.4.1 Flat-Face Distortion Correction

The theoretical basis for flat-face distortion is defined by the equation [3]:

$$\frac{d_a}{d_i} = \frac{R_a \sin\theta_a}{R_i \sin\theta_i}$$

This equation, the variables of which are specified in Figure 7.6, specifies the ratio of the true deflection to the deflection on an ideal surface that has its deflection center at the center of curvature. This distortion may be minimized through the use of special correction circuits. The departure from linearity (illustrated in Figure 7.7) leads to the correction equation:

$$d_a = d_i \left\{ 1 + \frac{K^2 I^2}{2} \right\}$$

Linearity correction can be achieved by using a circuit that compensates for the second term of the preceding equation (d_i). When rewritten by substituting KI for d_i the following statement results:

$$d_a = KI + \left\{ \frac{K^3 I^3}{2} \right\}$$

Because $d_i = KI$, the second term in the equation must be subtracted to achieve linearity, with the same type of correction applied to both the X and the Y axes. This is done by obtaining the current for each axis from the yoke winding and using the type of circuit shown in Figure 7.8. The cubic term is generated by means of a piecewise approximation, using as many diodes as necessary to achieve the desired accuracy of correction. Ten segments are usually sufficient, although only five are shown in the figure for simplicity. The circuit operates by biasing the diodes to different voltages so they will conduct only when the output of the first summing amplifier exceeds the voltages. The diode currents are then summed in the output amplifier. The output voltage is given by:

$$E_o = \frac{E_i R_f}{R_p}$$

Where:
R_p = the equivalent value of the summing resistors
E_i = uncorrected input voltage (see Figure 7.8)
E_o = corrected output voltage (Figure 7.8)

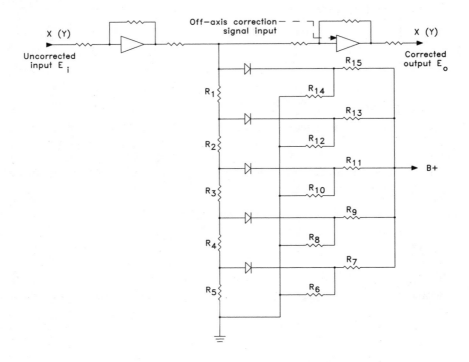

Figure 7.8 Basic linearity correction circuit for a CRT. (*After* [3].)

The similarity to a digital-to-analog (D/A) converter is apparent, but the resistors are not binary weighted, and E_i depends on the current in the yoke. The values for R_n in the resistor network may be determined by plotting:

$$E_o = K I^3$$

This plot is shown in Figure 7.9. From this plot a piecewise approximation can be made, reading the bias levels from the horizontal axis where a new segment begins, and deriving the summing resistor values from the slope of the segment. A fairly accurate result may be attained by choosing R1 to R5 (in Figure 7.8) with each increasing by a factor of 2 from the previous one so that the resistors are defined by $R_n/2_n$, where n takes on the values from zero to the maximum number of segments (minus one).

The network shown in Figure 7.8 corrects only for on-axis nonlinearity. However, when both *X* and *Y* deflection signals are present, they will affect each other. Therefore, it is necessary to cross couple the two inputs (as shown in Figure 7.10).

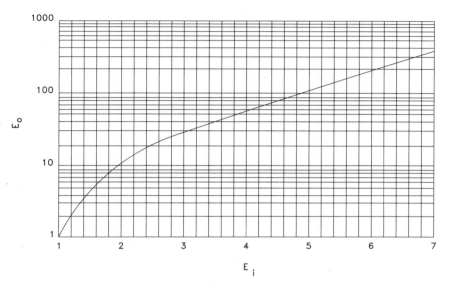

Figure 7.9 A plot of E_o versus E_i where $E_o = KI^3$. (*After* [3].)

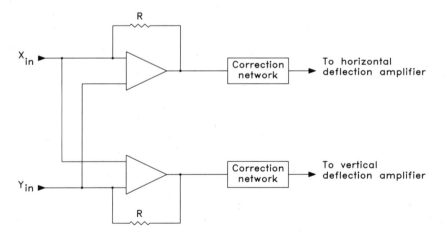

Figure 7.10 Block diagram of X/Y linearity correction circuit. (*After* [3].)

7.4.2 Dynamic Focusing

The effect of deflection on the focus of the electron beam can be minimized by means of a correction circuit. A typical approach is shown in Figure 7.11. The circuit operates on the

232 Video Display Engineering

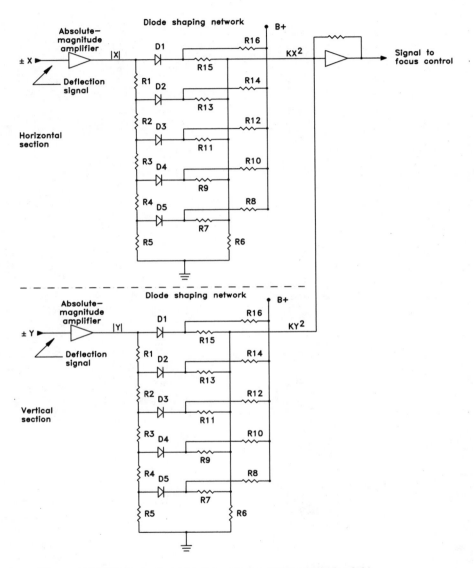

Figure 7.11 Basic design of a dynamic focus circuit. (*After* [3].)

basis that the diameter of the electron beam is affected by the deflection distance on the face of the CRT, given by:

$$r_s = K d^2$$

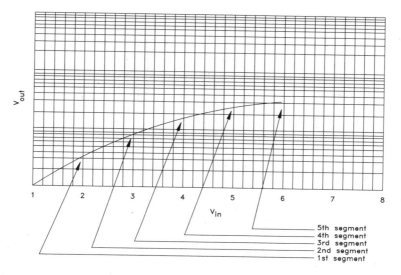

Figure 7.12 Plot of a dynamic focusing correction signal. (*After* [3].)

Where:
r_s = the change in spot size
d = deflection distance
K = a constant

The deflection from the center d consists of the X and Y components, and it is given by the following:

$$d^2 = X^2 + Y^2$$

The correction signal requires that the X and Y terms be generated and then summed. The square functions are produced by piecewise approximations using a diode network similar to one used for the cubic function described in the previous section. In the Figure 7.11 circuit, the diode break points and the values of the summing resistors may be determined by plotting the correction signal, as shown in Figure 7.12, and then converting the data into the number of segments required (usually not more than five). In the case illustrated, all resistors (except the first) may have the same value, which simplifies the circuit.

After the X^2 and Y^2 functions have been attained, they are added in the output summing amplifier and applied to the focusing circuit. It is possible through the use of this type of correction scheme to maintain a spot size ratio of less than 1.5:1 over 35° of deflection instead of a variation of 3:1, which is not uncommon without correction. Thus, the use of dynamic focus correction is a necessary part of any well-designed deflection system.

234 Video Display Engineering

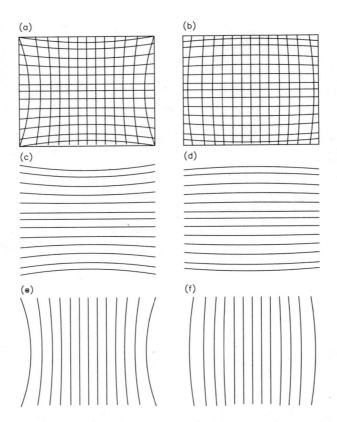

Figure 7.13 Pincushion distortion: (*a*) overall effect on the displayed image, (*b*) corresponding composite correction signal, (*c*) horizontal pincushion distortion component, (*d*) horizontal correction signal, (*e*) vertical pincushion distortion component, (*f*) vertical correction signal.

7.4.3 Pincushion Correction

Pincushion distortion is illustrated in Figure 7.13(*a*). Pincushion correction is used in all wide-angle deflection systems. This correction may be achieved by one of the following:

- A special yoke design that has controlled field distortion
- Predistorting the deflection current and applying it to a separate pincushion transformer that connects the correction current to the vertical yoke

The correction must compensate for the top lines shown in Figure 7.13 (*c*), which bow down, and the bottom lines, which bow up. The center line is straight. The corresponding correction to the raster is achieved by introducing a parabolic waveform to the vertical deflection signal, as shown in Figure 7.13(*d*). The vertical deflection signal is modulated by

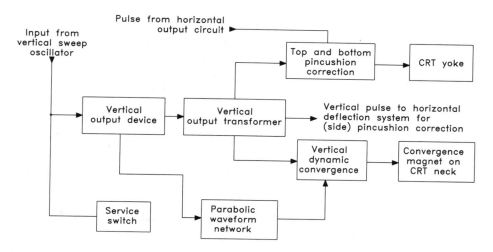

Figure 7.14 Block diagram of a pincushion correction circuit.

the parabolic correction waveform at the horizontal deflection rate. It is then combined with the vertical deflection system output to produce the corrected vertical deflection signal. A block diagram of such a system is shown in Figure 7.14.

Side correction is another form of pincushion compensation. The distortion to be corrected occurs in the vertical dimension, as shown in 7.13(e). In this case, the vertical deflection current modulates the horizontal deflection signal at the vertical scanning frequency. To accomplish this task, the vertical signal is modified by an RC network (or by an equivalent means) to produce the desired correction wave shape, which is then applied through a pincushion transformer to the horizontal yoke.

7.5 References

1. Whitaker, Jerry C., and K. Blair Benson (eds.): *Standard Handbook of Video and Television Engineering*, 3rd ed., McGraw-Hill, New York, N.Y., 2000.
2. Spangenberg, K. R., *Vacuum Tubes*, McGraw-Hill, New York, N.Y., 1948.
3. Sherr, S., *Fundamentals of Display Systems Design*, Wiley, New York, N.Y., 1970.
4. Cloz, R., et al.: "Mechanism of Thin Film electroluminescence," *Conference Record*, SID Proceedings, Society for Information Display, San Jose, Calif., vol. 20, no. 3, 1979.
5. Sherr, S., *Electronic Displays*, Wiley, New York, N.Y. 1979.
6. Popodi, A. E., "Linearity Correction for Magnetically Deflected Cathode Ray Tubes," *Elect. Design News*, vol. 9, no. 1, January 1964.
7. Fink, Donald, (ed.): *Television Engineering Handbook*, McGraw-Hill, New York, N.Y., 1957.

7.6 Bibliography

Aiken, W. R.: "A Thin Cathode Ray Tube," *Proc. IRE*, vol. 45, no. 12, pp. 1599–1604, December 1957.

Barkow, W. H., and J. Gross: "The RCA Large Screen 110° Precision In-Line System," ST-5015, RCA Entertainment, Lancaster, Pa.

Casteloano, Joseph A.: *Handbook of Display Technology*, Academic, New York, N.Y., 1992.

Fink, Donald, and Donald Christiansen (eds.): *Electronics Engineers Handbook*, 3rd ed., McGraw-Hill, New York, N.Y., 1989.

Hutter, Rudolph G. E., "The Deflection of Electron Beams," in *Advances in Image Pickup and Display*, B. Kazan (ed.), vol. 1, pp. 212–215, Academic, New York, N.Y., 1974.

Jordan, Edward C. (ed.): *Reference Data for Engineers: Radio, Electronics, Computer, and Communications*, 7th ed., Howard W. Sams, Indianapolis, IN, 1985.

Morell, A. M., et al.: "Color Television Picture Tubes," in *Advances in Image Pickup and Display*, vol. 1, B. Kazan (ed.), pg. 136, Academic, New York,N.Y., 1974.

Pender, H., and K. McIlwain (eds.), *Electrical Engineers Handbook*, Wiley, New York, N.Y., 1950.

Sinclair, Clive, "Small Flat Cathode Ray Tube," *SID Digest*, Society for Information Display, San Jose, Calif., pp. 138–139, 1981.

Zworykin, V. K., and G. Morton: *Television*, 2d ed., Wiley, New York, N.Y., 1954.

Chapter 8
Color CRT Display Devices

8.1 Introduction

Many types of color CRTs have been developed for video, data, and special display applications, including:

- Shadow-mask tube
- Parallel-stripe tube
- Voltage-penetration tube

The majority of color tubes used for consumer and professional applications fall into three size categories:

- 14-in diagonal
- 19-in diagonal
- 21-in diagonal

Within each category there are four primary grades of resolution, based on the center-to-center spacing between phosphor dots of the same color (*pitch*):

- *Low resolution*—dot pitch 0.44 to 0.47 mm
- *Medium resolution*—dot pitch 0.32 to 0.43 mm
- *High resolution*—dot pitch 0.28 to 0.31 mm
- *Ultrahigh resolution*—dot pitch 0.21 mm (or less) to 0.27 mm

8.1.1 Shadow-Mask CRT

The shadow-mask CRT is the most common type of color display device. As illustrated in Figure 8.1, it utilizes a cluster of three electron guns in a wide neck, one gun for each of the colors red, green, and blue. All the guns are aimed at the same point at the center of the shadow-mask, which is an iron-alloy grid with an array of perforations in triangular arrangement, generally spaced 0.025-in between centers for standard-definition entertain-

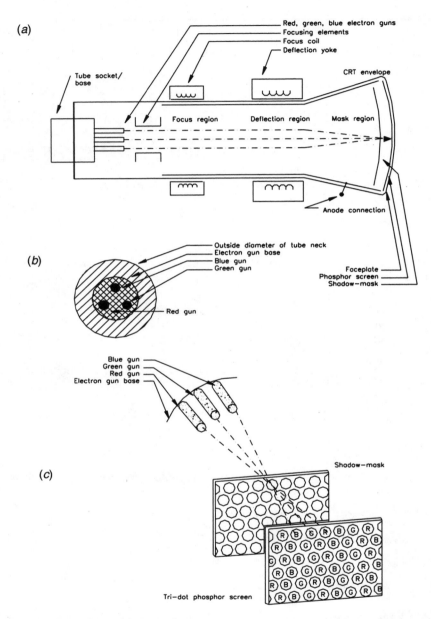

Figure 8.1 Basic concept of a shadow-mask color CRT: (*a*) overall mechanical configuration, (*b*) delta gun arrangement on tube base, (*c*) shadow-mask geometry.

ment television. For high-resolution studio monitor or computer graphic monitor applications, color CRTs with shadow-mask aperture spacing of 0.012-in center-to-center or less, are readily available. This triangular arrangement of electron guns and shadow-mask apertures is known as the *delta-gun configuration*. Phosphor dots on the faceplate just beyond the shadow-mask are arranged so that after passing through the perforations, the electron beam from each gun can strike only the dots emitting one color.

All three beams are deflected simultaneously by a single large diameter deflection yoke, which is usually permanently bonded to the CRT envelope by the tube manufacturer. The three phosphors together are designated P-22, individual phosphors of each color being denoted by the numbers P-22R, P-22G, and P-22B. Most modern color CRTs are constructed with rare-earth-element-activated phosphors, which offer superior color and brightness compared with previously used phosphors.

Because of the close proximity of the phosphor dots to each other, and the strict dependence on angle of penetration of the electrons through the apertures, tight control over electron optics must be maintained. Close attention is also paid to shielding the CRT from extraneous ambient magnetic fields and to degaussing of the shield and shadow-mask (usually carried out automatically when the equipment is switched on).

Even if perfect alignment of the mask and phosphor triads is assumed, the shadow-mask CRT is still subject to certain limitations, mainly with regard to resolution and brightness. Resolution restriction results from needing to align the mask apertures and the phosphor dot triads; the mask aperture size controls the resolution that can be attained by the device.

Electron beam efficiency in a shadow-mask tube is low, relative to a monochrome CRT. Typical beam efficiency is 9 percent; considering the three beams of the color tube, total efficiency is approximately 27 percent. By comparison, a monochrome tube may easily achieve 80 percent electron beam efficiency. This restriction leads to a significant reduction in brightness for a given input power to the shadow-mask CRT.

8.1.2 Parallel-Stripe Color CRT

The parallel-stripe class of CRT, such as the popular *Trinitron*, incorporates fine stripes of red-, green-, and blue-emitting phosphors deposited in continuous lines repetitively across the faceplate, generally in a vertical orientation. (*Trinitron* is a registered trademark of Sony.) The Trinitron is illustrated in Figure 8.2. This device, unlike a shadow-mask CRT, uses a single electron gun that emits three electron beams across a diameter perpendicular to the orientation of the phosphor stripes. This type of gun is called the *in-line gun*. Each beam is directed to the proper color stripe by the internal beam-aiming structure and a slitted aperture grille.

The Trinitron phosphor screen is built in parallel stripes of alternating red, green, and blue elements. A grid structure placed in front of the phosphors, relative to the CRT gun, is used to focus and deflect the beams to the appropriate color stripes. Because the grid spacing and stripe width can be made smaller than the shadow-mask apertures and phosphor dot triplets, higher resolutions may be attained with the Trinitron system.

Elimination of mask transmission loss, which reduces the electron beam-to-luminance efficiency of the shadow-mask tube, permits the Trinitron to operate with significantly greater luminance output for a given beam input power.

The in-line gun is directed through a single lens of large diameter. The tube geometry minimizes beam focus and deflection aberrations, greatly simplifying convergence of the red, green, and blue beams on the phosphor screen.

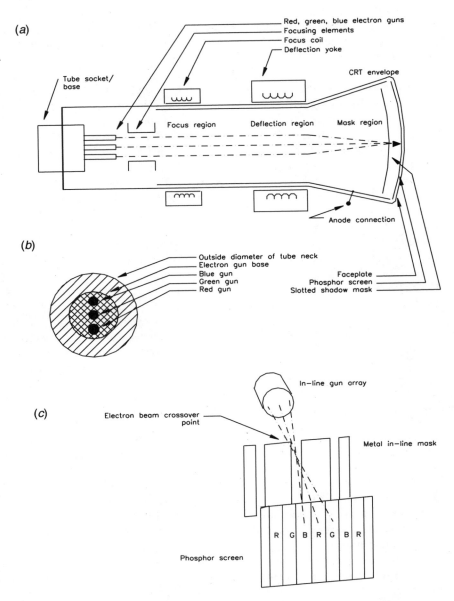

Figure 8.2 Basic concept of the Trinitron color CRT: (*a*) overall mechanical configuration, (*b*) in-line gun arrangement on tube base, (*c*) mask geometry.

The *Lawrence tube*, or *Chromatron*, is another example of the parallel-stripe-phosphor class of color CRT [1]. It employs a single electron beam. Color selection is accomplished by control voltages applied between interdigitated combs of wires, which constitute a steer-

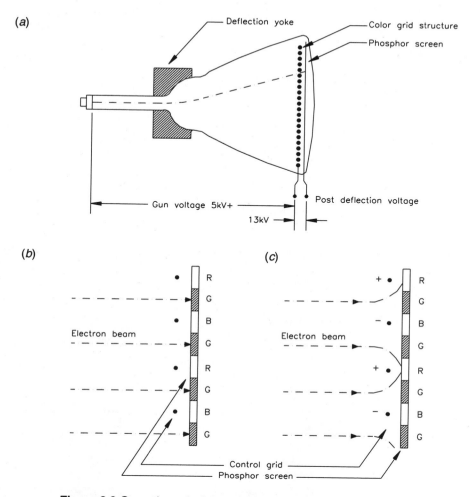

Figure 8.3 Operating principles of the Chromatron CRT: (*a*) overall mechanical configuration, (*b*) no deflection voltage applied to the grid and electrons strike the green phosphor strips, (*c*) deflection voltage applied to the grid with a polarity such that electrons are deflected to the blue phosphor strips (the opposite polarity produces red). (*After* [1].)

ing grid. The grid is placed in front of the phosphor plane to deflect the beam to the appropriate color strip during the scanning process. Figure 8.3 illustrates the theory of operation.

8.1.3 Beam Penetration Color CRT

The beam penetration color CRT is a mechanically simple design that uses a single electron beam, and does not require a mask or index stripes. Production of color is based on controlled depth of penetration of the electron beam into a multilayer phosphor screen.

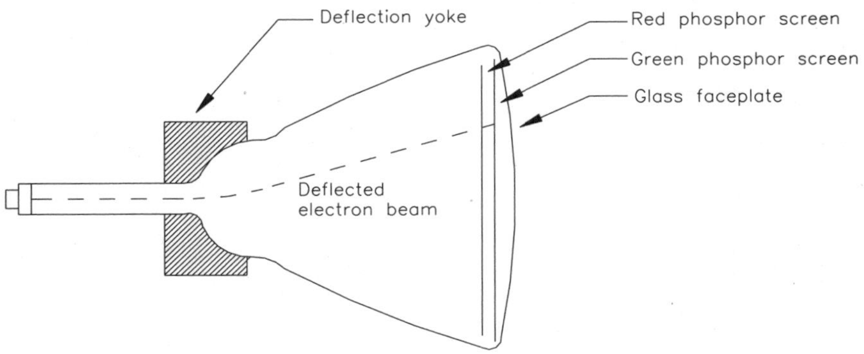

Figure 8.4 Mechanical layout of the beam penetration color CRT.

Figure 8.4 shows the basic mechanical structure. The beam-penetration CRT typically uses one of the following phosphor deposition methods:

- Two or three unstructured layers of phosphors emitting different colors are deposited upon each other, sometimes with a nonluminescing, transparent-barrier layer between them for better color differentiation.

- Individual phosphor grains are built-up in layers, called *onion skin phosphors*. The core phosphor is generally green-emitting. It is surrounded by a nonemitting layer, which in turn is surrounded by an outer red-emitting layer.

With both types of phosphor screens, a single electron beam is employed, and the resultant color of the screen is determined by preselected beam-accelerating voltages, which are varied to control the depth of beam penetration into the phosphor layers.

Intermediate colors are produced by the visual combination of the first color, resulting from the penetration of the first layer by the electron beam, with varying intensities of the second color (as more electrons penetrate into the second color-emitting layer). Therefore, the range of colors produced is limited by the hues of the two phosphor layers, which must lie on the same side of the CIE chromaticity diagram, as close to the spectral locus as possible. A major design challenge associated with the beam-penetration color CRT is the change in deflection sensitivity and focus of the electron beam as the screen voltage is varied to modify the color displayed. This usually dictates operation at only a few preset screen voltages (and therefore colors) where the deflection amplifications and focus voltages are also preset to correspond. Resolution of the beam penetration tube is limited primarily by electron beam size (as in a monochrome CRT).

8.2 Basics of Color CRT Design

The shadow-mask CRT is the workhorse of the video display industry. Used in the majority of color video displays since the introduction of color television in the early 1950s, the shadow-mask technique has been refined to yield excellent performance and low-manufacturing cost.

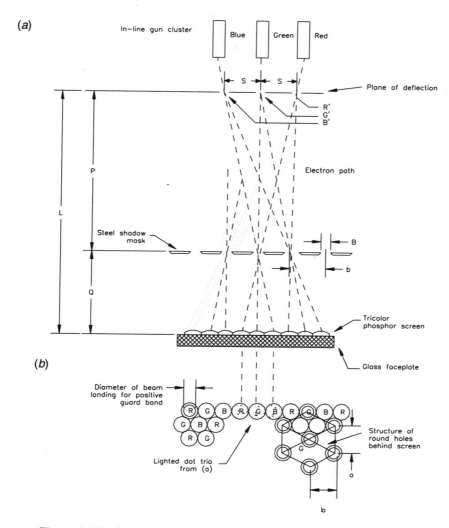

Figure 8.5 Shadow-mask CRT using in-line guns and round mask holes: (*a*) overall tube geometry, (*b*) detail of phosphor dot layout.

8.2.1 Tube Geometry

Figure 8.5 illustrates the shadow-mask geometry for a tube at face center using in-line guns and a shadow-mask of round holes. As an alternative, the shadow-mask may consist of vertical slots, as shown in Figure 8.6. The three guns and their undeflected beams lie in the horizontal plane, and the beams are shown converged at the mask surface. The beams can overlap more than one hole and the holes are encountered only as they happen to fall in the scan line. By convention, a beam in the figure is represented by a single straight line

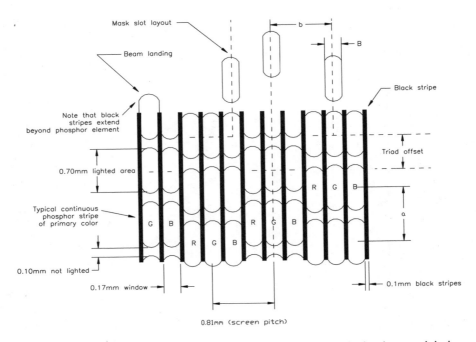

Figure 8.6 Shadow-mask CRT using in-line guns and vertical stripe mask holes.

projected backward at the incident angle from an aperture to an apparent *center of deflection* located in the *deflection plane*. In Figure 8.5, the points B', G', and R', lying in the deflection plane, represent such apparent *centers of deflections* for blue, green, and red beams striking an aperture under study. (These deflection centers move with varying deflection angles.) Extending the rays forward to the facepanel denotes the printing location for the respective colored dots (or stripes) of a tricolor group. Thus, centers of deflection become color centers with a spacing S in the deflection plane. The distance S projects in the ratio Q/P as the dot spacing within the trio. Figure 8.5 also shows how the mask hole horizontal pitch b projects as screen horizontal pitch in the ratio L/P. The same ratio applies for projection of mask vertical pitch a. The Q-space (mask-to-panel spacing) is optimized to obtain the largest possible theoretical dots without overlap. At panel center, the ideal screen geometry is then a mosaic of equally spaced dots (or stripes).

The stripe screen shown in Figure 8.6 is used extensively in color CRTs. One variation of this stripe (or line) screen uses a cylindrical faceplate with a vertically tensioned grill shadow-mask without tie bars. Prior to the stripe screen, the standard construction was a tri-dot screen with a delta gun cluster, as shown in Figure 8.7.

8.2.2 Guard Band

The use of *guard bands* is a common feature for aiding purity in a CRT. The guard band, where the lighted area is smaller than the theoretical tangency condition, may be either *positive* or *negative*. In Figure 8.5, the leftmost red phosphor exemplifies a positive guard

Color CRT Displays   245

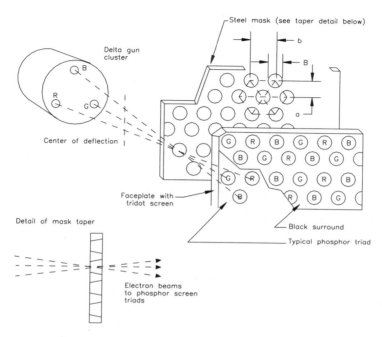

Figure 8.7 Delta-gun, round-hole mask negative guard band tri-dot screen. The taper on the mask holes is shown in the detail drawing only.

band; the lighted area is smaller than the actual phosphor segment, accomplished by mask hole diameter design. Figure 8.6, on the other hand, shows negative guard band (NGB) (or *window-limited*) construction for stripe screens. Vertical black stripes about 0.1 mm (0.004-in) wide separate the phosphor stripes forming windows to be lighted by a beam wider than the window opening by about 0.1 mm. Figure 8.7 shows NGB construction of a tridot screen.

8.2.3 Shadow-Mask Design

The shadow-mask for a color CRT is typically constructed of 0.13 mm low-carbon sheet steel chemically etched to the desired pattern of apertures using photoresist techniques. The photographic masters are made by a precision laser plotter. The completed flat mask is then press-formed to a contour approximately concentric to the faceplate. Mask-to-panel distance (*Q-spacing*) is locally modified to achieve optimum nesting of screen triplets.

Round-hole Mask

The array layout for a typical round-hole shadow-mask is shown in Figures 8.5 and 8.7. The round holes, numbering approximately 440,000 in a standard-definition television display, are placed at the vertices and centers of iterated regular hexagons (long axis vertical). The ratio of horizontal pitch to vertical pitch is given by:

$$\frac{b}{a} = \sqrt{3} = 1.732$$

Where:
a = spacing between holes in a vertical column
b = spacing between hole centers in a horizontal row

This ratio applies both to mask holes and screen triads. When references are made to mask or screen pitch without stating direction, the closest center-line spacing between like elements is assumed.

As illustrated in Figure 8.7, each aperture is tapered to present a more sharply-defined, limiting-aperture plane to an angled incident beam. This construction increases transmission efficiency and prevents color desaturation resulting from electrons being scattered by sidewalls of the apertures. Apertures are graded radially to smaller diameters at screen edge, because that is where the trio configuration and beam quality are less ideal, and registry is more critical. At the tube center, shadow-mask transmission is typically 16 percent for nonmatrix construction and 22 percent for *black matrix* shadow-mask construction.

The black matrix screen is designed to overcome the loss in luminance and resultant brightness caused by a neutral density filter as the faceplate. Such a filter can be used to increase the contrast ratio of the device. This condition arises as a result of the large amount of area in the viewing surface not covered by any phosphor. The need to align the phosphor dots with the mask holes to improve convergence can lead to 50 percent of the surface merely reflecting ambient light (in a medium-quality consumer television display). The black matrix screen covers this area with black, thus reducing back-scattered and reflected light by the same amount as the faceplate. This scheme reduces the loss in the faceplate without affecting contrast. A brighter displayed picture results.

The beam triad pattern can become distorted near the screen edges, resulting in poorer *packing factors* (*geometric nesting*) for the related phosphor dots. For delta-gun systems the beam triad triangles compress radially at all screen edges because of foreshortening. For in-line gun dot screens, on the other hand, the beam triad represents three circles in a horizontal line. Using compass points to designate axes and areas, there is no foreshortening at N and S; at E and W, the mask-panel spacing can be sufficiently increased to restore nesting quality. However, near the screen corners, there will be some rotation of the line trio, thereby demanding smaller holes, dots, and beam landings, which results in less efficient nesting.

Slot Mask

The relationship between a and b, horizontal and vertical pitch, can be chosen by the designer. Only the horizontal pitch will affect Q-space. The factors affecting the choice of horizontal pitch include:

- Display resolution
- Practical mask-panel spacing
- Attainable slot widths
- Ability to manufacture masks and screens with reasonable yields

The main considerations relating to the choice of vertical pitch and tie bars (or *bridges*) include:

- Avoidance of moiré pattern
- Strength of the tie bars
- Transmission efficiency

Vertical screen stripes can be regarded as lined-up oblong dots that have been merged vertically. As a result, there is no color purity or registration requirement in the vertical direction; this simplifies the design and manufacture of display devices.

8.2.4 Resolution and Moiré

Resolution is a measure of the definition or sharpness of detail in displayed image. Resolution may be measured vertically and horizontally. The layout and center-line spacings of the mask apertures are designed to provide sufficient horizontal and vertical lines in the pattern to ensure that they are not the limiting factor in resolution, compared with the number of raster lines. In addition, the number of lines running horizontally in the layout pattern is chosen to avoid moiré fringes resulting from a "beat" with the scan lines. (This latter criterion is not applicable to the cylindrical grill structure of continuous slits.)

For round-hole masks, the effective number of pattern lines running horizontally, allowing for the staggered pattern, is typically about 2.25 times the number of picture horizontal scan lines. In a conventional television display there are approximately 440,000 dot trios on the screen, compared with about 200,000 pixels in the raster. These criteria apply to tube sizes 19-in (48 cm) through 25-in (63 cm) screen diagonal. For smaller tubes, the pattern is made relatively coarser to avoid excessively small apertures and to avoid moiré pattern in the display. The hexagonal pattern used to lay out the centers of round holes is favorable for increasing the number of pattern lines, reducing the likelihood of moiré, and for nesting of screen dots. The zigzag, or staggering, of alternate columns effectively increases the number of columns and rows, as long as the pattern is significantly finer than the raster scan.

For slot apertures, spacings between columns and the resulting number of columns are chosen for resolution and cosmetic appearance. The tie bars or bridges, only about 0.13 mm (0.005-in) in width, are placed according to moiré and strength considerations and are not thought significant for resolution in most applications.

Application Example: Delta Gun Device

The EIA-registered 25VFLP22 90° color tube incorporates a delta gun and tri-dot NGB black matrix screen. The principal mechanical screen parameters are as follows:

- Nominal screen diagonal = 25-in (63 cm)
- Screen height (h) = 39.6 cm
- Screen width (w) = 52.8 cm
- Vertical spacing of trio centers (a) = 0.74 mm

Using the geometry of the hexagon pattern (but without correcting h or w for panel curvature) the following parameters can be determined:

$$N_v = \frac{2h}{a}$$

$$N_h = \frac{2w}{\sqrt{3}\,a}$$

$$N = \frac{(2)hw}{(\sqrt{3})a^2}$$

Where:
N_v = horizontal rows of trios in the pattern
N_h = vertical columns in pattern
N = number of dot trios

For the example device (25VFLP22):
N_v = 1070 horizontal rows of trios in pattern
N_h = 820 vertical columns in pattern
N = 441,417 dot trios

Note that the published value of dot trios (N) for the device in question is 439,000.

The values of N_v and N_h calculated above are not the actual vertical and horizontal resolution measurements, but instead the pattern structure values in comparison with the raster. The mask vertical pitch a_m for the above structure would be approximately 0.70 mm, and the center-hole diameter 0.5 a_m = 0.35 mm, which is smaller than the width of a television scan line. A scan line would then be expected to encompass at least a part of an aperture in both odd and even columns.

Assuming no more than 475 visible scan lines in a 525-line (NTSC) picture, the "2.25 rule" is validated by:

$$1070 \text{ horizontal lines} \div 470 \text{ scan lines} = 2.25$$

Application Example: In-line Gun Device

The EIA-registered 25VGDP22 90° color tube incorporates an in-line gun and NGB stripe screen. The principal mechanical screen parameters are as follows:

- Nominal screen diagonal = 25-in (63 cm)
- Screen height (h) = 39.6 cm
- Screen width (w) = 52.8 cm
- Horizontal center-line spacing of green stripes (b) = 0.82 mm

The number of columns is given by:

$$N_h = \frac{w}{b}$$

For this example, $N_h = 640$ vertical columns in the pattern.

This number is smaller than the comparable figure for the round-hole example, but it is still adequate compared with the nominal 525- or 625-line television raster.

8.2.5 Mask-Panel Temperature Compensation

Color CRTs for consumer, computer, and industrial applications are shifting toward larger, flatter physical dimensions, and higher resolution. Manufacturing such devices for good color purity and stability is a major engineering challenge.

Conventional printing of screen patterns requires that the mask assembly be removed and reinserted multiple times without loss of registry. In the most common mask suspension system, the interior skirt of the glass faceplate has three or four protruding metal studs that engage spring clips welded to the mask frame. Because the shadow-mask may intercept 75 percent (or more) of the beam current, it is subject to thermal expansion during operation. Misregistry would occur if thermal correction were not provided. This temperature compensation is accomplished by automatically shifting the mask slightly closer to the screen surface by thermal expansion of a bimetal structure incorporated into each spring support, or by a designed lever action resulting from the transverse expansion. The exact mechanism used varies from one tube manufacturer to another.

Because the shadow-mask is one of the most critical components in determining CRT performance, the material used to form the structure is an important design parameter. *Aluminum killed* (AK) steel has been extensively employed as mask material because of its etching characteristics, low cost, and relatively good magnetic properties. To build a wide aspect ratio (HDTV) CRT at a size of 30-in or above, mask thickness of at least 0.2 mm is required [2]. The pitch of the device must be 70-100 percent higher than a conventional CRT. Because of the practical limits of etching technology, mask transmission will be significantly lower than a conventional tube of similar size. The percentage of electron beam bombardment onto the mask will, thus, be significantly greater, requiring higher beam current to achieve sufficient brightness. This requirement contributes to color purity problems when the beam is mislanded onto the screen because of mask thermal expansion. This phenomenon is further pronounced on flat CRTs.

To overcome the effects of thermal expansion, an iron-nickel alloy has been introduced as a substitute to AK steel. *Invar*, a *Fe-36Ni* alloy, exhibits a thermal expansion coefficient 10 times lower than AK steel. The effects of misregistration resulting from thermal expansion, therefore, are significantly reduced, as illustrated in Figure 8.8. However, because of the relatively low modulus of elasticity of Invar (40 percent less than that of AK), manufacturing difficulties are encountered with the material, including mask forming and mask mounting to minimize springback effects and microphonics, respectively.

Tension Mask

Mask stability can also be improved by utilizing the *tension mask* method, in which the mask is tensioned during manufacture to a predetermined stress [3]. When the proper level

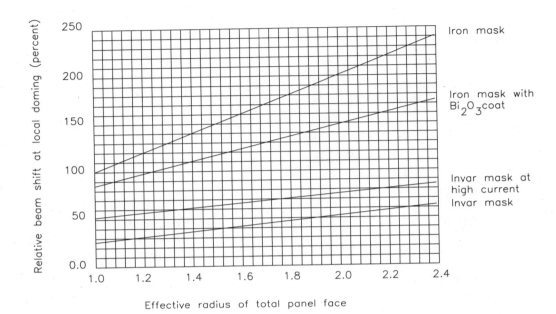

Figure 8.8 The relationship between beam shift at local doming and the effective radius of the faceplate with various mask materials. (*After* [2].)

of tension is applied to the mask, thermal impact on the structure reduces the tension to a certain extent, but will not cause displacement of the aperture features. The effects on color purity resulting from thermal expansion are, therefore, minimized, as shown in Figure 8.9.

The *taut shadow-mask* is a variation on the basic tension mask concept. This design departs from the conventional domed shadow-mask approach and, instead, stretches the shadow-mask into a perfectly flat contour, as illustrated in Figure 8.10 [5]. This design provides greater power handling capability and is less susceptible to Z axis vibration. The taut shadow-mask typically achieves a four-fold increase in maximum beam current for a specified color purity.

On a 5 × 5-in CRT, the taut mask is typically stretched to 50,000 psi. Relaxation occurs at between 1.8 and 2.0 W/in^2. The taut shadow-mask uses a nickel-chrome-titanium material (*Ni Span C*), which exhibits a higher-yield strength than steel and a lower *coefficient of thermal expansion* (CTE) as well. The lower CTE permits more power input before the mask loses its tension. The thickness of the Ni Span C mask is on the order of 0.001-in. Ceramic is used as the mask mounting ring because of its strength and manufacturing advantages.

8.2.6 Faceplate Screening

Screening of the CRT faceplate is typically accomplished using a three-step process. Three fields of phosphor color segments are sequentially printed on the faceplate by photoresist techniques using the tube's own shadow-mask as a master and a different light-

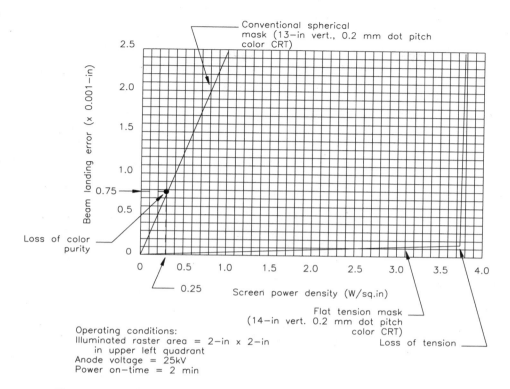

Figure 8.9 Comparison of small area mask doming for a conventional CRT and a flat-tension-mask CRT. (*After* [2].)

house setup for each color. Each lighthouse typically includes a mercury lamp for exposure and corrective optical lenses such that light rays passing through the mask land on the photoresist in a simulation of the deflected electron beam. The light intensity across the panel is controlled by a graded density pattern on a glass sheet. If black matrix is used, it is applied prior to phosphor screening using photoresist methods.

These exposure methods apply to biaxially stretched domed shadow-masks; their stretch-forming variability prohibits both interchangeability and the use of a dedicated corrective mask master for exposure instead of lenses. (The cylindrical-grill mask, which is not stretch-formed, is more amenable to modification of the foregoing methods using special procedures.)

The screened faceplate is aluminized (discussed in Chapter 6) and then low-temperature fitted to a glass funnel using fixturing, which essentially duplicates that used during the lighthouse step. A conductive graphite coating covers the interior wall of the funnel and some of the exterior surface according to a prescribed pattern.

The front faceplate of the CRT easily accumulates static charges because of the high anode voltages present during operation. This causes several inconveniences, including dust from the air that adheres to the outer surface of the device, and electrical shock to the user. In

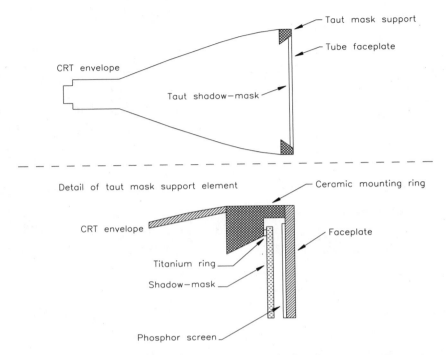

Figure 8.10 Mechanical structure of a taut shadow-mask CRT. (*After* [3].)

order to solve these problems, the outer surface the faceplate is usually covered with an antistatic/antiglare coating. Such coatings are typically based on silane compound systems. Work has also been reported [4] on using thermoplastic and thermosetting polymers for this purpose.

8.2.7 Magnetic Shielding

Because of the earth's magnetic effects, electron beam deflection will change—to some extent—depending on the location and orientation of the CRT. This effect also influences color purity. An external or internal magnetic metal shield is included, therefore, to compensate for the effect of the earth's magnetic field and stray magnetic fields. The low-carbon steel shadow-mask itself also contributes to the shielding. The shield does not extend into the yoke field because this would result in loading. Also, external shields must be clear of the anode button area. With the display orientation in the chosen compass direction, the shield and tube must be thoroughly degaussed to gain full purity. This treatment not only removes any residual magnetization of the shield and shadow-mask, but also induces a residual static magnetic field that bucks the ambient magnetic field.

Table 8.1 Color CRT Diagonal Dimension vs. Weight

Diagonal Visible (in)	Weight (kg)	Weight (lb)
19	12	26
25	23	51
30	40	88

8.2.8 X-radiation

The shadow-mask color tube operates with extremely high anode voltages (typically 20 to 30 kV). The possibility of x-rays must be considered for the safety of technicians and end-users. The color tube envelope is made from glass that has been formulated for x-ray absorbing characteristics. By closely controlling glass thickness, the glassmaker controls the extent of x-ray absorption. Also, the power levels at which the tube is operated must be controlled. Tube-type data sheets provide the relevant absolute maximum ratings (for voltages and currents) that will keep x-radiation within accepted safety levels.

8.2.9 Screen Size

As the consumer demands ever-larger picture sizes, the weight and depth of the CRT, and the higher power and voltage requirements become serious limitations. These are reflected in sharply increasing receiver cabinet costs to accommodate the larger tubes (a major share of the manufacturing costs) and in more complex circuitry for high-voltage operation.

To withstand the atmospheric pressures on the evacuated glass envelope, the CRT weight increases exponentially with the viewable diagonal. Typical figures for television receiver tubes designed for a 4:3 aspect ratio are shown in Table 8.1. (Figure 8.11 charts the relationship.) Nevertheless, manufacturers have continued to meet the demand for larger screen sizes with larger direct-view tubes. Examples include an in-line gun, 110° deflection tube with a 35-in diagonal screen. The device (Mitsubishi and Matsushita) was initially designed for computer displays, but is also suitable for HDTV. In the Trinitron configuration, 37-, 38-, and 43-in diagonal tubes have been produced (Sony).

The weight of 35-in and larger tubes and the depth of the receiver cabinet are of questionable practicality for home use. Consequently, a 27-in tube is the largest size suitable for the majority of home viewing situations.

CRTs designed for HDTV applications suffer an additional weight disadvantage because of the wide aspect ratio of the format (16:9). Furthermore, large-diameter electron guns and small deflection angles (90°) are often employed to improve resolution, making the tube necks fat, and giving the devices greater depth and weight.

8.2.10 Resolution Improvement Techniques

In a shadow-mask display device, higher resolution may be achieved by attention to the following parameters:

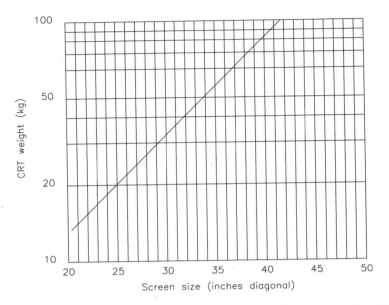

Figure 8.11 Relationship between screen size and CRT weight. (*After* [6].)

- Finer pitch
- Smaller-diameter apertures and screen dots
- Smaller Q-space
- Thinner mask material

Etching of mask holes becomes more demanding when diameters are smaller than the material thickness. The black matrix tri-dot system has a mask-manufacturing advantage because mask aperture diameters are larger than in comparable nonmatrix tri-dot tubes, or in comparable matrix (or nonmatrix) stripe tubes.

Table 8.2 shows comparative resolution capabilities of three 19-in (48 cm) visible (19V) devices with round-hole masks as screen pitch is reduced to increase resolution. The value N_r, a comparative measure of resolution, assumes that resolution is proportional to the square root of the number of trios in a square area with sides equal to screen height. To achieve higher resolution than the third design shown in the table (or for smaller tube sizes), it is necessary to use thinner mask material and smaller holes. The Q-space also reduces proportionally to pitch, becoming critically small for manufacturing.

Display devices for high-definition television of 1125 or 1250 lines include changing the aspect ratio from the conventional 4:3 standard to 16:9 for greater visual impact. Construction of a glass tube of the size required for HDTV presents several design and fabrication problems. The tensile stresses on the bulb must be minimized to avoid cracks resulting from temperature imbalance during and after manufacture, and the danger of subsequent implosion.

Table 8.2 Comparative Resolution of Shadow-mask Designs (After [7].)

19V (48cm) NGB Tube Type	Mask Material (mm)*	Vertical Pitch (mm)*	Center Hole dia. (mm)*	Screen Vertical Pitch (mm)	N_t, Trios in Screen	$N_r = \sqrt{NT}/1.33$
Conventional	0.15	0.56	0.27	0.60	400,000	500 lines
Monitor	0.15	0.40	0.19	0.43	800,000	775 lines
High resolution	0.13	0.30	0.15	0.32	1,400,000	1025 lines

* Flat shadow-mask
Data is based on round-hold mask, tri-dot screen (404 mm H × 303 mm V screen; aspect ratio = 1.33).

A 90° deflection angle is often chosen to provide accurate convergence of the three (RGB) beams and uniform white balance over the full width of the screen, as well as the height. A large neck diameter permits the use of a large-diameter electron gun, necessary to achieve a high degree of resolution at high values of beam current. Reduction of moiré to an unnoticeable level requires an increase in the frequency of the pattern, or a decrease in the amplitude. The pattern can be minimized through the use of certain specific values of triad pitch. However, selection of the pitch is limited by the thinner shadow-mask for a finer pitch, and by a sacrifice in resolution for a more coarse pitch. For a 40-in tube, a pitch of 0.46 mm provides a resolution of 1000 television lines and satisfies the other requirements of strength of the shadow-mask and reduction in moiré pattern.

High-Resolution Trinitron

One of the first color display devices intended for commercial production that satisfied the requirements for use as an HDTV monitor was a 37V-in Trinitron tube (Sony). The principal specifications of the tube included:

- A flat rectangular screen with a square-cornered, 16:9 aspect ratio viewing area.
- Resolution of more than 1000 television lines.
- Highlight luminous intensity of more than 95 cd/m^2

Tube specifications are listed in Table 8.3. An outline and principal dimensions of the device are shown in Figure 8.12.

The more stringent resolution requirements, particularly in the corners of the widescreen, were met through the use of an electron gun with a longer focus field than conventional tubes, and a customized magnetic deflection yoke. In addition, fine spot size and precise color tracking over the full range of luminance levels were achieved at high beam currents. The interaction of spot size with beam current is shown in Figure 8.13. Figure 8.14 illustrates the effects of beam current on tube luminance. The MTF at 1100 lines is reduced only by slightly less than 6 dB at a beam current of 500 µA. At this current, the spot size is approximately 1.25 mm.

Table 8.3 Specifications for a 38V-in High-Resolution Trinitron CRT (*After* [8].)

Parameter		Value
Maximum faceplate dimension (diagonal)		41.3-in (1048 mm)
Overall depth		31.4-in (798 mm)
Neck diameter		36.5-in (1.44 mm)
Weight		231 lb (105 kg)
Useful screen dimensions	Horizontal	33.5-in (852 mm)
	Vertical	18.8-in (477 mm)
	Diagonal	38.5-in (977 mm)
Number of vertical color-stripe trios		1861
Resolution of monitor (TV lines)	Horizontal	Over 1000
	Vertical	Over 750
Deflection angle		90°
Anode voltage		32 kV
Luminance		95 cd/m^2

8.2.11 Flat-Face Tubes

The production of CRTs with a flat faceplate has been a design goal of manufacturers for a number of years. One system (Zenith) uses a *flat tension mask* (FTM) configuration in which a tense, foil mask is supported in precise Q- spacing to a flat panel wherein the periphery of the mask is an integral part of the envelope structure. This monolithic assembly exhibits reduced susceptibility to shock damage, enabling an expanded range of possible applications.

Another system (Hitachi) utilizes a thin stretched-taught shadow-mask, *phototacky* phosphor coating process, and *quadrupole* electron gun. The stretched mask offers improved stability under changing shadow-mask temperatures. The phosphor coating facilitates the printing of fine phosphor dots during manufacture. The quadrupole gun electron geometry provides the optimum beam alignment for the flat mask, minimizing the need for distortion-correcting signals. The color display offers 2,000 × 1,500 pixel resolution (approximately 140 pixels/in).

Additional advantages of the flat face-type CRT device include:

- **Higher geometric brightness**. Full-raster power density is typically in excess of 100 mW/cm^2, and small-area power density is typically 700 mW/cm^2.

- **Higher resolution**. A perfectly flat screen and precision mask-to-panel registry and spacing simplify fabrication of high resolution devices.

- **Manufacturing cost advantages**. The flat face makes it possible to bond inexpensive flat window glass for implosion protection on the device. This attribute facilitates surface treatments, such as antireflective coatings.

(a)

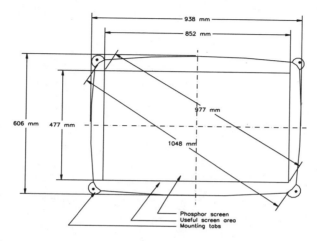

(b)

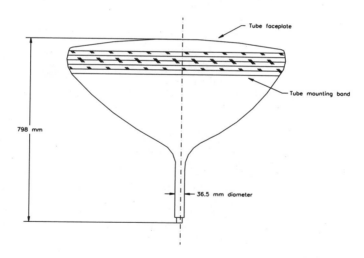

Figure 8.12 Principal dimensions of a 38V-in (965 mm) CRT: (*a*) front view, (*b*) side view. (*After* [8].)

- **Reduction of glare and EMI**. The absence of skirts on the inside of the faceplate permits the use of internal antiglare treatments. Electromagnetic interference shielding may also be incorporated into the bonding system.

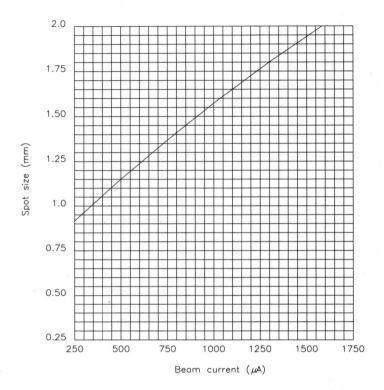

Figure 8.13 Increase in spot size as a function of beam current for a large-screen CRT. (*After* [8].)

- **User preference**. The look of video and data on a flat surface is preferable to many users.
- **Improved etching capabilities**. A thin tension mask is used with a flat face device, generally less than 1.5 mils thick. The thin mask permits the etching of apertures for very fine pitch (0.18 mm is common).

Flat face CRT technology is applicable to a wide range of tube sizes and display resolutions.

8.3 Electron Gun

Figure 8.15 illustrates the general electrode configuration for a shadow-mask color electron gun. The device can be subdivided into three major regions:

- *Beam-forming region*, which consists of the cathode, grid-1, and grid-2 electrodes
- *Prefocus lens region*, which consists of the grid-2 and lower grid-3 electrodes
- *Main lens region*, which consists of the grid-3 and grid-4 electrodes. These elements create a focusing field for the electron beam

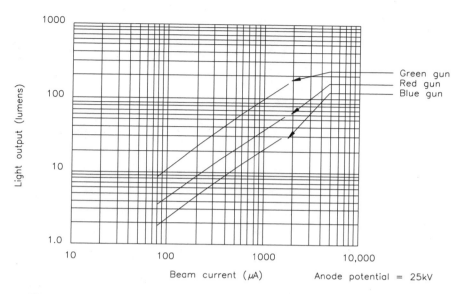

Figure 8.14 Beam current versus luminance characteristics for a large-screen CRT. (*After* [8].)

In more complicated lens systems, additional elements follow grid-4.

8.3.1 Operating Principles

The cathode current, I_k, under conditions of space-charge-limited emission at the cathode surface, can be calculated for grid drive by the empirically derived relation:

$$I_k = KV_D^{3.0} V_c^{-1.5} \; \mu A$$

Where:
K = modulation constant V_D-positive-going drive signal applied to negative grid-1 (grid drive), V
V_c = grid-1 spot cutoff voltage

This formula is valid until V_D reaches approximately one-half of V_c. Above that point, the formula is revised to the following:

$$I_k = KV_D^{3.5} V_c^{-2} \; \mu A$$

The modulation constant varies between 3.0 and 4.5, depending on geometry in the cathode grid-1/grid-2 region of the electron gun.

When a negative-going signal is applied to a positive cathode for cathode drive, the grid-2 accelerating voltage varies with respect to cathode and becomes a factor in the above equation. In this case, the formula must be modified to include the grid-2 effect. Grid-1

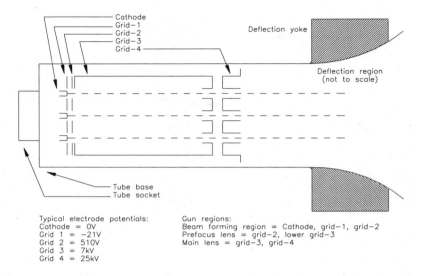

Figure 8.15 Simplified mechanical structure of a bipotential color electron gun.

spot-cutoff voltage in the formula is replaced with cathode spot cutoff, and the entire right-side of the equation is multiplied by the added factor:

$$1 + \left\{ \frac{\text{cathode drive signal}}{\text{grid-2 accelerating voltage}} \right\}^{1.5}$$

To yield the following:

$$I_k = K V_D^{3.5} V_k^{-2} \left\{ 1 + \left[\frac{K_d}{V_{g2}} \right]^{1.5} \right\} \mu A$$

Where:
V_k = cathode spot cutoff voltage
K_d = cathode drive signal
V_{g2} = grid-2 accelerating voltage

This equation is valid so long as the grid-2 voltage does not reach unusually small values (the cathode drive signal or less). Further, any grid-3 field penetrating through grid 2 will cause the effective accelerating voltage to be higher than the measured grid-2 voltage, requiring some adjustment to the grid-2 value in the formula.

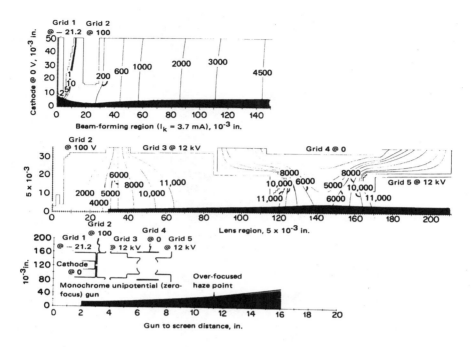

Figure 8.16 Bipotential electron gun configuration, shown with the corresponding field and beam plots. (*From* [7]. *Used with permission.*)

8.3.2 Electron Gun Classifications

Electron guns for color tubes can be classified according to the main lens configuration, which include:

- Unipotential
- Bipotential
- Tripotential
- Hybrid lenses

The unipotential gun (discussed in Chapter 6 as it applies to monochrome operation) is the simplest of all designs. This type of gun is rarely used for color applications, except for small screen sizes. The system suffers from a tendency toward arcing at high anode voltage, and relatively large, low-current spots.

The bipotential lens is the most commonly used gun in shadow-mask color tubes. The arrangement of gun electrodes is shown in Figure 8.16 along with a computer-generated plot of equipotential lines and electron trajectories. The main lens of the gun is formed in the gap between grid-3 and grid-4. When grid-3 operates at 18 to 22 percent of the grid-4 voltage,

262 Video Display Engineering

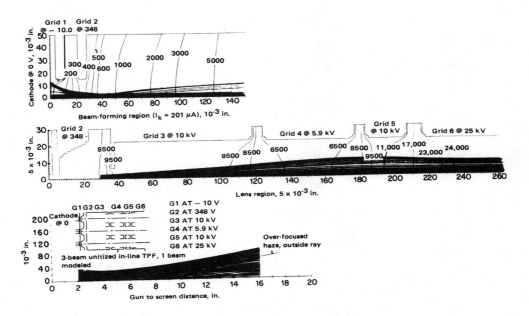

Figure 8.17 Tripotential electron gun configuration, shown with the corresponding field and beam plots. (*From* [7]. *Used with permission*.)

the lens is referred to as a *low bipotential* configuration, often called *LoBi* for short. When grid-3 operates at 26 to 30 percent of the grid-4 voltage, the lens is referred to as a *high bipotential* or *HiBi* configuration.

The LoBi configuration has the advantages of a short grid-3 and shorter overall length, with parts assembly generally less critical than the HiBi configuration. However, with its shorter object distance (grid-3 length), the lens suffers from somewhat larger spot size than the HiBi. The HiBi, on the other hand, with a longer grid-3 object distance, has improved spot size and resolution. The focus voltage supply for the LoBi also can be less expensive.

In the computer plot shown in Figure 8.16, it can be seen that the bipotential beam starts with a crossover of the beam near the cathode, rising to a maximum diameter in the lens. After a double bending action in the lens region, the rays depart in a convergent attitude toward the screen of the tube.

Further improvement in focus characteristics has been achieved with a tripotential lens. The electrode arrangement is shown in the computer model of Figure 8.17. The lens region has more than one gap and requires two focus supplies, one at 40 percent and the other at 24 percent of the anode potential. With this refinement the lens has lower spherical aberration. Together with a longer object distance (grids-3 to -5), the resulting spot size at the screen is smaller than the bipotential designs. Drawbacks of the tripotential gun include:

- The assembly is physically longer
- It requires two focus supplies

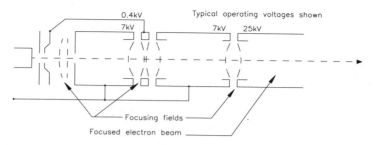

Figure 8.18 Electrode arrangement of a UniBi gun.

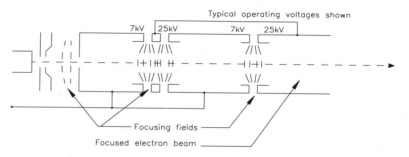

Figure 8.19 Electrode arrangement of a BiUni gun.

- It requires a special base to deliver the high focus voltage through the stem of the tube

Hybrid Lenses

Improved performance may be realized by combining elements of the unipotential and bipotential lenses in series. The more common of these configurations are known as *UniBi* and *BiUni*. The UniBi structure (sometimes referred to as *quadripotential focus*) combines the HiBi main lens gap with the unipotential type of lens structure to collimate the beam. As shown in Figure 8.18, the grid-2 voltage is tied to a grid-4 inserted in the object region of the gun, causing the beam bundle to collimate to a smaller diameter in the main lens. With this added focusing, the gun is slightly shorter than a bipotential gun having an equal focus voltage.

The BiUni structure, illustrated in Figure 8.19, achieves a similar beam collimation by tying the added element to the anode, rather than to grid-2. The gun structure is shorter because of the added focusing early in the device. With three high-gradient gaps, arcing can be a problem in the BiUni configuration.

264 Video Display Engineering

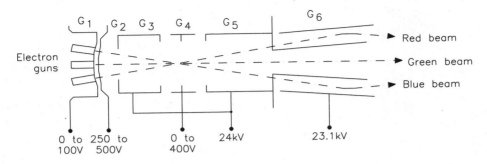

Figure 8.20 Electrode arrangement of the Trinitron gun. (*After* [9].)

Trinitron

The Trinitron gun consists of three beams focused through the use of a single, large, main-focus lens of unipotential design. Figure 8.20 shows the three in-line beams mechanically tilted to pass through the center of the lens, then reconverged toward a common point on the screen. The gun is somewhat longer than other color guns, and the mechanical structuring of the device requires unusual care and accuracy in assembly.

8.3.3 Gun Arrangements

Prior to 1970, most color gun clusters used a delta arrangement. Since that time the design trend has been toward the three guns in-line on the horizontal axis of the tube.

The electron optical performance of a delta gun is superior to in-line—in its basic form—by virtue of a larger lens diameter, resulting from more efficient use of the available area inside the neck. Larger neck diameters improve lens diameter, but at the expense of higher deflection power and more difficult convergence problems over the face of the tube. Typical delta gun lens diameters include:

- 12 mm (0.5-in) for a 51 mm (2-in) neck OD
- 9 mm (0.4 in) for a 36 mm (1.4-in) neck
- 7 mm (0.3 in) for a 29 mm (1.1-in) neck

Delta guns use individual cylinders to form the electron lenses, and the three guns are tilted toward a common point on the screen. The individual cylinders are subject to random errors in position that can cause misconvergence of the three undeflected spots. Furthermore, carefully tailored current waveforms are applied to magnetic pole pieces on the separate guns to dynamically converge the three beams over the full face of the tube.

In-line guns enjoy one major advantage that more than offsets their less efficient use of available neck area. That is, with three beams on the horizontal axis, a deflection yoke can be built that will maintain dynamic convergence over the full face of the tube, without the need for correcting waveforms. With this major simplification in needed circuitry almost all guns being built for commercial video applications are of the in-line design. In-line guns are available for 36 mm, 29 mm, and 22.5 mm neck diameters (among others). Typical lens diameters for these three guns are 7.5 mm, 5.5 mm, and 3.5 mm, respectively.

Unitized Construction

Most in-line guns are of *unitized construction* in which the three apertures or lens diameters are formed from a single piece of metal. With this arrangement, the apertures and lens diameters are accurately fixed with respect to each other so that beam landings at the screen are more predictable than with the cylinder guns. Also, self-converging deflection yokes need the more accurate positioning of the beams in the yoke area. The only trio of gun elements not electrically tied together are the cathodes. Therefore, all varying voltages controlling luminance and color must be applied to only the cathodes. Note that the three beams in unitized guns travel parallel until they reach the final lens gap. Here, an offset, or tilted, lens on the two outboard beams bends the two outer beams toward the center beam at the screen. Any change in the strength of this final lens gap, such as a focus voltage adjustment, will cause a slight change in the undeflected convergence pattern at the screen.

8.3.4 Guns for High-Resolution Applications

Improved versions of both the delta and in-line guns previously noted are used in high-resolution display applications. In both cases the guns are adjusted for the lower beam current and higher resolution needed in data and/or graphics display. The advantages and disadvantages noted in earlier sections for delta and in-line guns also apply here. For example, the use of a delta-type cylinder, or *barrel-type* gun, requires as many as 20 carefully tailored convergence waveforms to obtain near-perfect convergence over the full face of the tube. The in-line gun, with a self-converging yoke, avoids the need for these waveforms, at the expense of slightly larger spots, particularly in the corners where overfocused haze tails can cause problems.

Figure 8.21 compares spot sizes (at up to 1 mA of beam current) for high-resolution designs compared with commercial receiver-type devices. Both delta and in-line 13- and 19-in vertical (13V and 19V) devices are shown. Note the marked improvement in spot size at current levels below 500 µA for the high-resolution devices.

8.4 Deflecting Multiple Electron Beams

Deflection of the electron beams in a color CRT represents a difficult technical exercise. The main problems that occur when the three beams are deflected by a common deflection system are associated with spot distortions that occur in single-beam tubes (discussed in Chapter 7). However, the effect is intensified by the need for the three beams to cross over and combine as a spot on the shadow-mask. The two most significant effects are:

- Curvature of the field
- Astigmatism

Misalignment and misregistration of the three beams of the color CRT will lead to loss of purity for colors produced by combinations of the primary colors. Such reproduction distor-

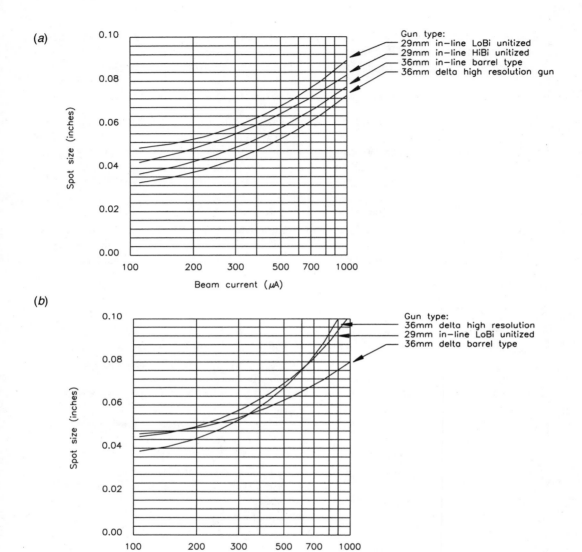

Figure 8.21 Spot-size comparison of high-resolution guns versus commercial television guns: (*a*) 13-in vertical display, (*b*) 19-in vertical display. (*After* [7].)

tions can also result in a reduction in luminance output because a smaller part of the beams are passing through the apertures. Additional errors that must be considered include:

- Deflection-angle changes in the yoke-deflection center
- Stray electromagnetic fields

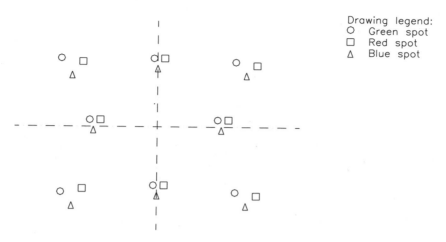

Figure 8.22 Astigmatism errors in a color CRT. (*After* [10].)

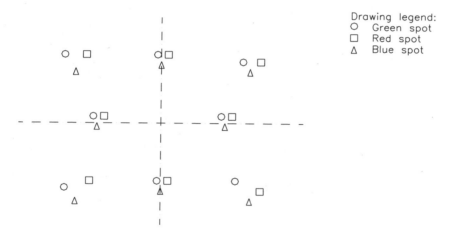

Figure 8.23 Astigmatism and coma errors. (*After* [10].)

8.4.1 Distortion Effects

Curvature of field and astigmatism distortions result in a misconvergence of the beam. Figure 8.22 illustrates distortion resulting from astigmatism (for a delta gun tube). Figure 8.23 illustrates misconvergence of the beam resulting from astigmatism and coma. The misconvergence that occurs in the four corners of the raster is shown in Figure 8.24. The result is that color rendition will not be true, particularly at the edges of the screen. This can be partially compensated by introducing quadripole fields, which cause the beams to be twisted and restore the equilateral nature of the triangle. The shapes these fields may take are illustrated in Figure 8.25 along with the currents required to produce the fields.

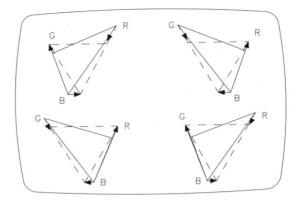

Figure 8.24 Misconvergence in the four corners of the raster in a color CRT. (*After* [10].)

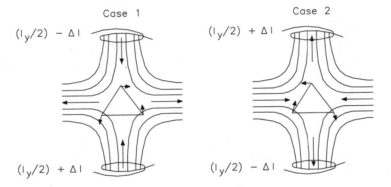

Figure 8.25 Field configurations suitable for correcting misconvergence in a color CRT. (*After* [10].)

Another technique for producing these fields involves placement of magnets around the neck of the CRT, or the use of internal pole pieces that are magnetized by external coils. Such a field is shown in Figure 8.26 for a four-pole field; a six-pole field is also commonly used. While these techniques have been used with some success, they require a rather cumbersome adjustment procedure to be effective. However, their adequacy is attested to by the success of the delta-gun shadow-mask tube in commercial television. The situation is less satisfactory for data display, where misregistration results in color errors that cannot be ignored, especially at the edges of the display.

The registration challenge is less severe in the case of the in-line gun; the three beams need only be converged into a vertical line, rather than the round spot required by the delta gun. For the in-line gun, a precision self-converging system is used where the yoke is designed to operate with one specific tube. This self-converging yoke causes the beams to diverge horizontally in the yoke, resulting in non-uniform fields that counteract the

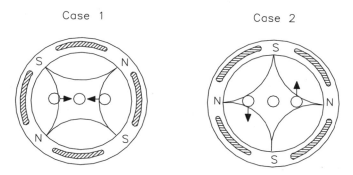

Figure 8.26 Four-pole field used for misconvergence correction in a color CRT. (*After* [10].)

overconvergence. The shape of the fields for horizontal and vertical deflection are shown in Figure 8.27. The horizontal yoke generates a pincushion field, while the vertical yoke generates a barrel-shaped field to accomplish these ends. However, complete self-convergence without the need for compensating adjustments is possible only with narrow-angle tubes; for 110° tubes it is necessary to add some convergence adjustments, although much less than is required for the delta gun. The drawback is that the yoke must be tailored to the specific tube with which it is used.

8.4.2 Deflection Amplifier Considerations

To establish requirements for the horizontal and vertical deflection amplifiers, it is necessary to identify the general characteristics of magnetic deflection for the type or class of CRT to be used. Of prime interest is the response of the amplifier/yoke system to the two most prevalent types of deflection signals:

- *Sawtooth*, used for the horizontal and vertical linear deflection periods
- *Step signal*, resulting from the flyback portion of the deflection signal

A general diagram of a deflection amplifier is shown in Figure 8.28. Note that the yoke is represented as a separate circuit element, and contribution of the yoke to the amplifier response is of great significance. The feedback resistor is another important circuit element because it contributes to both the small-signal and large-signal response of the amplifier.

8.4.3 Considerations for Flat Screen Devices

The trend toward flat faceplates, for both entertainment television and computer monitor applications, places additional demands on the deflection system, specifically the deflection amplifier and yoke. As the faceplate becomes flatter, the North/South (N-S) geometric distortion becomes increasingly pincushion-shaped [12]. Although misconvergence errors do not change significantly, any correction to the pincushion distortion by the yoke magnetic field results in a dramatic change in misconvergence. For a perfectly flat screen, this distortion can be as much as 6 percent, and corrrecting such a large amount of distor-

270 Video Display Engineering

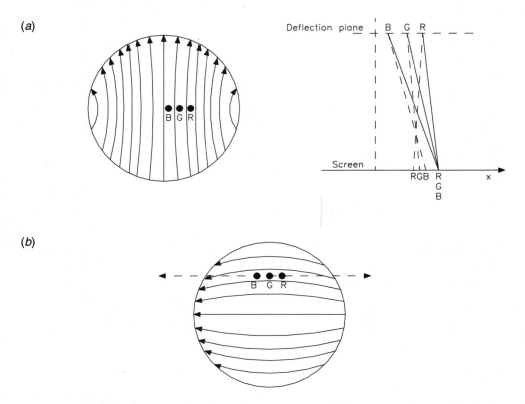

Figure 8.27 Self-converging deflection field in an in-line gun device: (*a*) horizontal deflection field, (*b*) vertical deflection field. (*After* [11].)

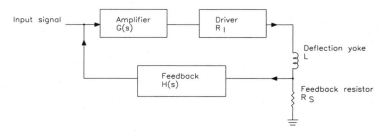

Figure 8.28 Simplified block diagram of a CRT magnetic deflection amplifier. (*After* [7].)

tion without adversely affecting self-convergence is quite difficult using conventional yoke design practices. A number of yoke and deflection amplifier enhancements, therefore, have been devised to accommodate the demanding requirements of the flat screen device.

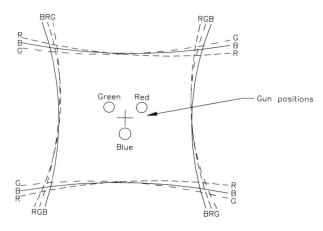

Figure 8.29 The effect of beam parallax. (*After* [13].)

A variety of solutions to this problem have been advanced, with varying degrees of success. Clearly, from a design point of view, the CRT yoke could be simplified if digital circuitry could be incorporated to carry the burden of correcting small residual misconvergence errors. Digital convergence, therefore, is an area of increasing interest by CRT manufacturers. The current selection of flat screen CRTs, however, demonstrate the value of simpler approaches to the problem (i.e., modified mechanical layout of the yoke).

One innovative—if perhaps a bit radical—proposal is the *in-neck yoke*, in which the yoke is included within the neck of the CRT itself. A less aggressive concept is the *integrated tube component* (ITC), where the yoke is aligned on the CRT and any residual errors corrected on a device-by-device basis. Although the ITC scheme requires a major rethinking of yoke design, it holds tremendous potential benefits insofar as labor assembly costs are concerned.

8.4.4 Dynamic Convergence

Dynamic convergence is an essential requirement for delta-gun CRT-based display systems. In-line tubes, which use predistortion in the deflection yokes for this purpose, do not require dynamic convergence of the type described in this section. However, it is useful to understand the effects that cause misconvergence when the beam is swept through angles of 70° and more. These distortions occur as the result of two effects:

- *Beam parallax*, where the beams are off axis when they arrive at the deflection yoke
- *Beam tilt*, where the beams contain a component of radial velocity as the result of being converged before deflection

The effect of beam parallax is shown in Figure 8.29 where, with the three guns arranged as illustrated, the three sweeps take on rhombic patterns. The effect of beam tilt is shown in Figure 8.30, and from the triangle A, D, Q it follows that:

$$\sin(\alpha+\beta) = K(i_d + i_0)$$

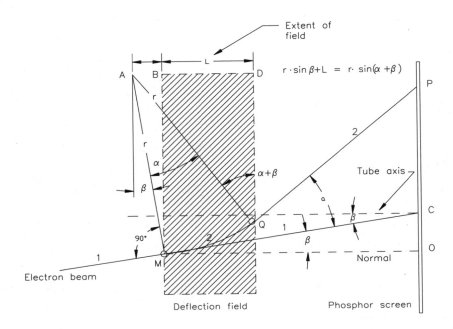

Figure 8.30 The effect of beam tilt. (*After* [12].)

Where:
K = a constant
i_d = deflection current
i_0 = direct current = $\sin \beta / K$

The dc term may be either positive or negative, depending on the sine and amplitude of the convergence angle β, and displaces the beams as shown in Figure 8.31. It is evident from the figure that the green beam is leading, and the red is trailing the blue beam, with the resultant shifts illustrated by the patterns shown in Figure 8.32. Correction can be achieved by generating parabolic waveforms of the type shown in Figure 8.33. The waveforms can then be applied to the deflection yokes to achieve dynamic convergence.

Alternatively, a special convergence yoke can be included. In either case, this discussion is primarily of historical interest, because in-line guns and predistorted yokes have essentially eliminated the need for separate dynamic-convergence circuits or yokes.

8.4.5 In-Line System Convergence

By providing nonuniform fields to overcome the basic overconvergence of in-line beams that are statically converged at the center, in-line display systems eliminate the need for special circuits and convergence yokes. These beams, when deflected, cross over before they reach the screen, as is shown in Figure 8.34. By using a self-converging yoke, nonuniform fields are generated that balance the overconvergence by causing the beams to di-

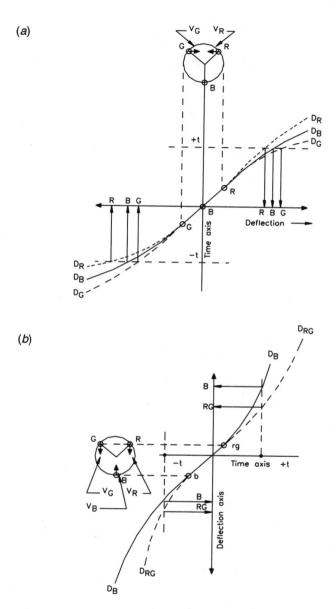

Figure 8.31 Color dot displacement: (*a*) horizontal deflection, (*b*) vertical deflection. (*After* [13].)

verge horizontally inside the yoke, so that the horizontal yoke generates a pincushion-shaped field with its intensity increasing as the horizontal distance from the axis increases. Similarly, the vertical yoke creates a barrel-shaped field that diverges the beams

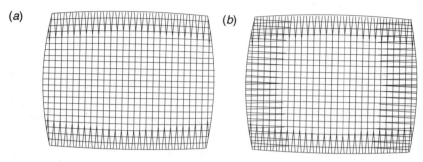

Figure 8.32 Color dot displacement: (*a*) horizontal shift of vertical green bars, (*b*) vertical shift of horizontal blue bars. (*After* [13].)

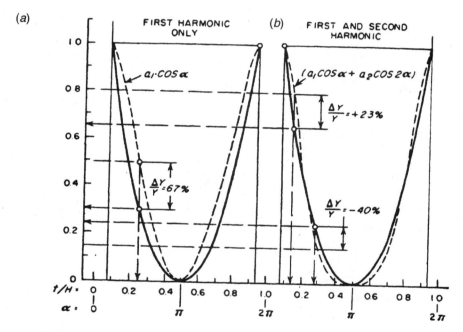

Figure 8.33 Parabolic correction waveforms: (*a*) first harmonic only, (*b*) first and second harmonic signals. (*From* [13]. *Used with permission.*)

horizontally. Thus, the self-converging yoke causes almost perfect vertical-line focus along the axis because of the horizontal-negative and vertical-positive isotropic astigmatism. In addition, the anisotropic astigmatism is removed and the coma resulting from the yoke is eliminated in the gun. However, this self-convergence without any dynamic convergence is effective only in small-angle (90°) systems. For 110° systems there is a small systematic convergence error that can be overcome by limiting it to the horizontal and using only one scanning frequency for correction. This may be achieved by either of the two

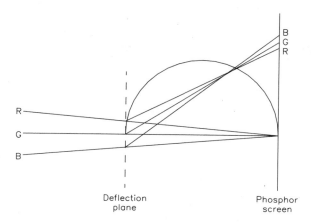

Figure 8.34 The image field showing beam crossover. (*After* [11].)

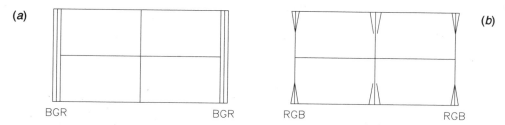

Figure 8.35 Self-converging in-line system: (*a*) vertical, (*b*) horizontal. (*After* [11].)

techniques shown in Figure 8.35. In either system, the horizontal lines converge over the whole raster, and the vertical lines converge along the horizontal axis. Convergence is achieved by means of quadripole windings on the yoke that are energized by one scanning frequency.

The in-line system with self-convergence along the horizontal axis results in better convergence and requires only vertical frequency correction current, which is less expensive and results in less deflection defocusing. The net result is improved convergence with only two preset dynamic controls. At the same time, the yoke is more compact and sensitive, and the CRT is shorter by 10 mm.

8.5 Flat CRT Devices

Although the CRT is the dominant display device for a wide variety of applications, the physical length of the component is a drawback for many users. Various attempts have been made to produce a thin—or even relatively flat—CRT. Some have been commercially successful, some have not.

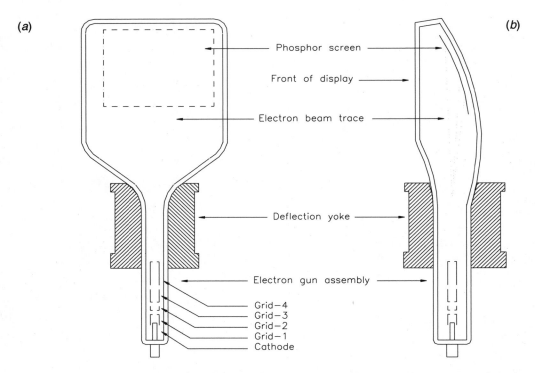

Figure 8.36 Design of a flat CRT (Sony): (*a*) front view, (*b*) side view. (*After* [15].)

8.5.1 Classic Device Designs

While flat tube development can be traced back to the 1950s, it was not until 1982 that a commercial product was offered to consumers in the form of a 2-in monochrome pocket television set (Sony). In this and similar designs, the electron source is positioned in the direction of the phosphor surface, rather than behind it, as illustrated in Figure 8.36. It can be seen from the drawing that large differences exist in electron travel distance from the gun to the phosphor screen. This situation results in deflection distortion and blurring of the focal point. Horizontal and vertical compensation circuits are required to correct for these distortions.

Improved versions of the original design include a 4-in diagonal flat tube, providing 600 lines of horizontal resolution. The video bandwidth of the display is 16 MHz. Larger devices are also possible.

Channel Multiplying CRT

The *channel multiplying* CRT (Philips) is a large format device intended primarily for military applications. The tube utilizes a box-type structure. As shown in Figure 8.37, the device is divided into two sections by a central plate. The electron gun is centered in the rear

Color CRT Displays 277

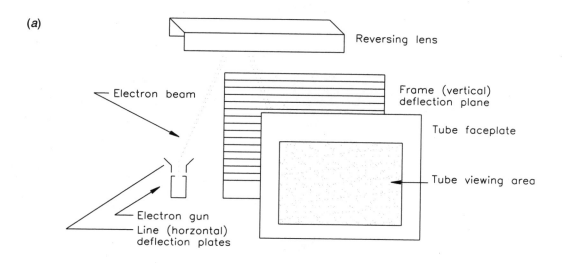

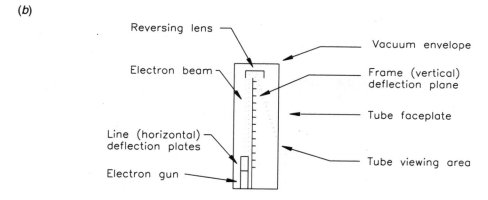

Figure 8.37 Channel multiplying flat CRT (Philips): (*a*) exploded front view, (*b*) side view. (After [16].)

section. The electron beam is projected toward the top of the tube, where a trough-shaped electrode at cathode potential is located. This element forces the beam to turn 180° and travel down the front section. Strip electrodes located on the central dividing plate are successively switched between cathode and anode potential (approximately 400 V) to permit the beam to trace a vertical line at the input of the *channel electron multiplier* (CEM) plate.

Line deflection is provided by a pair of electrostatic deflector plates at the gun. The CEM plate is constructed using shadow-mask technology. The plate is a laminated structure consisting of alternating layers of steel, which have openings spaced (in one design) 0.77 mm apart. The openings act as electrodes (*dynodes*) for secondary emission of electrons. The

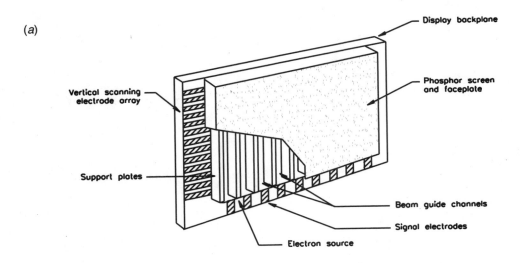

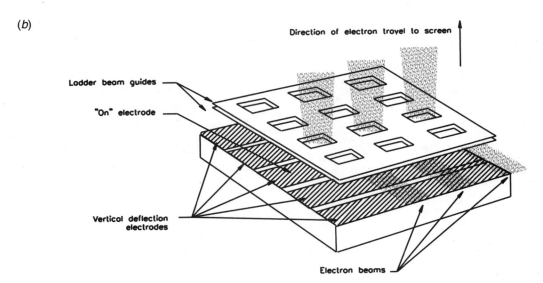

Figure 8.38 Beam guide flat CRT (RCA): (*a*) front view, (*b*) vertical beam deflection system, (*c*, next page) horizontal beam deflection system, (*After* [17].)

aligned holes form channels, each of which define one pixel. Multiplication is achieved by secondary emission from the wall of each hole, the secondary electrons being focused from one dynode to the next. Gain enhancement is achieved by coating the emitting surfaces with magnesium oxide.

This device has been produced in both monochrome and color versions.

(c)

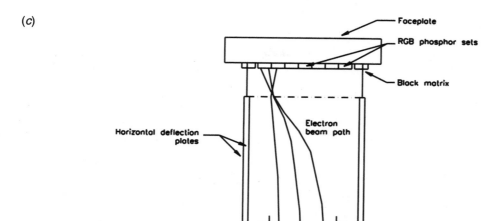

Figure 8.39 (continued)

Beam Guide CRT

The *beam guide* CRT (RCA) is a flat color device intended for consumer applications. Figure 8.38 illustrates the overall layout of the device. Pairs of grid plates, constituting ladder beam guides, form periodic electrostatic fields in which the electron beams are confined. The beam guides transfer the electron beam in a direction parallel to the display surface. In an early implementation of this CRT, 40 channels were provided at a width of 2.5 cm, giving a transverse screen width of 1 m. Three electron beams are guided in each channel toward a color phosphor layer. Scanning of each channel is accomplished by a scanning electrode on a support plate (as shown in the figure).

Matrix Drive and Deflection CRT

The *matrix drive and deflection system* (Matsushita) is another approach to flat color display. As illustrated in Figure 8.39, the device consists of an array of electron beams that pass through sandwiched deflection electrodes toward the phosphor screen. The prototype system utilized 3,000 controlled beams, generated from a matrix of 15 filament cathode and 200 beam control electrodes at right angles to the cathodes. Each beam is horizontally deflected in six steps (two steps of R, G, B), and vertically deflected in 32 steps (including interlace). The end result is an image consisting of 192,000 pixels on the phosphor panel. The displayed image is formed one line at a time in a scanning fashion.

Because no shadow-mask is used in this device, a finite electron beam is required (the same width as a phosphor stripe). Image brightness is controlled by varying the pulse width that drives the electron beams. This technique provided for 64 gray scale steps in the proto-

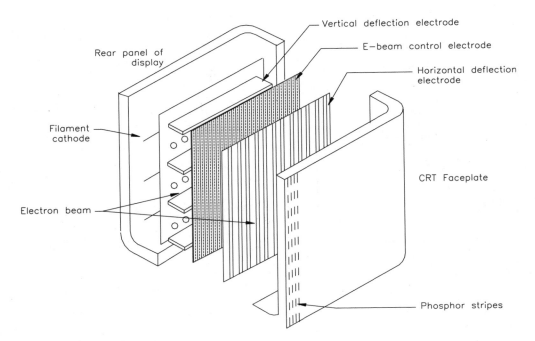

Figure 8.39 Matrix drive and deflection flat CRT (Matsushita). (*After* [18].)

type version. Control of color reproduction is accomplished by digitizing the input picture signal and alternately driving the red, green, and blue channels.

Horizontal Address Vertical Deflection CRT

The *horizontal address vertical deflection* (HAVD) device uses a hot line cathode whose vertical deflection is accomplished by modulating the electron currents for one line simultaneously by means of control electrodes. Figure 8.40 illustrates the operation of the device.

The HAVD CRT is relatively simple in design. A sawtooth voltage (typically +400 V to –400 V) is applied to the upper and lower pair of deflecting electrodes, providing address control in the vertical direction. Barrier electrodes are provided (maintained at a negative potential) to prevent the electron currents in one set of elements from affecting the neighboring elements. The function of these two types of electrodes are interchanged at each horizontal interval to increase the displayed resolution. This technology is applicable to small format monochrome devices.

Practical Considerations

The flat CRT devices discussed in this section have found varying degrees of commercial success (or lack thereof). Although the market potential for a high resolution flat color CRT is great, cost considerations are often the determining factor. While scientists have

Color CRT Displays 281

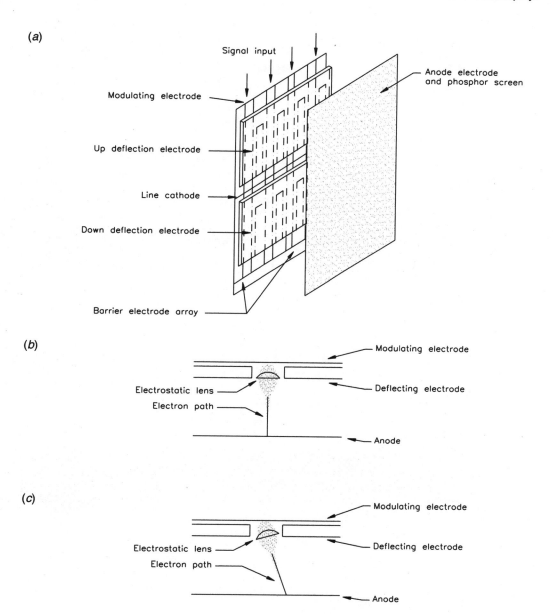

Figure 8.40 Horizontal address vertical deflection flat CRT: (*a*) exploded view, (*b*) detail of electron path under no-deflection condition, (*c*) detail of electron path under deflection condition. (*After* [19].)

worked to develop new and improved methods of producing flat devices, other scientists have successfully improved conventional (long) CRT designs. The end result is that conventional CRTs continue to be the mainstay of most image display applications.

As noted, cost is a key point in the practical use of a display device. Consumer television is a critically important factor in this regard. The production volumes associated with television and—more recently computer—applications are enormous. Flat panel designs will become common only when they are cost-competitive with conventional designs for large-volume applications.

8.6 New Consumer Devices

Progress continues to be made in mass-market CRT technology. Numerous improvements have been made over the years, and considerable new efforts have accompanied the move to digital television broadcasting (DTV). A case in point is the *FD Trinitron Wega* (Sony), the first consumer TV with a flat face CRT [14].

Glass is stronger when it is curved. For this reason, detailed modeling and strength analysis were necessary to make the face of the tube capable of withstanding the applied atmospheric pressure without failing or deforming. Because of the pressure gradient on the glass, it was necessary to develop a new tempered glass based upon the principles of automobile windshield glass. For additional implosion protection, a 188 mm thick film was laminated on the panel surface. The film also serves as an anti-reflective coating to improve display contrast, a common practice in computer monitors.

To realize a flat *aperture grille* (AG) with high tension, there were two major problems to overcome: *mask wrinkle* and AG vibration. Mask wrinkle is caused by localized heating during the welding process. It was found that the AG wires and even the frame itself were susceptible to sympathetic vibrations from the receiver speakers. The answer was to double the springs that attach the frame to the CRT panel.

As the faceplate gets flatter, it becomes more difficult for the electron gun to make a round spot towards the edges of the screen. As a solution to this problem, the focal length of the electron gun was elongated by 20 percent compared with a conventional Trinitron gun. The result was improved focus at the edges of the flat CRT. In order to keep the overall length of the TV set the same, other parts of the gun were widened and shortened.

A new deflection yoke was developed using a horizontal coil with a large square-front-bend to eliminate the usual pincushion distortion introduced by the flat CRT. A modulation coil was added to the vertical winding that used the horizontal frequency to correct vertical misconvergence.

Additional improvements included the following:

- Dynamic focus to vary the voltage across scan lines for sharp focus from edge-to-edge
- Velocity modulation to change the speed of the beam as it scans horizontally across the screen and sharpens the focus

This device is representative of the new generation of CRT displays now finding their way to consumers. The trend to higher quality, lower priced displays will likely accelerate as DTV penetration in the home increases.

8.7 References

1. Dressler, R.: "The PDF Chromatron—A Single or Multi-Gun CRT," *Proc. IRE*, vol. 41, no. 7, July 1953.

2. Tong, Hua-Sou: "HDTV Display—A CRT Approach," in *Display Technologies*, Shu-Hsia Chen and Shin-Tson Wu (eds.), Proc. SPIE 1815, SPIE, Bellingham, Wash., pp. 2–8, 1992.
3. Hockenbrock, Richard: "New Technology Advances for Brighter Color CRT Displays," in *Display System Optics II*, Harry M. Assenheim (ed.), Proc. SPIE 1117, SPIE, Bellingham, Wash., pp. 219–226, 1989.
4. Robinder R., D. Bates and P. Green: "A High Brightness Shadow Mask Color CRT for Cockpit Displays," *SID Digest*, Society for Information Display, San Jose, Calif., vol. 14, pp. 72–73, 1983.
5. Hu, C., Y. Yu and K. Wang: "Antiglare/Antistatic Coatings for Cathode Ray Tube Based on Polymer System, in *Display Technologies*, Shu-Hsia Chen and Shin-Tson Wu (eds)., Proc. SPIE 1815, SPIE, Bellingham, Wash., pp.42–48, 1992.
6. Mitsuhashi, Tetsuo: "HDTV and Large Screen Display," *Large-Screen Projection Displays II*, William P. Bleha, Jr., (ed.), Proc. SPIE 1255, SPIE, Bellingham, Wash., pp. 2–12, 1990.
7. Whitaker, Jerry C., and K. Blair Benson (eds.): *Standard Handbook of Video and Television Engineering*, McGraw-Hill, New York, N.Y., 2000.
8. Benson, K. B., and D. G. Fink: *HDTV: Advanced Television for the 1990s*, McGraw-Hill, New York, N.Y., 1990.
9. Morrell, A., et al.: *Color Television Picture Tubes*, Academic, New York, N.Y., 1974.
10. Hutter, Rudolph G. E.: "The Deflection of Electron Beams," in *Advances in Image Pickup and Display*, B. Kazen (ed.), Academic, New York, N.Y., vol. 1, pp. 212–215, 1974.
11. Barkow, W. H., and J. Gross: "The RCA Large Screen 110° Precision In-line System," ST-5015, RCA Entertainment, Lancaster, Penn.
12. Dasgupta, B. B.: "Recent Advances in Deflection Yoke Design," *SID International Symposium Digest of Technical Papers*, Society for Information Display, San Jose, Calif., pp. 248–252, May 1999.
13. Fink, Donald, (ed.): *Television Engineering Handbook*, McGraw-Hill, New York, N.Y., 1957.
14. Eccles, D. A., and Y. Zhang: "Digital-Television Signal Processing and Display Technology," SID 99 Digest, Society for Information Display, San Jose, Calif., pp. 108–111, 1999.
15. Maeda, M.: *Japan Display '83*, pg. 2, 1971.
16. Woodhead, A., et al.: *1982 SID Digest*, Society for Information Display, San Jose, Calif., pg. 206, 1982.
17. Credelle, T. L., et al.: *Japan Display '83*, pg. 26, 1983.
18. Casteloano, Joseph A.: *Handbook of Display Technology*, Academic, New York, N.Y., 1992.
19. Sakamoto, Y.: and E. Miyazaki, *Japan Display '83*, pg. 30, 1983.

8.8 Bibliography

Aiken, J. A.: "A Thin Cathode Ray Tube," *Proc. IRE*, vol. 45, pg. 1599, 1957.

Ashizaki, S., Y. Suzuki, K. Mitsuda, and H. Omae: "Direct-View and Projection CRTs for HDTV," *IEEE Transactions on Consumer Electronics*, vol. 34, no. 1, pp. 91–98, February 1988.

Barbin, R., and R. Hughes: "New Color Picture Tube System for Portable TV Receivers," *IEEE Trans. Broadcast TV Receivers*, vol. BTR-18, no. 3, pp. 193–200, August 1972.

Blacker, A., et al.: "A New Form of Extended Field Lens for Use in Color Television Picture Tube Guns," *IEEE Trans. Consumer Electronics*, pp. 238–246, August 1966.

Blaha, R.: "Degaussing Circuits for Color TV Receivers," *IEEE Trans. Broadcast TV Receivers*, vol. BTR-18, no. 1, pp. 7–10, February 1972.

Carpenter, C.: et al., "An Analysis of Focusing and Deflection in the Post-Deflection Focus Color Kinescope," *IRE Trans. Electron Devices*, vol. 2, pp. 1–7, 1955.

Chang, I.: "Recent Advances in Display Technologies," *Proc. SID*, Society for Information Display, San Jose, Calif., vol. 21, no. 2, pg. 45, 1980.

Chen, H., and R. Hughes: "A High Performance Color CRT Gun with an Asymmetrical Beam Forming Region," *IEEE Trans. Consumer Electronics*, vol. CE-26, pp. 459–465, August, 1980.

Chen, K. C., W. Y. Ho and C. H. Tseng: "Invar Mask for Color Cathode Ray Tubes, in *Display Technologies*, Shu-Hsia Chen and Shin-Tson Wu (eds.), Proc. SPIE 1815, SPIE, Bellingham, Wash., pp.42–48, 1992.

Clapp, R., et al.: "A New Beam Indexing Color Television Display System," *Proc. IRE*, vol. 44, no. 9, pp. 1108–1114, September 1956.

Cohen, C.: "Sony's Pocket TV Slims Down CRT Technology," *Electronics*, pg. 81, February 10, 1982.

Credelle, T. L.: "Modular Flat Display Device with Beam Convergence," U.S. Patent 4,131,823.

Credelle, T. L., et al.: "Cathodoluminescent Flat Panel TV Using Electron Beam Guides," *SID Int. Symp. Digest*, Society for Information Display, San Jose, Calif., pg. 26, 1980.

"CRT Control Grid Having Orthogonal Openings on Opposite Sides," U.S. Patent 4,242,613, Dec. 30, 1980.

"CRTs: Glossary of Terms and Definitions," Publication TEP92, Electronic Industries Association, Washington, 1975.

Davis, C., and D. Say: "High Performance Guns for Color TV—A Comparison of Recent Designs," *IEEE Trans. Consumer Electronics*, vol. CE-25, August 1979.

Donofrio, R.: "Image Sharpness of a Color Picture Tube by MTF Techniques," *IEEE Trans. Broadcast TV Receivers*, vol. BTR-18, no. 1, pp. 1–6, February 1972.

"Electron Gun with Astigmatic Flare-Reducing Beam Forming Region," U.S. Patent 4,234,814, Nov. 18, 1980.

Fink, Donald, and Donald Christiansen (eds.): *Electronics Engineers Handbook*, 3rd ed., McGraw-Hill, New York, N.Y., 1989.

Fiore, J., and S. Kaplin: "A Second Generation Color Tube Providing More Than Twice the Brightness and Improved Contrast," *IEEE Trans. Consumer Electronics*, vol. CE-28, no. 1, pp. 65–73, February 1982.

Flechsig, W.: "CRT for the Production of Multicolored Pictures on a Luminescent Screen," French Patent 866,065, 1939.

Godfrey, R., et al.: "Development of the Permachrome Color Picture Tube," *IEEE Trans. Broadcast TV Receivers*, vol. BTR-14, no. 1, 1968.

Gow, J., and R. Door: "Compatible Color Picture Presentation with the Single-Gun Tri Color Chromatron," *Proc. IRE*, vol. 42, no. 1, pp. 308–314, January 1954.

Hasker, J.: "Astigmatic Electron Gun for the Beam Indexing Color TV Display," *IEEE Trans. Electron Devices*, vol. ED-18, no. 9, pg. 703, September 1971.

Herold, E.: "A History of Color TV Displays," *Proc. IEEE*, vol. 64, no. 9, pp. 1331–1337, September 1976.

Hoskoshi, K., et al.: "A New Approach to a High Performance Electron Gun Design for Color Picture Tubes," 1980 IEEE Chicago Spring Conf. Consumer Electronics.

Johnson, A.: "Color Tubes for Data Display—A System Study," Philips ECG, Electronic Tube Division.

Law, H.: "A Three-Gun Shadowmask Color Kinescope," *Proc. IRE*, vol. 39, pp. 1186–1194, October 1951.

Lucchesi, B., and M. Carpenter: "Pictures of Deflected Electron Spots from a Computer," *IEEE Trans. Consumer Electronics*, vol. CE-25, no. 4, pp. 468–474, 1979.

Masterson, W., and R. Barbin: "Designing Out the Problems of Wide-Angle Color TV Tube," *Electronics*, pp. 60–63, April 26, 1971.

Mears, N., "Method and Apparatus for Producing Perforated Metal Webs," U.S. Patent 2,762,149, 1956.

Mokhoff, N.: "A Step Toward Perfect Resolution," *IEEE Spectrum*, IEEE, New York, N.Y., vol. 18, no. 7, pp. 56–58, July 1981.

Morrell, A.: "Color Picture Tube Design Trends," *Proc. SID*, Society for Information Display, San Jose, Calif., vol. 22, no. 1, pp. 3–9, 1981.

Moss, H.: *Narrow Angle Electron Guns and Cathode Ray Tubes*, Academic, New York, N.Y., 1968.

Oess, F.: "CRT Considerations for Raster Dot Alpha Numeric Presentations," *Proc. SID*, Society for Information Display, San Jose, Calif., vol. 20, no. 2, pp. 81–88, second quarter, 1979.

Ohkoshi, A., et al.: "A New 30V" Beam Index Color Cathode Ray Tube," *IEEE Trans. Consumer Electronics*, vol. CE-27, p. 433, August 1981.

Palac, K.: Method for Manufacturing a Color CRT Using Mask and Screen Masters, U.S. Patent 3,989,524, 1976.

Pitts, K., and N. Hurst: "How Much do People Prefer Widescreen (16×9) to Standard NTSC (4×3)?," *IEEE Transactions on Consumer Electronics*, vol. 35, no. 3, pp. 160–169, August 1989.

"Recommended Practice for Measurement of X-Radiation from Direct View TV Picture Tubes," Publication TEP 164, Electronics Industries Association, Washington, D.C., 1981.

Robbins, J., and D. Mackey: "Moire Pattern in Color TV," *IEEE Trans. Consumer Electronics*, vol. CE-28, no. 1, pp. 44–55, February 1982.

Rublack, W.: "In-Line Plural Beam CRT with an Aspherical Mask," U.S. Patent 3,435,668, 1969.

Say, D.: "Picture Tube Spot Analysis Using Direct Photography," *IEEE Trans. Consumer Electronics*, vol. CE-23, pp. 32–37, February 1977.

Say, D.: "The High Voltage Bipotential Approach to Enhanced Color Tube Performance," *IEEE Trans. Consumer Electronics*, vol. CE-24, no. 1, pg. 75, February 1978.

Schwartz, J.: "Electron Beam Cathodoluminescent Panel Display," U.S. Patent 4,137,486.

Sherr, S.: *Electronic Displays*, Wiley, New York, N.Y., 1979.

Sinclair, C.: "Small Flat Cathode Ray Tube," *SID Digest*, Society for Information Display, San Jose, Calif., pg. 138, 1981.

Stanley, T.: "Flat Cathode Ray Tube," U.S. Patent 4,031,427.

Swartz, J.: "Beam Index Tube Technology," *SID Proceedings*, Society for Information Display, San Jose, Calif., vol. 20, no. 2, p. 45, 1979.

Uba, T., K. Omae, R. Ashiya, and K. Saita: "16:9 Aspect Ratio 38V-High Resolution Trinitron for HDTV," *IEEE Transactions on Consumer Electronics*, vol. 34, no. 1., pp. 85–89, February 1988.

Yoshida, S., et al.: "25-V Inch 114-Degree Trinitron Color Picture Tube and Associated New Development," *Trans. BTR*, pp. 193-200, August 1974.

Yoshida, S., et al.: "A Wide Deflection Angle (114°) Trinitron Color Picture Tube," *IEEE Trans. Electron Devices*, vol. 19, no. 4, pp. 231–238, 1973.

Yoshida. S.: et al., "The Trinitron—A New Color Tube," *IEEE Trans. Consumer Electronics*, vol. CE-28, no. 1, pp. 56–64, February 1982.

Chapter 9
Projection Display Systems

9.1 Introduction

As the need to present high-resolution video and graphics information steadily increases, the use of large screen projection displays continues to grow rapidly. High-definition television requires large screens (greater than 40-in diagonal) to provide effective presentation. Some form of projection is, therefore, the only practical solution. The role of HDTV and film in future theaters is also being explored, along with performance criteria for effective large screen video presentations.

The number of specific applications for large screen displays continues to grow. Military command centers provide a stringent test where high resolution, brightness, and reliability are critical. Flight simulation presents some of the greatest challenges, because large images must be presented on domes or other surfaces with high brightness, contrast, and resolution. Success in this application may lead to new entertainment vehicles, where domed entertainment theaters will combine traditional presentations with dynamic, high-definition video enhancements.

Extensive developmental efforts are being conducted into large screen display systems. Much of this research is aimed at advancing new technologies such as plasma, electroluminescent and LCD, lasers, and new varieties of CRTs. Currently, CRT systems lead the way for applications requiring full-color and high resolution. LCD systems, which are advancing rapidly, may capture a sizable portion of the video-only marketplace.

9.1.1 Display Types

Large screen projectors fall into four broad classes, or grades:

- *Graphics*. Graphics projectors are the highest quality—and generally most expensive—projectors. These systems are capable of the highest operating frequency and resolution. They can genlock to almost any computer or image source, and offer resolutions of $2,000 \times 2,000$ pixels (or more) with horizontal sweep rates of 89 kHz (and higher).

- *Data*. Data projectors are less expensive than graphics projectors and are suitable for use with common computer image generators, such as PCs equipped with standard graphics cards. Data projectors offer resolutions of at least 1,024 × 780 pixels, and horizontal sweep rates of 49 kHz and higher.
- *HDTV*. HDTV projectors provide the quality level necessary to take full advantage of high-definition imaging systems. Resolution on the order of 1,125 TV lines is provided at an aspect ratio of 16:9.
- *Video*. Video projectors provide resolution suitable for NTSC-level images. Display performance of 380 to 480 TV lines is typical.

Computer signal sources can follow many different—and sometimes incompatible—standards. It is not a trivial matter to connect any given projector to any given computer. Interface kits are available that make the job easier. Also, multisync projectors have become commonplace. Generally speaking, projectors are *downward compatible*. In other words, most graphic projectors can function as data projectors, HDTV projectors, and video projectors; most data projectors can also display HDTV and video, and so on.

9.1.2 Displays for HDTV Applications

The most significant difference between a conventional display and one designed for high-definition video is the increased resolution and wider aspect ratio of HDTV. Three basic technologies are practical for viewing high-definition images of 40-in diagonal and larger:

- *Light valve projection display*. Capable of modulating high-power external light sources, light valves are mainly used for large screen displays measuring greater than 200-in.
- *CRT projection displays*. Widely used for middle-sized screen displays of 45 to 200-in, CRT projection systems are popular because of their relative ease of manufacturing (and, hence, competitive cost) and good performance. These displays are likely to remain the mainstay technology for HDTV in the near-term future.
- *Flat panel PDP* (plasma display panel) and *LCD* (liquid crystal display). Flat panel PDP displays have been produced in 40-in and larger sizes. Similar LCD systems have also marketed. Both hold great promise for future HDTV applications as the underlying technologies mature.

9.1.3 Displays for Military Applications

Full-color, large screen display systems can enhance military applications that require group presentation, coordinated decisions, or interaction between decision makers. Projection display technology already plays an important role in operations centers, simulation facilities, conference rooms, and training centers. Each application requires unique values of luminance, resolution, response time, reliability, and video interface. The majority of large screen applications involve fixed environments where commanders and their staff interact as a group to track an operation and make mission-critical decisions. As sensor technology improves and military culture drives toward *scene presence*, future applica-

Table 9.1 Performance Requirements for Large Screen Military Display Applications (*After* [1].)

Parameter		Performance Requirement
Output luminance[1]	Small conference room	50–200 lumens
	Large conference room	200–600 lumens
	Control center	600–2000 lumens
Addressable resolution		1280 × 1024 pixels
Visual resolution		15% absolute minimum modulation depth to an alternate pixel input
Video interface		15.75 to 80 kHz horizontal scan, 60 to 120 Hz noninterlaced refresh
Video sources		10–20 different sources, autosynchronous lock
Response to new update		Less than 1 s, no smear to dynamic response
Reliability		1000 hours MBTF, excluding consumables
Operating cost		Less than $10 per hour per projector, including ac power
Maintenance		*Level 5* maintenance technician

[1] Ambient lighting of 20–40 footcandles is assumed. The values shown are modulated luminance, spatially averaged over the full screen area.

tions will require large screen display systems that offer greater realism and real-time response. Table 9.1 lists the basic requirements for military display applications. The available technologies meeting these demands are illustrated in Figure 9.1.

9.2 Geometric Optics

Geometric optics[1] deals with image formation using geometric methods. It is based on two postulates:

- That light travels in straight lines in a homogeneous medium
- That two rays may intersect without affecting the subsequent path of either

The fundamental laws of geometric optics can be developed from general principles, such as Maxwell's electromagnetic equations or Fermat's principle of least time. However, the laws of reflection and refraction can also be determined in a simple way by means of Huygen's principle, which states that every point of a wave front may be considered as a source of small waves spreading out in all directions from their centers, to form the new wave front along their envelope.

9.2.1 Laws of Reflection and Refraction

The laws of reflection and refraction for optics may be stated as follows:

---

1 Portions of this section were adapted from: Jerry C. Whitaker and K. Blair Benson (eds.): *Standard Handbook of Video and Television Engineering,* 3rd ed., McGraw-Hill, New York, N.Y., 2000.

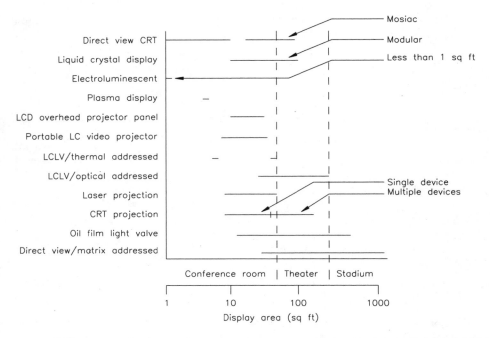

Figure 9.1 Basic grading of full-color display technologies vs. display area for military applications. (*After* [1].)

- *Law of Reflection*. The angle of the reflected ray is equal to the angle of the incident ray.
- *Law of Refraction*. A ray entering a medium in which the velocity of light is different is refracted so that $n \sin i = n' \sin r$, where i is the angle of incidence, r is the angle of refraction, and n and n' are the indexes of refraction of the two media.

A *ray* is an imaginary line normal to the wave front. The angle the advancing ray forms with the line normal to the surface in question, is the *angle of incidence* and is equal to the angle the wave front forms with the surface.

The *index of refraction* is the ratio of the velocity of light c in a vacuum to the velocity v in the medium:

$$n = \frac{c}{v} \quad n' = \frac{c}{v'}$$

For air, the velocity of light is generally considered equal to the velocity *in vacuo* so $n = 1.0$ and the equation may be simplified to the following:

$$\frac{\sin i}{\sin r} = n'$$

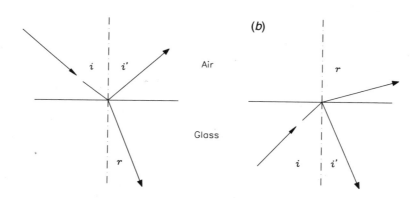

Figure 9.2 Refraction and reflection at air-glass surface: (*a*) beam incident upon glass from air, (*b*) beam incident upon air from glass. (*After* [2].)

When a ray passes from a medium of smaller index into one of larger index, as from air to glass, the angle of refraction is less than the angle of incidence, and the ray is bent toward the normal. In passing from glass to air, the ray is bent away from the normal, as illustrated in Figure 9.2. The incident ray, reflected ray, refracted ray, and the normal to the surface at the point of incidence all lie in the same plane.

A ray passing from a medium of higher index to one of lower index may be totally internally reflected. The following relationship applies:

$$n \sin i = n' \sin r$$

The value of $\sin r$ is always greater than $\sin i$ when n is greater than n'. The maximum value for $\sin r$ is unity $r = 90°$) and occurs for some value of i, called the *critical angle*, which is determined by the refractive indexes of the two media. For a water-air surface:

$$\frac{n}{n'} = 1.33$$

and the critical angle is 48.5°.

When the angle of incidence exceeds the critical angle, the ray is not refracted into the medium of lower index but is totally reflected, as illustrated in Figure 9.3. For angles smaller than the critical angle the rays are partially reflected.

Application of the *sine law* to two parallel surfaces, such as a glass plate, shows that the ray emerges parallel to the entering ray, but is displaced. The most important applications of the laws of reflection and refraction relate to the formation of images by means of spherical surfaces, such as mirrors and lenses.

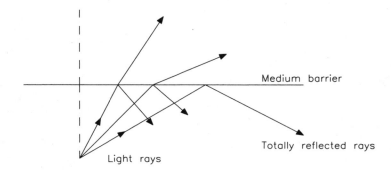

Figure 9.3 The result when the angle of refraction exceeds the critical angle and the ray is totally reflected. (*After* [2].)

9.2.2 Refraction at a Spherical Surface in a Thin Lens

It can be shown by tracing a ray through a single refracting surface that:

$$\frac{n}{s} + \frac{n'}{s'} = \frac{n'-n}{R}$$

Where:
s = object distance to refracting surface
s' = image distance to refracting surface
R = radius of curvature of surface
n = index of refraction of object medium
n' = index of refraction of image medium

A ray traversing two refractive surfaces, as in a lens in air, has a path whose image distance and object distance are found by applying the foregoing equation to each of the two surfaces. For a lens whose thickness may be considered negligible relative to the image distance, the following applies:

$$\frac{1}{s} + \frac{1}{s'} = (n-1)\left\{\frac{1}{R_1} - \frac{1}{R_2}\right\}$$

Where:
n = the index of refraction of the lens
R_1 = the radii of curvature of the first surface
R_2 = the radii of curvature of the second surface

The right side of the equation contains quantities that are characteristic of the lens, called the *power* of the lens. The reciprocal of this expression is referred to as the *focal length f*:

$$\frac{1}{f} = (n-1)\left\{\frac{1}{R_1} - \frac{1}{R_2}\right\}$$

For a thin lens in air the object distance, image distance, and focal length are related as follows:

$$\frac{1}{s} + \frac{1}{s'} = \frac{1}{f}$$

Certain conventions of algebraic sign must be observed in the use of this and previous equations. The conventions may be summarized as follows:

- All figures are drawn with the light incident on the reflecting or refracting surface from the left.

- The object distance s is considered positive where the object lies at the left of the *vertex*. The vertex is the intersection of the reflecting or refracting surface with the axis through the center of curvature of the surface.

- The image distance s' is considered positive when the image lies at the right of the vertex.

- The radii of curvature is considered positive when the center of curvature lies at the right of the vertex.

- Angles are considered positive when the slope of the ray with respect to the axis is positive.

- Dimensions, such as image height, are considered positive when measured upward from the axis.

In general, after observing the first two conventions, the others follow the rules of coordinate geometry with the vertex as the origin.

From the previous equations and the foregoing sign conventions, it is apparent that the sign of the focal length may be negative or positive. For a lens in air, and parallel incident rays, the focal length is positive when the transmitted rays converge and negative when they diverge. Cross sections of simple converging and diverging lenses are shown in Figure 9.4.

There are two focal points of a lens, located on the lens axis. All incident rays parallel to the lens axis are refracted to pass through the second focal point; all incident rays from the first focal point emerge parallel to the lens axis, as illustrated in Figure 9.5. For a thin lens, the distances from the two focal points to the lens are equal and denote the focal length.

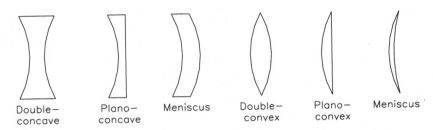

Figure 9.4 Various forms of simple converging and diverging lenses. (*After* [2].)

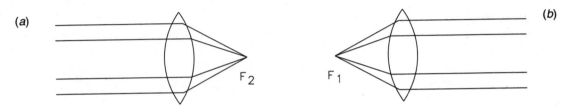

Figure 9.5 Lens effects: (*a*) parallel rays incident upon the lens pass through the second focal point, (*b*) rays passing through the first focal point incident upon the lens emerge parallel. (*After* [2].)

The magnification m provided by a lens is defined as the ratio of the image height (y') to the object height (y):

$$m = \frac{y'}{y}$$

The principles of magnification are illustrated in Figure 9.6. From the similar triangles *ABC* and *CDE*, it follows:

$$m = \frac{y'}{y} = \frac{s'}{s}$$

9.2.3 Reflection at a Spherical Surface

By considering reflection as a special case of refraction, many of the previous equations can be applied to reflection by a spherical mirror if the convention is adopted that $n' = -n$. This yields:

$$\frac{1}{s} - \frac{1}{s'} = \frac{2}{R}$$

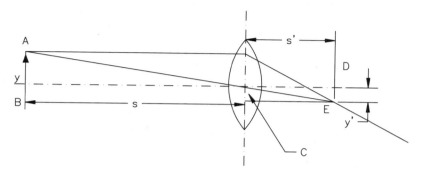

Figure 9.6 Magnification of a simple lens. (*After* [2].)

The focal point is the axial point that is imaged at infinity by the mirror; its distance from the mirror is the focal length. Hence:

$$f = \frac{R}{2}$$

For a concave mirror, R is negative and the focal point lies at the left of the mirror, following the convention of signs described previously. For any spherical mirror:

$$\frac{1}{s} - \frac{1}{s'} = \frac{1}{f}$$

It also follows that:

$$m = \frac{y'}{y} = \frac{s'}{s}$$

Such mirrors are subject to spherical aberration as in lenses, such as the failure of the centrally reflected rays to converge at the same axial point as the rays reflected from the mirror edge. Aspherical surfaces formed by a *paraboloid of revolution* have the property that rays from infinity incident on the surface are all imaged at the same point on the axis. Thus, for the focal point and infinity, spherical aberration is eliminated. This is a useful device in projection components where the light source is placed at the focal point to secure a beam of nearly parallel rays.

Spherical aberration in mirrors can be eliminated by inserting lenses before the mirror. The *Schmidt corrector* is an aspherical lens, with one surface convex in the central region and concave in the outer region. The other surface is plane. A Schmidt system of spherical mirror and corrector plate can be made with a high relative aperture; $f/0.6$ is a typical value. Because of the efficiency and low cost of these systems, compared with projection lens systems, they have been used to obtain enlarged images from CRT devices for large screen display (among other applications).

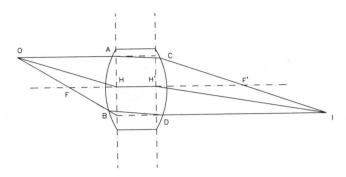

Figure 9.7 Focal points and principal planes for a thick lens. (*After* [2].)

Another type of corrector for a spherical mirror is a *meniscus lens* having no aspherical surfaces, known as a *Maksutov corrector*. The spherical surfaces of the meniscus lens permits easy manufacturing.

9.2.4 Thick (Compound) Lenses

The equations given previously apply to thin lenses. When the thickness of the lens cannot be ignored, measurements must be made from reference points other than the lens surface, such as the focal points—which have already been defined—or from the *principal points*. The principal points are located as follows (see Figure 9.7):

- Consider ray OA proceeding from the object parallel to the lens axis. This will be refracted to pass through the focal point F'.

- The ray OB, which passes through the focal point F, will emerge along DI parallel to the lens axis.

- If OA and $F'I$ are extended, their point of intersection lies in the second principal plane.

- The point H', where this plane intersects the axis, is called the *second principal point*. Similarly, the intersection of OF and DI extended lies in the first principal plane, and H is the *first principal point*.

- The distances FH and $F'H'$ are the first and second focal lengths, respectively.

When the index of the medium on both sides of the lens is the same, as for a lens in air, the first and second focal lengths are equal.

If the direction of the light ray is reversed (the object is placed at the image position), the ray retraces its path and the image is formed at the former object position. Any two corresponding object and image points are said to be conjugate to each other, and hence are *conjugate points*.

The equation

$$\frac{1}{s} + \frac{1}{s'} = \frac{1}{f}$$

given previously for a thin lens, continues to hold for a thick lens, but s and s' are measured from their respective principal points, as is the focal length f. The object distance, image distance, and focal length are related in another form, known as the *Newtonian form* of the lens equation. If x is the distance of the object from its focal point, and x' the image distance from its focal point, then:

$$xx' = f^2$$

9.2.5 Lens Aberrations

Up to this point, optical images have been considered to be faithful reproductions of the object. The equations given have been derived from the general expressions for the refraction of a ray at a spherical surface when the angle between the ray and the axis is small so that $\sin\theta = \theta$. This approximation is known as *first-order theory*. The departures of the actual image from the predictions of first-order theory are called aberrations. von Seidel extended the first-order theory by including the third-order terms of the expanded sine function. The *third-order theory* contains five terms to be applied to the first-order theory. When no aberrations are present, and monochromatic light is passed through the optical system, the sum of the five terms is zero. Thus von Seidel's sums provide a logical classification for the five monochromatic aberrations. In addition, two forms of chromatic aberration can occur because of variation of index with wavelength. The five monochromatic aberrations are:

- Spherical aberration
- Coma
- Astigmatism
- Curvature of field
- Distortion of field

These terms were described in chapter 6 as they relate to the electron optics of a cathode ray tube.

Spherical aberration may be described as the failure of rays from an axial point to form a point image in the direction along the axis. In general, spherical aberration can be minimized if the deviation of the rays is equally divided between the front and rear surfaces of the lens. In a system of two or more lenses, spherical aberration can be eliminated by making the contribution of the negative elements equal and opposite to that of the positive elements.

Coma relates to failure of the rays from an off-axis point to converge at the same point in the plane perpendicular to the axis. Coma can be eliminated for a given object and image distance in a single lens by proper choice of radii of curvature.

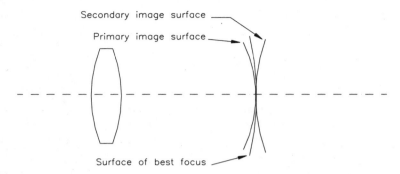

Figure 9.8 Surfaces of best focus, illustrating lens astigmatism. (*After* [2].)

Astigmatism contains aspects of both spherical aberration and coma. It resembles coma in that the off-axis points are affected, but—like spherical aberration—results from spreading of the image in a direction along the axis. The rays from a point converge on the other side of the lens to form a *line image*, actually the axis of a degenerate ellipse; continuing, the rays join with other rays to form a circle, and then at a still further distance form a second image crossed perpendicularly to the first. The best focus occurs when a circular image is formed. The locus of inner line images—the primary images—is a surface of revolution about the lens axis, called the *primary image surface*, shown in Figure 9.8. The locus of outer line images forms the secondary image surface. The locus of *circles of least confusion* forms the *surface of best focus*. As shown in the figure, these surfaces are tangent to one another at the lens axis.

Astigmatism is the failure of the primary and secondary image surfaces to coincide. The surface of best focus is usually not a plane but a curved surface; this type of aberration is known as *curvature of field*. It is not possible to eliminate both astigmatism and curvature of field in a single lens.

All rays passing through a lens from the center to the edge should result in equal magnification of the image. Distortion of the image occurs when the magnification varies with axial distance. If the magnification increases with axial distance, the effect is known as *pincushion distortion*, and the opposite effect is known as *barrel distortion*. The types of distortion are illustrated in Chapter 6.

The five types of lens aberration described in this section can occur in uncorrected lenses even though light of a single wavelength forms the image. When the image is formed by light from different regions of the spectrum, two types of *chromatic aberration* can occur:

- *Axial* or *longitudinal chromatism*

- *Lateral chromatism*

Axial chromatism results from the convergence of rays of different wavelength at different points along the axis; the lens focal length varies with wavelength. Because magnification depends upon the focal length, the images are also of different size, producing lateral chromatism. In many instances, lenses are corrected so that the focal points coincide for two

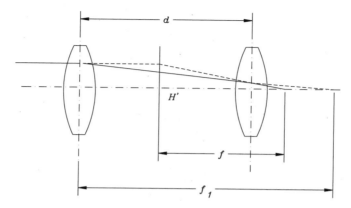

Figure 9.9 Lens system treated as a single thick lens.

or three colors, thus eliminating longitudinal chromatism. However, unless the focal lengths are also made to coincide, the images will be of slightly different size. This defect results in color fringing in the outer portions of the field.

9.2.6 Lens Systems

A combination of lenses may be treated as a thick lens. Consider two thin lenses of focal lengths f_1 and f_2 separated by a distance d. (See Figure 9.9.) The second principal plane is found in the same manner as for a single thick lens. The focal length of the combination f is the distance from the focal point to the principal plane. It is related to the focal lengths of the two thin lenses by the following:

Lenses are shaped to have *spherical properties* (uniform properties about the center of the lens) or *cylindrical properties* (uniform properties about the horizontal or vertical axis of the lens). An *anamorphic lens* is designed to produce different magnification of an image in the horizontal and vertical axes.

9.2.7 Color Beam-Splitting Systems

Many color projection display systems require spatial separation of the red, green, and blue source light. These beams may also need to be filtered to eliminate spurious or undesired wavelengths. The design goal of a color beam-splitting system is to reflect all the light of one primary color and to transmit the remaining visible radiation. *Dichroic mirrors* and prisms are used with supplemental trimming filters to accomplish this end. The most efficient systems utilize dichroic mirrors.

A dichroic mirror is made by coating glass with alternate layers of two materials having high and low indices of refraction. The material must have a thickness of 1/4-wavelength at the center of the band to be reflected. Figure 9.10 shows a typical mirror arrangement. The blue light is reflected by the first mirror, and the red and green light is transmitted. The red light is reflected by the second mirror, and the green light is passed. The curves of Figure

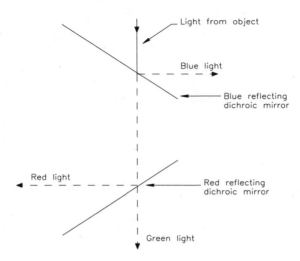

Figure 9.10 Arrangement of a dichroic mirror beam-splitting system. (*After* [2].)

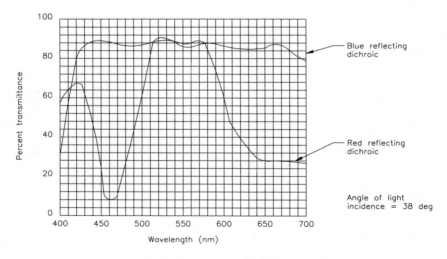

Figure 9.11 Transmission characteristics of typical dichroic mirrors.

9.11 show typical transmittance versus wavelength characteristics. The blue reflecting mirror transmits about 90 percent of the green and red light, and the red reflecting mirror transmits nearly 90 percent of the blue and green light.

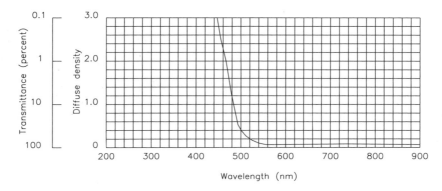

Figure 9.12 Response curve of a spectral trim filter (yellow). (*Source: Eastman Kodak Company.*)

Dichroic Prism

It can be seen in Figure 9.3 that when the angle of incidence of a light ray exceeds the critical angle, the ray is totally reflected. The critical angle for an air-glass surface is 42° for a typical index of refraction for glass of 1.50. Hence, a 45-45-90° glass prism offers a totally reflecting surface. Other designs permit partial reflection and refraction. Coatings at the prism surface, as for dichroic mirrors, will selectively pass or reflect different colors.

Spectral Trim Filters

A typical dichroic mirror does not abruptly change spectral reflection at some specific wavelength. This property can be observed in Figure 9.11 Instead, there is a gradual transition over a wide band. This transition must be eliminated to maintain the purity of the red, green, and blue color signals. The spectral reflectance transmittance bands are trimmed by inserting filters having abrupt divisions between high and low transmittance. These filters are constructed of glass, plastic, or gelatin containing light-absorbing substances.

The *neutral density filter* is another type useful in beam splitter applications. The filter absorbs equally, or nearly so, all wavelengths in the visible spectrum. These filters are available in different densities so that color beams may be balanced for equal signal output.

Figure 9.12 shows the response curve of a spectral trim filter (yellow). Figure 9.13 illustrates the response of a neutral density filter (D = 1.0).

9.2.8 Interference Effects

Huygen's principle was mentioned earlier in this chapter as forming the basis for the laws of reflection and refraction. To this concept should now be added the *principle of superposition*, which states that the resultant effect of the superposition of two or more waves at a point may be found by adding the instantaneous displacements that would be produced at the point by the individual waves if each were present alone.

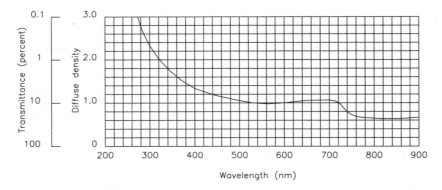

Figure 9.13 Response curve of a neutral density filter (visible spectrum). (*Source: Eastman Kodak Company.*)

If the wave path is thought of as a sinuous path consisting of alternate crests and troughs, the maximum height of the crests or depth of the troughs is called the *maximum amplitude*. Starting at zero amplitude and progressing through a crest back to zero and through a trough to zero constitutes one cycle; the distance traveled through the medium is one wavelength. The number of cycles per unit time is the frequency. The distance traveled per unit time is the velocity. Therefore, the velocity u is the product of the wavelength λ and the frequency v:

$$u = v\lambda$$

If two waves meet in such a manner that the crests reinforce each other to produce the maximum possible amplitude, the waves are said to be in phase, but if the crest meets the trough to produce the minimum possible amplitude, they are said to be 180° out of phase. Thus, the phase expresses the distance between the crests of the waves. If two waves of the same amplitude, traveling in the same or opposite directions, are 180° out of phase, they are said to *completely interfere* and no disturbance is noted. If the phase, amplitude, or frequency of the waves is not the same, then the waves can reinforce at certain points and destroy at other points to produce an interference pattern.

A commonly observed example of interference is the array of colors seen in a thin film of oil on a wet pavement. Light waves are reflected from the front and rear surfaces of the film. When the thickness of the film is an odd number of quarter wavelengths, the light of that wavelength reflected from the front and back surfaces of the film reinforces itself and light is strongly reflected (assuming normal incidence). For an even number of quarter wavelengths, destructive interference occurs. Thus, from one area of film only blue-green light may be reflected while from another area of different thickness red light can be observed.

If—instead of a thin film of oil—two reflecting surfaces, such as two glass surfaces, are placed together but not in complete optical contact, an interference pattern is formed by light from the front and rear surfaces. Frequently the pattern takes the form of concentric rings, called *Newton's rings*.

Interference patterns are useful in grinding optical surfaces. The new surface may be tested by bringing it in contact with a surface of known curvature and noting the shape and separation of the fringes. By repeating the test at intervals, the new surface may be gradually worked to the desired precision.

9.2.9 Diffraction Effects

If an obstacle such as a slit or straight edge is placed in a beam of light—according to Huygen's principle—each point along the slit becomes a source for new wavelets. It can be shown that as these wavelets fan out beyond the obstacle they tend to reinforce or destroy each other in various regions, forming an interference pattern. As the wavelets fan out beyond the obstacle, the light "bends around" it, producing light areas in regions that would be dark if the light traveled only in straight lines. The effects produced by blocking part of a wave front to form interference patterns are called *diffraction effects*. If a wave front is incident on a circular opening such as a lens aperture, the diffraction pattern consists of a bright central disk surrounded by alternate dark and bright rings. The angle α formed at the lens by the diffraction circle is dependent upon the diameter of the lens opening D as follows:

$$\alpha = \frac{2.4\lambda}{D}$$

The *diffraction grating* is an important device utilizing diffraction principles. This element consists essentially of a large number of parallel slits of the same width spaced at regular intervals. Light passing through the slits is diffracted to form interference patterns. The waves will reinforce to form a maximum when the following condition is met:

$$\sin\theta = \frac{n\lambda}{d}$$

Where:
θ = angle of deviation from the direction of incident light
d = distance between successive grating slits
n = an integer denoting order of the maximum

Some light will pass directly through the grating. This is called the *zero order*. The first maximum (assuming monochromatic light) lies beyond the zero order and is called the *first order*. The next maximum is the second order and so on. If white light is incident on the grating, the zero order is a white image followed by a first-order spectrum, then second-order, and so on. By proper ruling of the grating lines, a large proportion of the incident light can be directed into one of the first-order spectra. These gratings are used in many spectral-analysis instruments because of their high efficiency.

9.2.10 Polarization Effects

Because light is a series of electromagnetic waves, each wave can be separated into its electric (E) and magnetic (H) vectors, vibrating in planes at right angles to each other. A series of electromagnetic waves will have E vectors, for example, vibrating in all possible planes perpendicular to the direction of travel. By means of reflection, double refraction, or scattering, the waves can be sorted into two resultant components with their E vectors at right angles to each other. Each ray is said to be *plane-polarized*, that is, made up of waves vibrating in a single plane. If two rays with waves of equal amplitude are brought together, they can form elliptically, plane, or circularly polarized light depending upon whether:

- The phase difference between the vibrating waves lies between 0 and $\pi/2$ for elliptical polarization
- The phase is at 0 or π for plane polarization
- The phase is at $\pi/2$ for circular polarization.

The angle of incidence at which light reflected from a polished surface will be completely polarized is given by the equation known as *Brewster's law*:

$$\tan\theta = \frac{n'}{n}$$

Where n' and n are the indexes of refraction of the two media.

For glass and air, $n' = 1.5$ and $n = 1$, the polarizing angle is 56°. Of the natural light incident at the polarizing angle, about 7.5 percent is reflected and is polarized with its vibration plane perpendicular to the plane of incidence. The rest of the light is transmitted and consists of a mixture of the light with a vibration plane parallel to the plane of incidence and the balance of the perpendicular component. By passing the mixture through successive sheets of glass stacked in a pile, more of the perpendicular component is removed at each reflection and the transmitted fraction consists of the parallel component.

The velocity of a light wave through many transparent crystalline materials is not the same in all directions. Because the ratio of the velocity of light in a medium to the velocity in a vacuum is the index of refraction, these materials have more than one index of refraction. When oriented in one position with respect to the direction of the incident ray, the crystal behaves normally and that direction is called the *optic axis* of the crystal. A ray incident on the crystal to form an angle with the optic axis is broken into two rays, one of which obeys the ordinary laws of refraction and is called the *ordinary ray*; the second ray is called the *extraordinary ray*. The two rays are plane-polarized in mutually perpendicular planes. By eliminating one of the rays, such doubly refracting materials can be used to obtain plane-polarized light. In some materials, one of the components is more strongly absorbed than the other. Crystals of iodoquinine sulfate are an example. The parallel orientation of layers of such crystals in plastic has been used to form polarizing filters.

Kerr discovered that some liquids become doubly refracting when an electric field is applied. The *Kerr effect* makes it possible to control the transmission of light by an electric field. A Kerr cell consists of a transparent cell containing a liquid such as nitrobenzene. The

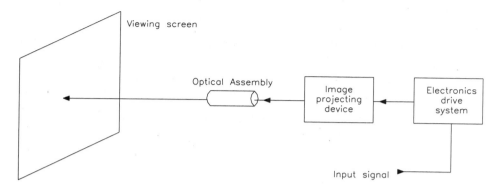

Figure 9.14 Principal elements of a video projection system.

cell is placed between crossed polarizers. When an electric field is applied light is transmitted; it is cut off when the field is removed.

9.3 Projection System Fundamentals

Video projection systems provide a method of displaying a much larger image than can be generated on a direct-view cathode-ray picture tube. Optical magnification and other techniques are employed to throw an expanded image on a passive viewing surface that may have a diagonal dimension of 75-in or more. CRTs are generally restricted to less than 39-in diagonal. A primary factor limiting CRT size is glass strength. In order to withstand atmospheric pressure on the evacuated envelope, CRT weight increases exponentially with linear dimension (discussed in Chapter 8).

The basic elements of a projection system, illustrated in Figure 9.14, include:

- Viewing screen
- Optical elements
- Image source
- Drive electronics

The major differences of projection systems from direct-view displays are embodied in the first three areas, while the electronics assembly is (typically) essentially the same as for direct-view systems.

9.3.1 Projection Requirements

To provide an acceptable image, a projection system must approach or equal the performance of a direct-view device in terms of brightness, contrast, and resolution. Whereas the first two parameters may be compromised to some extent, large displays must excel in resolution because of the tendency of viewers to be positioned less than the normal relative distance of four to eight times the picture height from the viewing surface. Table 9.2 pro-

Table 9.2 Performance Levels of Video and Theater Displays

Display System	Luminous Output (brightness), nits (ft·L)	Contrast Ratio at Ambient Illumination (fc)	Resolution (TVL)
Television receiver	200–400, 60–120	30:1 at 5	275
Theater (film projector)	34–69, 10–20	100:1 at 0.1[1]	4800
[1] Limited by lens flare			

vides performance levels achieved by direct-view standard-definition video displays and conventional film theater equipment.

Evaluation of overall projection system brightness B, as a function of its optical components, can be calculated using the following equation:

$$B = \frac{L_G \, G \, T \, R^M \, D}{4W_G \, (f/N)^2 \, (1+m)^2}$$

Where:
L_G = luminance of the green source (CRT or other device)
G = screen gain
T = lens transmission
R = mirror reflectance
M = number of mirrors
D = dichroic efficiency
W_G = green contribution to desired white output (percent)
f/N = lens f-number
m = magnification

For systems in which dichroics or mirrors are not employed, those terms drop out.

Two basic categories of viewing screens are employed for projection video displays. As illustrated in Figure 9.15, the systems are:

- *Front projection*, where the image is viewed from the same side of the screen as that on which it is projected.

- *Rear projection*, where the image is viewed from the opposite side of the screen as that on which it is projected.

Front projection depends upon reflectivity to provide a bright image, while rear projection requires high transmission to achieve that characteristic. In either case, screen size influences display brightness inversely as follows:

$$B = \frac{L}{A}$$

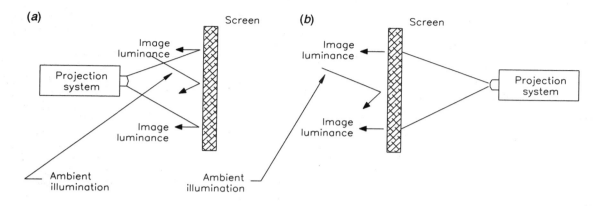

Figure 9.15 Projection system screen characteristics: (a) front projection, (b) rear projection.

Where:
B = apparent brightness (cd/m^2)
L = projector light output (lm)
A = screen viewing area (m^2)

Thus, for a given projector luminance output, viewed brightness varies in proportion to the reciprocal of the square of any screen linear dimension (width, height, or diagonal). An increase in screen width from the conventional aspect ratio of 4:3 (1.33) to an HDTV ratio of 16:9 (1.777) requires an increase in projector light output of approximately 33 percent for the same screen brightness.

To improve apparent brightness, directional characteristics can be designed into viewing screens. This property is termed *screen gain* (G), and the preceding equation becomes:

$$B = G \times \frac{L}{A}$$

Gain is expressed as screen brightness relative to a *lambertian surface*. Table 9.3 lists some typical front-projection screens and their associated gains.

Screen contrast is a function of the manner in which ambient illumination is treated. Figure 9.15 illustrates that a highly reflective screen (used in front projection) reflects ambient illumination as well as the projected illumination (image). The reflected light thus tends to dilute contrast, although highly directional screens diminish this effect. A rear-projection screen depends upon high transmission for brightness but can capitalize on low reflectance to improve contrast. A scheme for achieving this is equivalent to the black matrix utilized in tricolor CRTs. Illustrated in Figure 9.16, the technique focuses projected light through lenticular lens segments onto strips of the viewing surface, allowing intervening areas to be coated with a black (nonreflective) material. The lenticular segments and black stripes normally are oriented in the vertical dimension to broaden the horizontal viewing angle. The overall result is a screen that transmits most of the light (typically 60 percent) incident from

Table 9.3 Screen Gain Characteristics for Various Materials

Screen Type	Gain
Lambertian (flat-white paint, magnesium oxide)	1.0
White semigloss	1.5
White pearlescent	1.5–2.5
Aluminized	1–12
Lenticular	1.5–2
Beaded	1.5–3
Ektalite (Kodak)	10–15
Scotch-light (3M)	Up to 200

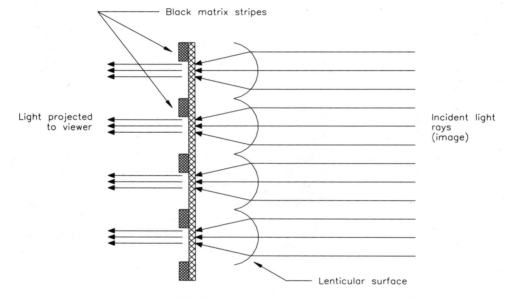

Figure 9.16 High-contrast rear-projection system.

the rear while absorbing a large percentage of the light (typically 90 percent) incident from the viewing side, thus providing high contrast.

Rear-projection screens usually employ extra elements, including diffusers and directional correctors, to maximize brightness and contrast in the viewing area.

As with direct-view CRT screens, resolution can be affected by screen construction. This is not usually a problem with front-projection screens, although granularity or lenticular patterns can limit image detail. In general, any screen element, such as the matrix arrangement described previously, that quantizes the image (breaks it into discrete segments) limits resolution. For 525- and 625-line video, this factor does not provide the limiting aperture. High-resolution applications, however, may require attention to this parameter.

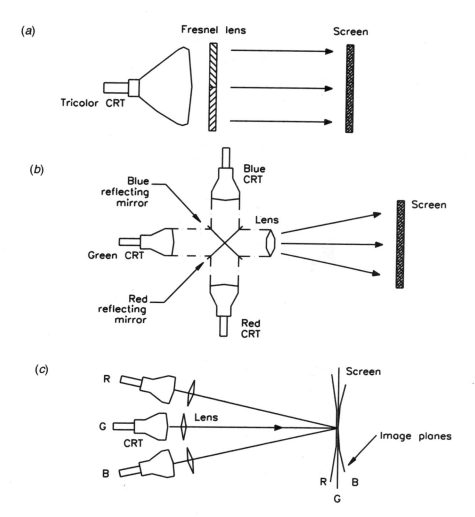

Figure 9.17 Optical projection configurations: (*a*) single tube, single lens rear projection system; (*b*) crossed-mirror trinescope rear projection; (*c*) three tube, three lens rear projection system; (*d*, next page) folded optics, front projection with three tubes in-line and a dichroic mirror; (*e*, next page) folded optics, rear projection. (*After* [6].)

9.3.2 Optical Projection Systems

Both refractive and reflective lens configurations have been used for the display of a CRT raster on screens 40-in or more in diagonal width. The first attempts merely placed a lenticular Fresnel lens, or an inefficient $f/1.6$ projection lens in front of a shadow mask direct view tube, as shown in Figure 9.17*a*. The resulting brightness of no greater than two or three ft·L was suitable for viewing only in a darkened room. Figure 9.17*b* shows a varia-

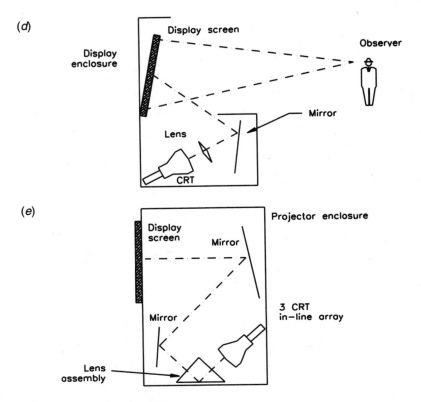

Figure 9.17 (Continued)

tion on this basic theme. Three individual CRTs are combined with cross reflecting mirrors and then focused onto the screen. The in-line projection layout is shown in 9.17c using three tubes, each with its own lens. This is the most common system used for multi-tube displays. Typical packaging to reduce cabinet size for front or rear projection is shown in 9.17d and e, respectively.

Because of the off-center positioning of the outboard color channels, the optical paths differ from the center channel, and keystone scanning height modulation is necessary to correct for differences in optical throw from left to right. The problem, illustrated in Figure 9.18, is more severe for wide-screen formats.

Variables to be evaluated in choosing among the many schemes include the following:

- Source luminance
- Source area
- Image magnification
- Optical-path transmission

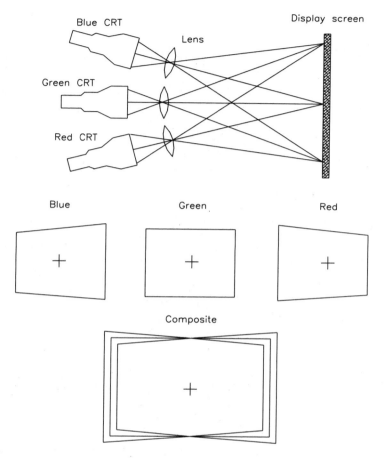

Figure 9.18 Three tube in-line array. The optical axis of the outboard (red and blue) tubes intersect the axis of the green tube at screen center. Red and blue rasters show trapezoidal (keystone) distortion and result in mis-convergence when superimposed on the green raster, thus requiring electrical correction of scanning geometry and focus. (*After* [6].)

- Light collection efficiency (of the lens)
- Cost, weight, and complexity of components and corrective circuitry

The lens package is a critical factor in rendering a projection system cost-effective. The package must possess good luminance collection efficiency (small *f*-number), high transmission, good *modulation transfer function* (MTF), light weight, and low cost. Table 9.4 compares the characteristics of some available lens complements.

The total light incident upon a projection screen is equal to the total light emerging from the projection optical system, neglecting losses in the intervening medium. The distribution of this light generally is not uniform. Its intensity is less at screen edges than at the center in

Table 9.4 Performance of Various Types of Projection Lenses

Lens	Aperture	Image Diagonal (mm)	Focal Length (mm)	Magnification	Response at 300 TVL (percent)
Refractive, glass	f/1.6	196	170	8	33
Refractive, acrylic	f/1.0	127	127	10	13
Schmidt	f/0.7	76	87	30	15
Fresnel	f/1.7	305	300	5	6

most projection systems as a result of light ray obliquity through the lens ($\cos^4 \theta$ law) and *vignetting effects* [3].

Light output from a lens is determined by collection efficiency and transmittance, as well as source luminance. Typical figures for these characteristics are:

- Collection efficiency: 15–25 percent

- Transmittance: 75–90 percent

Collection efficiency is partially a function of the light source, and the figure given is typical for a lambertian source (CRT) and lens having a half-field angle of approximately 25°.

9.3.3 Optical Distortions

Optical distortions are important to image geometry and resolution. Geometry is generally corrected electronically, both for pincushion/barrel effects and keystoning, which result from the fact that the three image sources are not coaxially disposed in the common in-line array.

Resolution, however, is affected by lens astigmation, coma, spherical aberration, and chromatic aberration. The first three factors are dependent upon the excellence of the lens, but chromatic aberration can be minimized by using line emitters (monochromatic) or narrow band emitters for each of the three image sources. Because a specific lens design possesses different magnification for each of the three primary colors (the index of refraction varies with wavelength), *throw distance* for each must also be adjusted independently to attain perfect registration.

In determining final display luminance, transmission, reflectance, and scattering by additional optical elements such as dichroic filters, optical-path folding mirrors, or corrective lenses must also be accounted for. Dichroics exhibit light attenuations of 5 to 30 percent, and mirrors can reduce light transmission by as much as 5 percent each. Front-surface mirrors exhibit minimum absorption and scattering but are susceptible to damage during cleaning. Contrast is also affected by the number and nature of optical elements employed in the projection system. Each optical interface generates internal reflections and scattering, which dilute contrast and reduce MTF amplitude response. Optical coatings can be utilized to minimize these effects, but their contribution must be balanced against their cost.

9.3.4 Image Devices

CRTs and light valves are the two most common devices for creating images to be optically projected. Each is available in a multitude of variations, the most successful of which are described in Section 9.4. Projection CRTs have historically ranged in size from 1-in (2.5 cm) to 13-in (33 cm) diagonal (diameter for round envelope types). Because screen power must increase in proportion to the square of the magnification ratio, it is clear that faceplate dissipation for CRTs used in projection systems must be extremely high. Electrical-to-luminance conversion efficiency for common video phosphors is on the order of 15 percent [4]. A 50-in (1.3 m) diagonal screen at 60 ft·L requires a 5-in (12.7 cm) CRT to emit 6000 ft·L, exclusive of system optical losses, resulting in a faceplate dissipation of approximately 20 W in a 3-in (7.6 cm) by 4-in (10.2 cm) raster. A practical limitation for ambient air-cooled glass envelopes (to minimize thermal breakage) is 1 mW/mm^2 or 7.74 W for this size display. Accommodation of this incompatibility must be achieved in the form of improved phosphor efficiency or reduced strain on the envelope via cooling. Because phosphor development is a mature science, maximum benefits are found in the latter course of action with liquid cooling assemblies employed to equalize differential strain on the CRT faceplate. Such implementations produce an added benefit through reduction of phosphor thermal quenching and thereby supply up to 25 percent more luminance output than is attainable in an uncooled device at equal screen dissipation [5].

A liquid-cooled CRT assembly, shown in Figure 9.19, depends upon a large heat sink to carry away and dissipate a substantial portion of the heat generated in the CRT. Large screen projectors using such assemblies commonly operate CRTs at four to five times their rated thermal capacities. Economic constraints mitigate against the added cost of cooling assemblies, however, and methods to improve phosphor conversion efficiency and optical coupling/transmission efficiencies continue to be investigated.

Concomitant to high power are high voltage and/or high beam current. Each has benefits and penalties. Resolution, dependent on spot diameter, is improved by increased anode voltage and reduced beam current. For a 525- or 625-line display, spot diameter should be 0.006-in (0.16 mm) on the 3 by 4-in (7.6 by 10.2-cm) raster discussed previously. Higher resolving-power displays require yet smaller spot diameters, but a practical maximum anode-voltage limit is 30 to 32 kV when x-radiation, arcing, and stray emission are considered. Exceeding 32 kV typically requires special shielding and CRT processing during assembly.

One form of the in-line array benefits from the relatively large aperture and light transmission efficiency of Schmidt reflective optics by combining the electron optics and phosphor screen with the projection optics in a single tube. The principal components of an integral Schmidt system are shown in Figure 9.20. Electrons emitted from the electron gun pass through the center opening in the spherical mirror of the reflective optical system to scan a metal-backed phosphor screen. Light from the phosphor (red, green, or blue, depending on the color channel) is reflected from the spherical mirror through an *aspheric corrector lens*, which serves as the face of the projection tube. Schmidt reflective optical systems are significantly more efficient than refractive systems because of the lower *f* characteristic and the reduced attenuation by glass in the optical path.

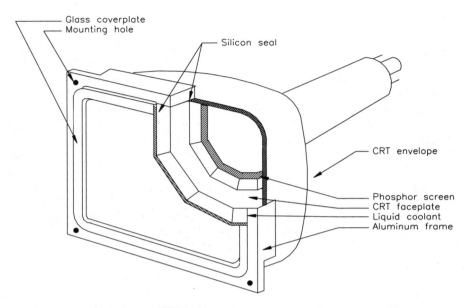

Figure 9.19 Mechanical construction of a liquid-cooled projection CRT.

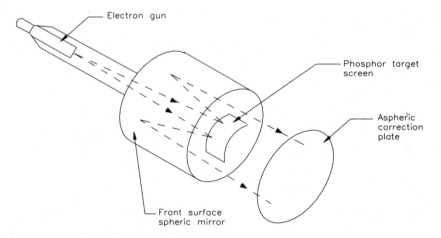

Figure 9.20 Projection CRT with integral Schmidt optics.

Advanced CRTs for Projection Display

Among the promising technologies for improved CRT devices intended for HDTV applications is the YAG (*yttrium aluminum garnet*) luminescent screen. Prepared using a *liq-*

uid-phase epitaxy (LPE) process, such devices can obtain high luminance and high definition. However, the tube envelop—made from high-density sintered alumina—is costly, and the sealing process between the YAG faceplate and tube envelope is complicated. Screening is likewise difficult, and efficiency is a concern [6].

Additional work has resulted in a 2–3 in. diagonal YAG phosphor screen for HDTV [7, 8]. The substrate material (i.e. the CRT faceplate), replaces conventional glass, offering high thermal conductivity and optical clarity. The phosphor screen is prepared by depositing phosphor on YAG using improved centrifugal sedimentation screening techniques.

Further development has resulted in a practical device [9]. The YAG phosphor screen faceplate has significantly higher thermal conductivity ($\lambda = 0.12$ W-cm^{-1}-K^{-1} at room temperature) than a conventional glass faceplate ($\lambda = 0.01$ W-cm^{-1}-K^{-1} at room temperature). Moreover, the new device has high mechanical strength, excellent insulation characteristics, and is much thinner ($D = 2$ mm) than glass ($D = 10$ mm, approximately).

Common rules of thermodynamics dictate that when the phosphor screen is in thermal equilibrium, the conducted thermal energy Q per unit area of phosphor screen faceplate per unit time is equal to:

$$Q = \frac{\lambda \Delta T}{D}$$

Where:
λ = thermal conductivity of the faceplate
ΔT = temperature difference between the inside and outside of faceplate
D = thickness of faceplate

Using this equation, it can be determined that for the same electron excitation power density, the YAG phosphor screen temperature difference is only 1/60 that of the glass phosphor screen. By means of calculation and experiment, the temperature difference between the YAG phosphor screen inside and outside surfaces is only 1.8°C, compared to 110°C for a typical glass phosphor screen at 1-W/cm^2 electron excitation power density [9].

Through the use of liquid cooling, the YAG phosphor screen avoids to a great extent phosphor thermal quenching and burn out caused by high-density electron bombardment, which is a serious problem with conventional glass projection phosphor screens. Therefore, a device based on the YAG phosphor screen can obtain higher luminance and longer life than one based on a glass phosphor screen.

In order to further improve the luminance, chromaticity, and resolution on the YAG projection system, a multilayer interference filter an be applied between the phosphor layer and the YAG faceplate, comprising a number of layers that are alternately of high and low refractive index [9]. This filter has the transmittance characteristics of a shortwave pass for the central wavelength of light emitted by the phosphor. Therefore, large-angle light is reflected by the filter to the phosphor and is scattered with an angular redistribution, resulting in part of the light being re-emitted at small angles. This changes the angular distribution of the emitted light from Lambertian to a more peaked emission in the forward direction with a gain at the smaller angles. Using this approach, significantly improved collection of light by the lens is achieved, which increases the luminance on the projection screen. Furthermore, because the filter is a narrow-bandpass type, it changes the spectral distribution of the emitted light and results in an attenuation of unwanted long-wave components. This improves the chromaticity and reduces chromatic aberration.

9.4 Projection Display Systems

The use of large screen video displays has been increasing for a wide variety of applications. Because of the greater resolution provided by HDTV and advanced computer graphics systems, higher image quality is being demanded from large screen display systems. Available basic technologies include the following:

- CRT-based display systems
- *Eidophor* reflective optical system
- *Talaria* transmissive color system
- Various forms of light valve systems
- Laser beam projection scanning system
- LCD projection system
- Digital mirror device-based system

HDTV applications present additional, more stringent requirements than conventional 525/625-line color displays, including:

- Increased light output for wide screen presentation
- Increased horizontal and vertical resolution
- Broader gamut of color response to meet system specifications
- Projection image aspect ratio of 16:9 rather than 4:3

9.4.1 CRT Projection Systems

Large screen color projection systems typically employ three monochrome CRT assemblies (red, green, blue) and the necessary optics to project full-color images. CRT systems based on 5 and 7-in CRT technology are popular because of their low cost, however, they suffer from low luminance levels (less than 50-60 average lumens). Nine-inch CRTs provide higher luminance levels (160 lumens is typical). The larger CRT surface area permits increased beam energy without sacrificing resolution. Thirteen-inch CRT projection systems offer still higher luminance and resolution.

Static and dynamic convergence circuitry is used to align the three beams over the display area. Digital convergence circuits permit convergence files tailored for each video source to be stored within a dedicated microprocessor. Advanced operating capabilities include auto-sync to a variety of input sources, built-in diagnostics, and automatic setup.

With CRT technology, high luminance and high resolution are conflicting requirements. For a given acceleration voltage, increased beam current will provide increased luminance, but reduced resolution because the spot size tends to increase. High luminance levels also raise the operating temperature of the device, which may shorten its expected lifetime.

Table 9.5 Operating Characteristics of Basic Light Valve Projection Systems

Trade Name	Company	Addressing Method	Control Layer	Optical Effect
Eidophor	Gretag	Electron beam	Oil film on spherical mirror	Diffraction in reflection
Talaria and MLV	General Electric	Electron beam	Oil film on glass disk	Diffraction in transmission
SVS (TITUS)	SODERN (Philips)	Electron beam	Electro-optic crystal	Pockels effect in reflection
Super-projector	GM/Hughes	CRT + photo-conductor	ECB/liquid-crystal cell	Birefringence in reflection
HDTV LCLV	Sharp	Active TFT matrix in amorphous Si	TN-liquid crystal cell	Polarization rotation in transmission

Multiple sets of CRT-based projection systems can be linked to increase luminance and resolution for a given application. The multiple beams overlay each other to yield the improved performance. Convergence of the 6 to 12 CRT assemblies, however, can become complex. Multiple CRT systems are satisfactory for NTSC video sources, but are usually inadequate for the display of computer-generated graphics data with single pixel characters.

The *mosaic* approach to improved luminance and resolution subdivides the screen into segments, each allocated to an individual projection system. As a result, larger images are delivered with higher overall performance. This scheme has been used effectively in simulation training to provide seamless, panoramic displays. Real-time processors divide the input image(s) into the proper display subchannels. Careful edge correction is necessary to prevent brightness falloff and geometric linearity distortions.

9.4.2 Light Valve Systems

Light valves can be defined as devices that, like film projectors, employ a fixed light source modulated by an optical-valve intervening source and projection optics. Light valve display technology is a rapidly developing discipline. Light valve systems offer high brightness, variable image size, and high resolution. Table 9.5 lists some of the more common light valve systems and their primary operating parameters. The first four systems are based on *electron-beam addressing* in a CRT. The last system is based on *liquid crystal light valve* (LCLV) technology. Progress in light valve technology for HDTV depends upon developments in two key areas:

- Materials and technologies for light control
- Integrated electronic driving circuits for addressing picture elements

Eidophor Reflective Optical System

Light valve systems are capable of producing images of substantially higher resolution than are required for 525/625-line systems. They are ideally suited to large screen theater displays of HDTV. The *Eidophor* system (Gretag) is in common usage.

318   Video Display Engineering

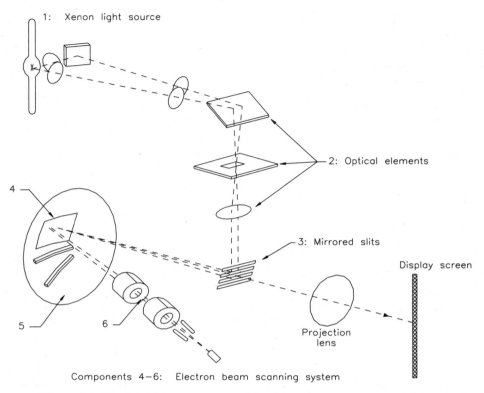

Figure 9.21 Mechanical configuration of the Eidophor projector optical system. (*After*[10].)

In a manner similar to film projectors, a fixed light source is modulated by an optical valve system (Schlieren optics) located between the light source and the projection optics (see Figure 9.21). In the basic Eidophor system, collimated light typically from a 2 kW xenon source (component 1 in the figure) is directed by a mirror to a viscous oil surface in a vacuum by a grill of mirrored slits (component 3).

The slits are positioned relative to the oil-coated reflective surface so that when the surface is flat, no light is reflected back through the slits. An electron beam scanning the surface of the oil with a video picture raster (components 4–6) deforms the surface in varying amounts, depending upon the video modulation of the scanning beam. Where the oil is deformed by the modulated electron scanning beam, light rays from the mirrored slits are reflected at an angle that permits them to pass through the slits to the projection lens. The viscosity of the liquid is high enough to retain the deformation over a period slightly greater than a television field.

Projection of color signals is accomplished through the use of three units, one for each of the red, green, and blue primary colors converged on a screen.

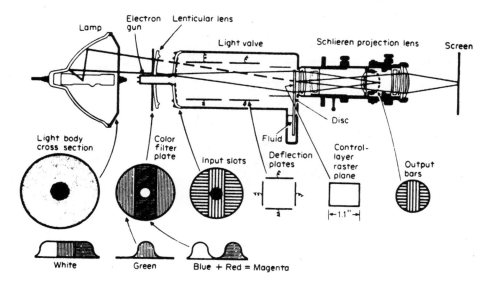

Figure 9.22 Functional operation of the General Electric single-gun light valve system. (*After* [10].)

Talaria Transmissive Color System

The Talaria system (General Electric) also uses the principle of deformation of an oil film to modulate light rays with video information. However, the oil film is transmissive rather than reflective. In addition, for full-color displays, only one gun is used to produce red, green, and blue colors. This is accomplished in a single light valve by the more complex Schlieren optical system shown in Figure 9.22.

Colors are created by writing *diffraction grating*, or grooves, for each pixel on the fluid by modulating the electron beam with video information. These gratings break up the transmitted light into its spectral colors, which appear at the output bars where they are spatially filtered to permit only the desired color to be projected onto the screen.

Green light is passed through the horizontal slots and is controlled by modulating the width of the raster scan lines. This is done by means of a high frequency carrier, modulated by the green information, applied to the vertical deflection plates. Magenta light, composed of red and blue primaries, is passed through the vertical slots and is modulated by diffraction gratings created at right angles (*orthogonal diffraction*) to the raster lines by velocity-modulating the electron beam in the horizontal direction. This is done by applying 16 MHz and 12 MHz carrier signals for red and blue, respectively, to the horizontal deflection plates and modulating them with the red and blue video signals. The grooves created by the 16 MHz carrier have the proper spacing to diffract the red portion of the spectrum through the output slots while the blue light is blocked. For the 12 MHz carrier, the blue light is diffracted onto

320 Video Display Engineering

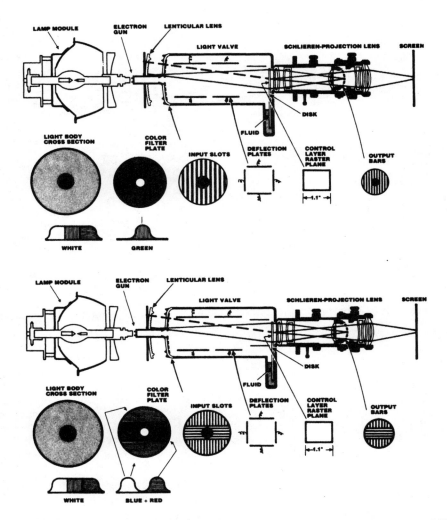

Figure 9.23 Functional operation of the two channel HDTV light valve system. (*After*[10].)

the screen while the red light is blocked. The three primary colors are projected simultaneously onto the screen in registry as a full-color picture.

To meet the requirements of HDTV, the basic Talaria system can be modified as shown in Figure 9.23. In the system, known as Talaria MLV-HDTV, one monochromatic unit with green dichroic filters produces the green spectrum. Because of the high scan rate for HDTV (on the order of 33.75 kHz), the green video is modulated onto a 30 MHz carrier instead of the 12 or 15 MHz used for 525- or 625-line displays. Adequate brightness levels are produced using a 700 W xenon lamp for the green light valve and a 1.3 kW lamp for the magenta (red and blue) light valve.

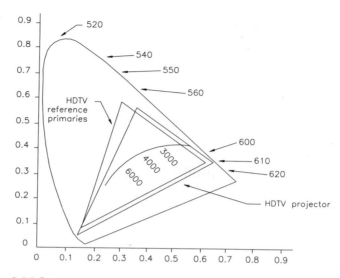

Figure 9.24 Color characteristics of the GE HDTV light valve system. (*After* [10].)

A second light valve with red and blue dichroic filters produces the red and blue primary colors. The red and blue colors are separated through the use of orthogonal diffraction axes. Red is produced when the writing surface diffracts light vertically. This is accomplished by negative amplitude modulation of a 120 MHz carrier, which is applied to the vertical diffraction plates of the light valve. Blue is produced when the writing surface diffracts light horizontally. This is accomplished by modulating a 30 MHz carrier with the blue video signal and applying it to the horizontal plate, as is done in the green light valve.

The input slots and the output bar system of the conventional light valve are used, but with wider spacing of the bars. Therefore, the resolution limit is increased. The wider bar spacing is achievable because the red and blue colors do not have to be separated on the same diffraction axis as in the single light valve system. This arrangement eliminates the cross-color artifact present with the single light valve system, and therefore improves the overall colorimetric characteristics, as shown in Figure 9.24.

High performance electron guns help provide the required resolution and modulation efficiency for HDTV systems of up to 1,250 lines. The video carriers have been optimized to increase the signal bandwidth capability to 30 MHz.

In the three element system, all three devices are monochrome light valves with red, blue, and green dichroic filters. The use of three independent light valves improves color brightness, resolution, and colorimetry. Typically, the three light valves are individually illuminated by xenon arc lamps operating at 1 kW for the green and at 1.3 kW for the red and blue light valves.

The contrast ratio is an important parameter in light valve operation. The amount of light available from the arc lamp is basically constant; the oil film modulates the light in response to picture information. The key parameter is the amount of light that is blocked during picture conditions when totally dark scenes are being displayed (the *darkfield performance*).

322   Video Display Engineering

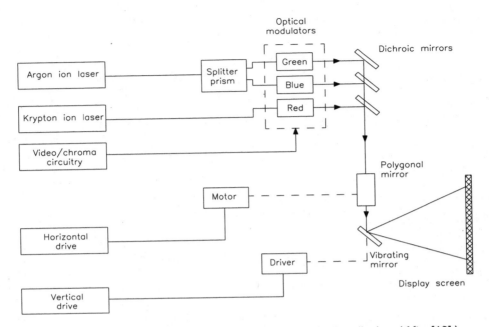

Figure 9.25 Block diagram of a laser-scanning projection display. (*After* [10].)

Another important factor is the ability of the display device to maintain a linear relationship between the original scene luminance and the reproduced picture. The amount of pre-distortion introduced by the camera must be compensated for by an opposite amount of distortion at the display device.

Laser Beam Projection Scanning System

Several approaches to laser projection displays have been implemented. The most successful employs three optical laser light sources whose coherent beams are modulated electro-optically and deflected by electromechanical means to project a raster display on a screen. The scanning functions are typically provided by a rotating polygon mirror and a separate vibrating mirror. A block diagram of the basic system is shown in Figure 9.25.

The flying spots of light used in this approach are one scan line (or less) in height and a small number of pixels wide. This means that any part of the screen may only be illuminated for a few nanoseconds. The scanned laser light projector is capable of high contrast ratios (as high as 1000:1) in a darkened environment. A laser projector may, however, be subject to a brightness variable referred to as *speckle*. Speckle is a sparkling effect resulting from interference patterns in coherent light. This effect causes a flat, dull projection surface to look as though it had a beaded texture. This tends to increase the perception of brightness, at the expense of image quality.

Figure 9.26 shows the configuration of a laser projector using continuous wave lasers and mechanical scanners. The requirements for the light wavelengths of the lasers are criti-

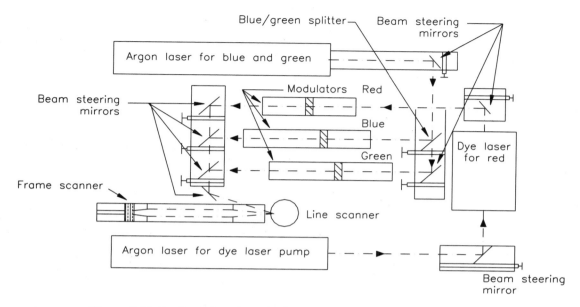

Figure 9.26 Configuration of a color laser projector. (*After* [11].)

cal. The blue wavelength must be shorter than 477 nm, but as long as possible. The red wavelength should be longer than 595 nm, but as short as possible. Greens having wavelengths of 510, 514, and 532 nm have been used with success. Because laser projectors display intense colors, small errors in color balance can result in significant distortions in gray scale or skin tone. The requirement for several watts of continuous-wave power further limits the usable laser devices. While several alternatives have been considered, most color laser projectors use argon ion lasers for blue and green, and an argon ion laser pumped dye laser for red.

To produce a conventional video signal, a modulator is required with a minimum operating frequency of 6 MHz. Several approaches are available. The bulk acousto-optic modulator is well suited to this task in high power laser projection. A directly-modulated laser diode is used for some low power operations. The modulator does not absorb the laser beam, but rather deflects it as needed. The angle of deflection is determined by the frequency of the acoustic wave.

The scanner is the component of the laser projector that moves the point of modulated light across and down the image plane. Several types of scanners can be used including mechanical, acousto-optic, and electro-optic. Two categories of scanning devices are used in the system:

- *Line scanner*, which scans the beam in horizontal lines across the screen. Lines are traced at 15,000 to 70,000 times per second. The rotating polygon scanner is commonly used, featuring 24 to 60 mirrored facets. Figure 9.27 illustrates a mechanical rotating polygon scanner.

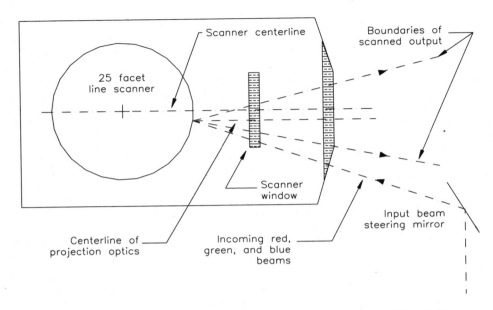

Figure 9.27 Rotating polygon line scanner for a laser projector. (*After* [11].)

- *Frame scanner*, which scans the beam vertically and forms frames of horizontal scan lines. The frame scanner cycles 50 to 120 times per second. A galvanometer-based scanner is typically used in conjunction with a mirrored surface. Because the mirror must fly back to the top of the screen in less than a millisecond, the device must be small and light. Optical elements are used to force the horizontally scanned beam into a small spot on the frame scanner mirror.

The operating speed of the deflection components may be controlled either from an internal clock, or locked to an external timebase.

The laser projector is said to have *infinite focus*. To accomplish this, optics are necessary to keep the scanned laser beam thin. Such optics allow a video image to remain in focus from approximately four feet to infinity. For high resolution applications, the focus is more critical.

Heat is an undesirable byproduct of laser operation. Most devices are water-cooled and use a heat exchanger to dump the waste heat.

Another approach large screen display employs an electron beam pumped monocrystalline screen in a CRT to produce a 1-in raster. The raster screen image is projected by conventional optics, as shown in Figure 9.28. This technology offers three optical benefits:

- High luminance in the image plane
- Highly directional luminance output for efficient optical coupling
- Compact, lightweight, and inexpensive projection optics

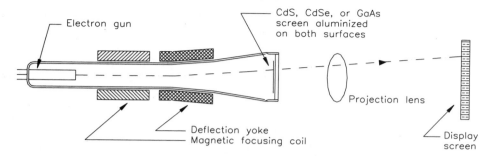

Figure 9.28 Laser-screen projection CRT. (*After* [12].)

Full-color operations is accomplished using three such devices, one for each primary color

LCD Projection Systems

Liquid crystal displays have been widely employed in high information content systems and high density projection devices. The steady progress in active-matrix liquid crystal displays, in which each pixel is addressed by means of a thin film transistor (TFT), has led to the development of full-color video TFT-addressed liquid crystal light valve (LCLV) projectors. Compared with conventional CRT-based projectors, LCLV systems have a number of advantages, including:

- Compact size
- Light weight
- Low cost
- Accurate color registration

Total light output (brightness) and contrast ratio, however, remain an issue in some applications. Improvements in the transmittance of polarizer elements have helped increase display brightness. By arrangement of the direction of the polarizers, the LCD can be placed in either the *normally black* (NB) or *normally white* (NW) mode. In the NW mode, light will be transmitted through the cell at V_{off} and will be blocked at V_{on}. The opposite situation applies for the NB mode. A high contrast display cannot be obtained without a satisfactory dark state.

Cooling the LC panels represents a significant technical challenge as the light output of projection systems increases. The contrast ratio of the displayed image will decrease as the temperature of the LC panels rise. Furthermore, long-term operation under conditions of elevated temperature will result in shortened life for the panel elements.

The conventional approach to cooling has been to circulate forced air over the LC panels. This approach is simple and works well for low- to medium-light-output systems. Higher power operation, however, may require liquid cooling. One system (Sanyo) incorporates a liquid-cooled heat exchanger to maintain acceptable contrast ratio on a high-output HDTV projector. The structure of the cooling system is shown in Figure 9.29. The cooling unit,

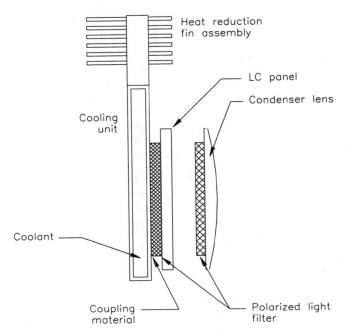

Figure 9.29 Structure of a liquid-cooled LC panel. (*After* [12].)

mounted behind the LC panel, is made of frame glass and two glass panels. The coolant is a mixture of water and ethylene glycol, which prevents the substance from freezing at low temperatures.

Figure 9.30 compares the temperature in the center of the panel for conventional air-cooling and for liquid-cooling. Because LC panel temperature is a function of the brightness distribution of the light source, the highest temperature is at the center of the panel, where most of the light is usually concentrated. As the amount of light from the lamp increases, the cooling activity of the liquid-based system accelerates. Waste heat is carried away from the projector by directing cooling air across the heat reduction fins.

Application Example

The numerous recent improvements in LC display technology have permitted significant advances in LC projection techniques. One system (Sharp) incorporates three 5.5-in, 1.2 million pixel TFT active matrix LCD panels as light valves to project onto a 110-in (or larger) screen. A simplified block diagram of the system is shown in Figure 9.31. A short-arc, single jacket type high intensity metal halide lamp is used as the light source. A parabolic cold mirror is utilized that reflects only the visible spectrum of the light beam to the front of the projector, allowing the infrared and ultraviolet spectrum rays to pass behind the mirror. Such rays have a harmful effect on the LC panels. The white light from the

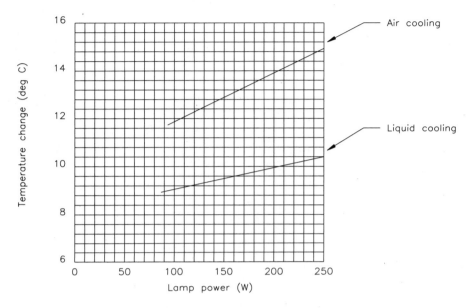

Figure 9.30 LC panel temperature dependence on lamp power in a liquid crystal projector. (*After* [12].)

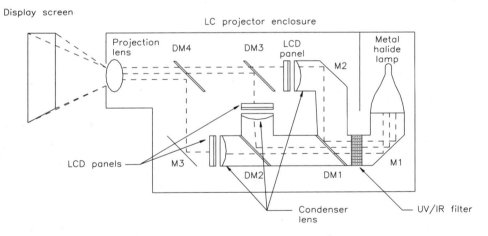

Figure 9.31 LC projection system for HDTV applications. (*After* [13].)

lamp is separated into the primary colors by dichroic mirrors. Each of the three primary color beams are regulated as they pass through their respective LC panels.

The LC system offers good signal clarity, compared to other projection systems. A single lens optical system is used; the primary colors are combined before entering the projection lens. The convergence of the picture projected, therefore, is unaffected by changes in the dis-

328 Video Display Engineering

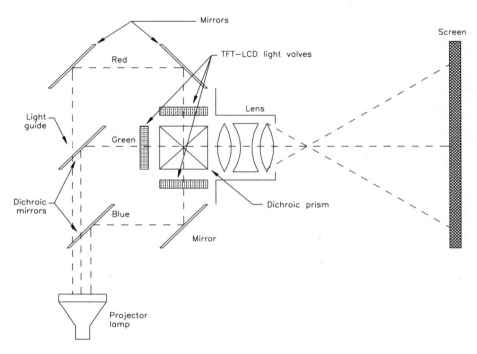

Figure 9.32 Optical System for an NTSC LCD projector. (*After* 14].)

tance between the projector and the screen. Furthermore, changes in the viewing position relative to the center axis of the display screen will not adversely affect color purity.

Figure 9.32 illustrates the optical system of an LC projector (JVC) designed for NTSC applications. The TFT active matrix LC panels contain 211,000 pixels. Vertical resolution of 440 TV lines is achieved through the use of noninterlaced progressive scanning. Interpolative picture processing techniques render the scanning lines essentially invisible. As shown in the figure, white light is separated into red, green, and blue light by dichroic mirrors. Light of each color is directed to its corresponding RGB LC imaging panel. The LC panels are mounted to a dichroic prism for fixed convergence. The RGB images are combined into a single full-color image in the dichroic prism and projected by the lens onto the screen.

Homeotropic LCLV

The *homeotropic* (perpendicular alignment) LCLV (Hughes) is based on the optical switching element shown in Figure 9.33[2]. Sandwiched between two transparent idium-tin

---

[2] This section is based on: Victor J. Fritz, "Full-Color Liquid Crystal Light Valve Projector for Shipboard Use," *Large Screen Projection Displays II*, William P. Bleha, Jr., (ed.), Proc. SPIE 1255, SPIE, Bellingham, Wash., pp. 59–68, 1990

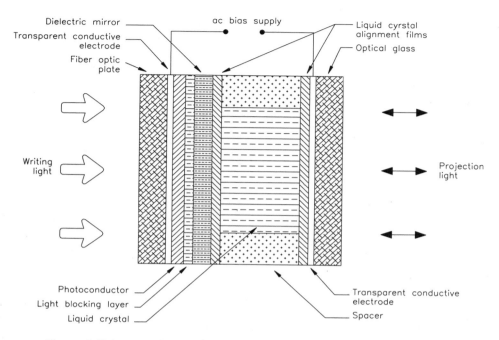

Figure 9.33 Layout of a homeotropic liquid crystal light valve. (*After* [15].)

oxide electrodes are a layer of cadmium sulfide photoconductor, a cadmium telluride light blocking layer, a dielectric mirror, and a 6 micron thick layer of liquid crystal. An ac bias voltage connects the transparent electrodes. The light blocking layer and the dielectric mirror are thin and have high dielectric constants, and so the ac field is primarily across the photoconductor and the liquid crystal layer. When there is no writing light impinging on the photoconductor, the field or voltage drop is primarily across the photoconductor and not across the liquid crystal. When a point on the photoconductor is activated by a point of light from an external writing CRT, the impedance at that point drops and the ac field is applied to the corresponding point in the liquid crystal layer. The lateral impedances of the layers are high, so the photocarriers generated at the point of exposure do not spread to adjacent points. Thus, the image from the CRT that is exposing the photoconductor is reproduced as a voltage image across the liquid crystal. The electric field causes the liquid crystal molecules to rotate the plane of polarization of the projected light that passes through the liquid crystal layer.

The homeotropic LCLV provides a perpendicular alignment in the *off* state (no voltage across the liquid crystal) and a pure optical birefringence effect of the liquid crystal in the *on* state (voltage applied across the liquid crystal). To implement this electro-optical effect, the liquid crystal layer is fabricated in a perpendicular alignment configuration; the liquid crystal molecules at the electrodes are aligned with their long axes perpendicular to the electrode surfaces. In addition, they are aligned with their long axes parallel to each other along a preferred direction that is fabricated into the surface of the electrodes.

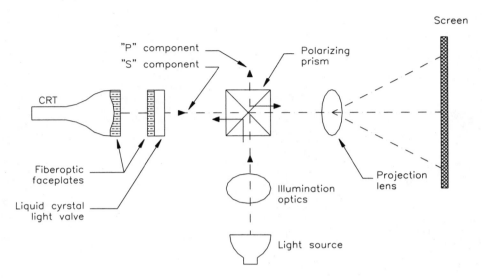

Figure 9.34 Block diagram of an LCLV system. (*After* [15].)

A conceptual drawing of an LCLV projector using this technology is shown in Figure 9.34. Light from a xenon arc lamp is polarized by a polarizing beamsplitting prism. The prism is designed to reflect *S* polarized light (the polarization axis, the E-field vector, is perpendicular to the plane of incidence) and transmit *P* polarized light (the polarization axis, the E-field vector, is parallel to the plane of incidence). The *S* polarization component of the light is reflected to the light valve. When the light valve is activated by an image from the CRT, the reflecting polarized light is rotated 90° and becomes *P* polarized. The *P* polarization component transmits through the prism to the projection lens. In this type of system the prism functions as both a polarizer and an analyzer. Both monochrome and full-color versions of this design have been implemented. Figure 9.35 shows a block diagram of the color system.

The light valve is coupled to the CRT assembly by a fiber optic backplate on the light valve and a fiber optic faceplate on the CRT.

The heart of this system is the fluid filled prism assembly. This device contains all of the filters and beamsplitter elements that separate the light into different polarizations and colors (for the full-color model). Figure 9.36 shows the position of the beamsplitter elements inside the prism assembly.

The light that exits the arc lamp housing enters the prism through the ultraviolet (UV) filter. The UV filter is a combination of a UV reflectance dichroic on a UV absorption filter. The UV energy below 400 nm can be destructive to the life of the LCLV. Two separate beamsplitters *prepolarize* the light: the green prepolarizer and the red/blue prepolarizer. Prepolarizers are used to increase the extinction ratio of the two polarization states. The green prepolarizer is designed to reflect only the *S* polarization component of green light. The *P* polarization component of green and both the *S* and *P* polarizations of red and blue are transmitted through the green prepolarizer. The green prepolarizer is oriented 90° with re-

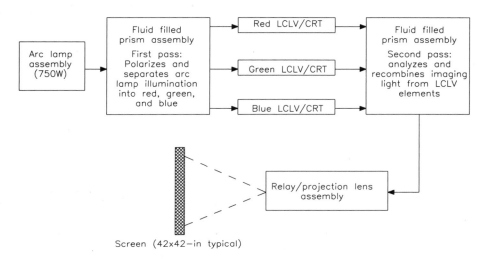

Figure 9.35 Functional optical system diagram of an LCLV projector. (*After* [15].)

spect to the other beamsplitters in the prism. This change of orientation means that the transmitted P component of the green light appears to be S polarized relative to the other beamsplitters in the prism assembly. The red/blue prepolarizer is designed to reflect the S polarization component of red and blue. The remaining green that is now S polarized and the red and blue P polarized components are transmitted to the main beamsplitter.

The main beamsplitter is a broadband polarizer designed to reflect all visible S polarized light and transmit all visible P polarized light. The S polarized green light is reflected by the main beamsplitter to the green mirror. The green mirror reflects the light to the green LCLV. The P polarized component of red and blue transmits through the main polarizer. The red/blue separator reflects the blue light and transmits the red to the respective LCLVs.

Image light is rotated by the light valve and is reflected to the prism. The rotation of the light or phase change (P polarization to S polarization) allows the image light from the red and blue LCLVs to reflect off the main beamsplitter toward the projection lens. When the green LCLV is activated, the green image light is transmitted through the main beamsplitter.

The fluid filled prism assembly is constructed with thin film coatings deposited on quartz plates and immersed in an index matching fluid. Because polarized light is used, it is necessary to eliminate stress birefringence in the prism, which—if present—would cause contrast degradation and background non-uniformity in the projected image. Using a fluid filled prism effectively eliminates mechanically and thermally induced stress birefringence as well as any residual stress birefringence. Expansion of the fluid with temperature is compensated for by using a bellows arrangement.

Laser-Addressed LCLV

The laser-addressed liquid crystal light valve is a variation on the CRT-addressed LCLV discussed in the previous section. Instead of using a CRT to write the video information on

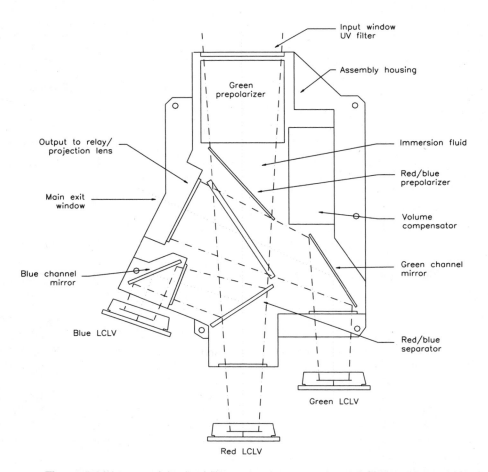

Figure 9.36 Layout of the fluid filled prism assembly for an LCLV projector. (*After* [15].)

the LC element, a laser is used. The principle advantages of this approach is the small spot size possible with laser technology, thereby increasing the resolution of the displayed image. In addition, a single beam from a laser scanner can be directed to three separate RGB light valves, reducing potential convergence problems by having only one source for the image.

A block diagram of the laser-addressed LCLV is shown in Figure 9.37. Major components of the system include the following:

- Laser
- Video raster scanner
- Polarizing multiplexing switches
- LCLV assemblies

Projection Display Systems

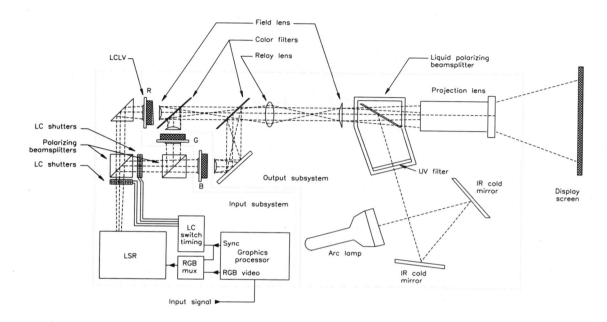

Figure 9.37 Block diagram of a color laser-addressed light valve display system. (*After*[16].)

- Projection optics

Projector operation can be divided into two basic subsystems:

- Input subsystem, which includes the laser, scanning mechanism, and polarizing beamsplitters
- Output subsystem, which includes the LCLV assemblies, projection optics, and light source

The output subsystem is similar in nature to the homeotropic LCLV discussed in the previous section.

The heart of the input subsystem is the *laser raster scanner* (LRS). The X-Y deflection system incorporated into the LRS uses four acousto-optic devices to achieve all solid-state modulation and scanning. A diode-pumped Nd:YAG, doubled to 535 nm (18 mW at 40:1 polarization ratio), is used as the laser source for the raster scan system.

Because only a single-channel scanner is used to drive the LCLV assemblies, a method is required to sequence the RGB video fields to their respective light valves. This sequencing is accomplished through the use of LC polarization switches. These devices act in a manner similar to the LCLV in that they change the polarization state of incident light. In this case, the object is to rotate the polarization by 0° or 90° and exploit the beam-steering capabilities of polarizing beamsplitters to direct the image field to the proper light valve at the proper time.

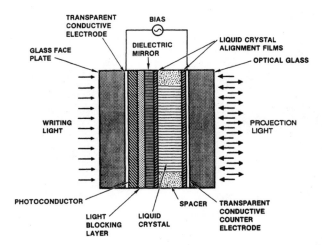

Figure 9.38 Cross-sectional diagram of the Image Light Amplifier light valve. (*After* [17]. *Courtesy of Hughes-JVC Technology.*)

As shown in Figure 9.38, each switch location actually consists of a pair of individual switches. This was incorporated to achieve the necessary fall time required by the LRS to avoid overlapping of the RGB fields. Although these devices have a short rise time (approximately 100 μs), their decay time is relatively long (approximately 1 ms). By using two switches in tandem, compensation for the relatively slow relaxation time is accomplished by biasing the second switch to relax in an equal, but opposite, polarization sense. This causes the intermediate polarization state that is present during the switch-off cycle to be nulled out, allowing light to pass through the LC switch during this phase of the switching operation.

Image Light Amplifier (ILA)

The Image Light Amplifier (Hughes, JVC) combines the advantages of a CRT-type projector with the high-brightness capability of a xenon arc lamp projection source. The enabling technology of the projector is the ILA modulator, which accepts a low-intensity image from a CRT and replicates the image on a high-intensity white xenon arc lamp beam. The ILA uses solid-state thin-film and liquid crystal technologies to produce a system with up to 12,000 lm output, at a contrast ratio of 1000:1. Resolutions of 2800×1500 TV lines (and higher) have been produced [17]. The ILA projector incorporates three modulators for the RGB channels of a color system.

The ILA light valve (ILA-LV) is a spatial light modulator that accepts a low-intensity input image and converts it, in real time, into an output image with light from another source. [18]. In the system, the image light sources are high-resolution CRTs, and the output light source is a xenon arc lamp. The ILA-LV is designed to operate in a reflective mode so that the input CRT and output xenon light beam are incident on opposite faces of the device. A

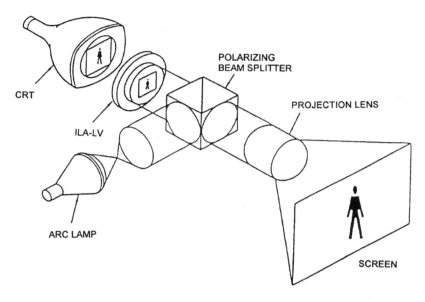

Figure 9.39 Optical channel of the basic ILA projector system. (*After* [17]. *Courtesy of Hughes-JVC Technology.*)

cross-sectional diagram of the ILA-LV is shown in Figure 9.38. Note the similarities to the homeotropic LCLV shown in Figure 9.33.

The basic ILA projector optical channel is shown in Figure 9.39. This is the building block for the full-color system, which is shown in Figure 9.40. The input image, provided by the high-resolution CRT, is imaged on the ILA-LV through a relay lens. The xenon arc lamp and condensing optics provide the output projection light beam, which is linearly polarized by a McNeille-type *polarizing beam-splitter* (PBS) before reaching the ILA-LV. The PBS polarizes the light to a high extinction ratio without absorption. The projected beam passes through the liquid crystal, reflects from the dielectric mirror, and passes through the liquid crystal again before returning to the polarizing beam-splitter.

As the beam passes through the liquid crystal, the direction of the linearly polarized light is rotated in direct response to the level of input image modulation of the liquid crystal birefringence. The PBS then operates on the output image from the ILA-LV, passing rotated polarized light to the projection lens and returning nonrotated light toward the lamp. Finally, the projection lens focuses and magnifies the ILA-LV image onto a screen.

Among the benefits of this system are:

- High brightness
- Undefined pixel structure (which permits a variety of input signal types and formats)
- High resolution
- Good contrast ratio

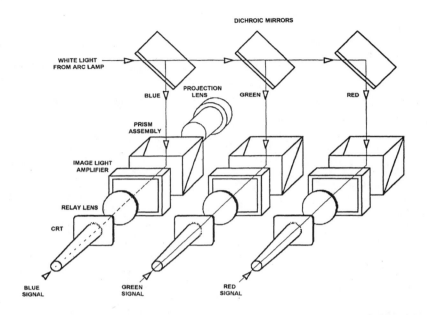

Figure 9.40 Full-color optical system for the ILA projector (Series 300/400). (*After*[17]. *Courtesy of Hughes-JVC Technology.*)

- Excellent color rendition, as illustrated in Figure 9.41.

Digital Micromirror Device

DMD is a semiconductor-based array of fast, reflective digital light switches that precisely control a light source using a binary pulse-width modulation (PWM) technique [18]. Individual DMD elements can be combined with image processing, memory, a light source, and optics to form a Digital Light Processing (DLP, Texas Instruments) system capable of projecting high-resolution color images. DLP-based projection displays are well suited to high-brightness and high-resolution applications. Attributes of the technology include the following:

- The digital light switch is reflective and has a high fill factor, resulting in high optical efficiency at the pixel level and low pixelation effects in the projected image.

- As the resolution and size of the DMD array increases, the overall system optical efficiency grows because of higher lamp-coupling efficiency.

- The DMD operates with conventional CMOS voltage levels (5 V), so integrated row and column drivers are readily employed to minimize the complexity and cost of scaling to higher resolutions.

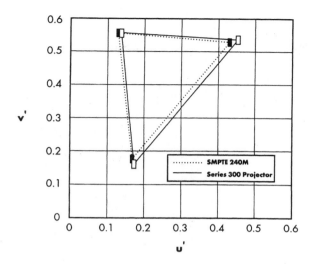

Figure 9.41 Color gamut of the ILA projector (Series 300/400) compared with SMPTE 240M. (*After* [17]. *Courtesy of Hughes-JVC Technology.*)

- Because the DMD is a reflective technology, the DMD chip can be effectively cooled through the chip substrate, thus facilitating the use of high-power projection lamps without thermal degradation of the DMD.

The DMD light switch, shown in Figure 9.42, is a member of a class of devices known as *microelectromechanical systems* (MEMS). Other MEMS devices include pressure sensors, accelerometers, and microactuators. The DMD is monolithically fabricated by CMOS-like processes over a CMOS memory element. Each light switch has a 16-mm-square aluminum mirror that can reflect light in one of two directions, depending on the state of the underlying memory cell. Rotation of the mirror is accomplished through electrostatic attraction produced by voltage differences developed between the mirror and the underlying memory. With the memory cell in the on (1) state, the mirror rotates to $+10°$; with the memory cell in the off (0) state, the mirror rotates to $-10°$.

When the DMD is combined with a suitable light source and projection optics, as illustrated in Figure 9.43, the mirror reflects incident light either into or out of the pupil of the projection lens by a simple beam-steering process. As a result, the (1) state of the mirror appears bright, and the (0) state of the mirror appears dark. Compared with diffraction-based light switches, the beam-steering action of the DMD light switch provides a favorable trade-off between contrast ratio and the overall brightness efficiency of the system.

Image gray scale is achieved through pulse-width modulation of the incident light. Color is achieved by using color filters, either stationary or rotating, in combination with one, two, or three DMD chips. A detailed photo of a DMD element is shown in Figure 9.44.

The simultaneous update of all mirrors produces an inherently low flicker display; there is no line-to-line temporal phase shift. Furthermore, a bit-splitting PWM algorithm pro-

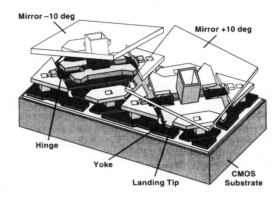

Figure 9.42 A pair of DMD pixels with one mirror shown at −10° and the other at +10°. (*After* [18]. *Courtesy of Texas Instruments.*)

duces short-duration light pulses that are distributed uniformly throughout the video field time, eliminating a temporal decay in brightness.

DLP optical systems have been designed in a variety of configurations, distinguished by the number of DMD chip arrays in the system [19]. The 1- and 2-chip schemes rely on a rotating color disk to time-multiplex the colors. The 1-chip configuration is used for lower brightness applications and is the most compact. The 2-chip systems yield higher brightness performance, but are primarily intended to compensate for the color deficiencies resulting from spectrally imbalanced lamps (e.g., the red deficiency in many metal halide lamps). For applications requiring the highest brightness, 3-chip systems are used.

A 3-chip DLP optical system is shown in Figure 9.45. Because the DMD is a simple array of reflective light switches, no polarizers are required. Light from a metal halide or xenon lamp is collected by a condenser lens. For proper operation of the DMD light switch, this light must be directed at 20° relative to the normal of the DMD chip. To accomplish this in a method that eliminates mechanical interference between the illuminating and projecting optics, a *total internal reflection* (TIR) prism is interposed between the projection lens and the DMD color-splitting/color-combining prisms.

The color-splitting/color-combining prisms use dichroic interference filters deposited on their surfaces to split the light by reflection and transmission into red, green, and blue components. The red and blue prisms require an additional reflection from a TIR surface of the prism in order to direct the light at the correct angle to the red and blue DMDs. Light reflected from the on-state mirrors of the three DMDs is directed back through the prisms, and the color components are recombined. The combined light then passes through the TIR prism and into the projection lens because its angle has been reduced to below the critical angle for total internal reflection in the prism air gap.

As the DMD resolution is increased, the pixel pitch is held constant and the chip diagonal is allowed to increase, as detailed in Figure 9.46. This approach to display design has several advantages:

Projection Display Systems 339

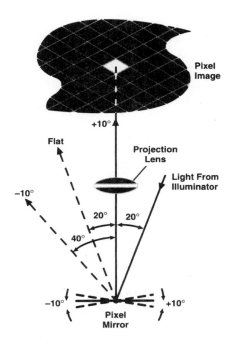

Figure 9.43 The basics of DMD optical switching. (*After* [18]. *Courtesy of Texas Instruments.*)

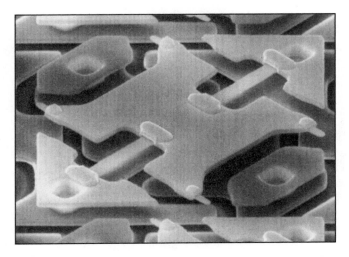

Figure 9.44 A scanning electron microscope view of the DMD yoke and spring tips (the mirror has been removed). (*After* [18]. *Courtesy of Texas Instruments.*)

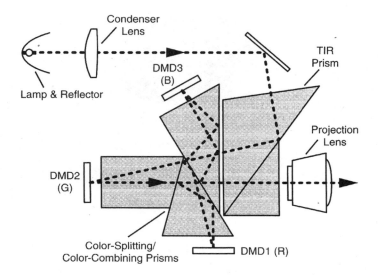

Figure 9.45 Optical system of the DLP 3-chip implementation. (*After* [18]. *Courtesy of Texas Instruments.*)

- The high optical efficiency and contrast ratio of the pixel are maintained at all resolutions.
- Pixel timing is common to all designs, and high address margins are maintained.
- The chip diagonal increases with resolution, which improves the DMD system optical efficiency.

DLP projection systems have been demonstrated at a variety of resolutions (and aspect ratios), including a 16:9 aspect ratio high-definition (1920 × 1080) system [20, 21, 22].

Grating Light Valve Display

The *Grating Light Valve* (GLV) technology is a means for creating a high-performance spatial light modulator on the surface of a silicon chip [23]. It is based on simple optical principles that leverage the wavelike behavior of light, using diffractive interference as the basis for discriminating between *on* and *off* pixel states. A GLV array is fabricated using conventional CMOS materials and equipment, adopting techniques from the emerging field of micro-electromechanical systems (MEMS). Pixels are comprised of a series of identical mechanical structures, fabricated using a relatively small number of masks and processing steps.

A GLV pixel is an addressable diffraction grating created by moving microscopic planar structures. A typical GLV pixel consists of an even number of parallel, dual-supported "ribbons" formed of silicon nitride and coated with a reflective aluminum top-layer, as illustrated in Figure 9.47. These ribbons are suspended above a thin air gap, allowing them to move vertically relative to the plane of the surface. The ribbons are held in tension, such that

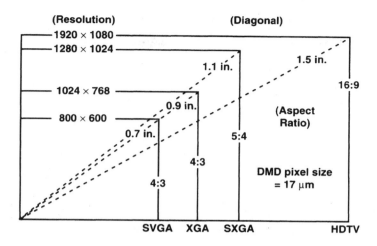

Figure 9.46 DMD display resolution as a function of chip diagonal. (*After* [18]. *Courtesy of Texas Instruments.*)

Figure 9.47 A typical GLV pixel showing alternate ribbons being addressed. (*From* [23]. *Courtesy of Silicon Light Machines.*)

in their unaddressed state, the surfaces of the ribbons collectively function as a mirror. A GLV pixel is addressed by inducing a voltage potential between the top of the ribbons and the substrate, thereby deflecting alternate ribbons. Viewed in cross-section (as in Figure 9.48), the up/down pattern of reflective surfaces creates a *square-well* diffraction grating [24]. The GLV device uses digitally-generated signals to produce specific drive voltages for each pixel, achieving precise variable control over the proportion of light that is reflected or diffracted.

Scanned Linear GLV Architecture

In contrast to a more conventional 2-D array made up of physical pixels that correspond to each bright or dark spot in the final image, the *Scanned Linear* GLV architecture consists of a linear array of physical GLV pixels oriented along a vertical column of image data [25]. Once per image refresh, this linear array is optically scanned across the screen to produce a complete two-dimensional image. During one scan, each pixel writes successive values corresponding to one entire row of image data. In this way, a single scan of the linear GLV array creates the complete image.

The GLV architecture imposes considerable performance requirements on the spatial light modulator used in the system, specifically:

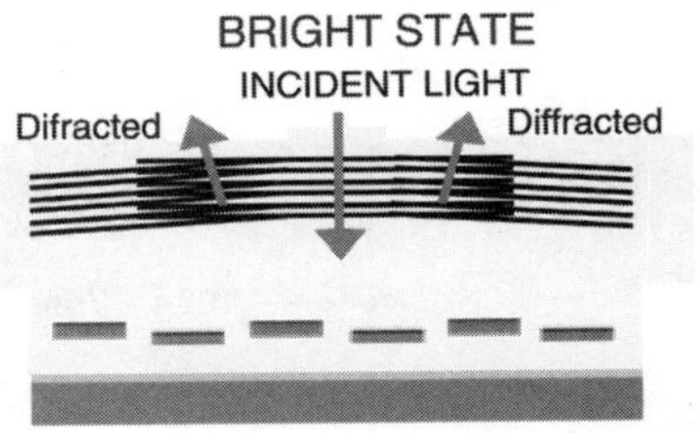

Figure 9.48 The diffractive (bright) state of a GLV pixel. (*After* [23]. *Courtesy of Silicon Light Machines.*)

- It must be capable of extremely fast switching speeds
- Capable of withstanding very high optical power density
- Capable of rendering continuous tone gray scale

GLV Video Processing Architecture

A 1080P projection display prototype based on the Scanned Linear GLV architecture has been developed [23]. (See Figure 9.49.) The system accepts 1080P video data at 24 or 30 f/s via a standard SMPTE 292M serial digital interface. The electronics architecture supports the following system parameters:

- 1920×1080 resolution
- Up to 120 Hz refresh, progressive scan
- 10 bits/channel R, G, B

The SMPTE 292M serial digital input contains luminance for all pixels and chrominance for odd pixels. The even pixel chroma values are generated by FIR filtering the red and blue chroma inputs. The luma and chroma are decoded into red, green, and blue with gamma correction using multipliers and adders. The decoder can support any 10-bit standard such as ITU-Rec. 601 or SMPTE 240M.

Gamma removal and color transformation are combined to allow a $9K \times 16$-bit table to take the place of three nonlinear mappings and nine multiply steps. This approach supports an arbitrary change of color primaries and any 10-bit gamma standard, such as ITU-Rec. BT.709 or SMPTE 240M. The table outputs are added and rounded to yield linear RGB for the system primaries. The prototype system uses solid-state RGB lasers as illumination sources and is capable of displaying a considerably wider color gamut than SMPTE RP 145. The table entries can be modified to change the whitepoint and/or to interpret highlights, depending on whether the input is standard video for HDTV or a custom transfer intended to produce a more film-like appearance.

The display mapping/gamma adjustment step maps RGB intensity to the GLV intensity-voltage characteristic. Conventional spatial light modulators that create grayscale values through pulse width modulation have an inherently linear optical response. However, the inherent GLV electro-optic response creates a natural, continuous grayscale with wide dynamic range that is well matched to the human visual system. (See Figure 9.50.)

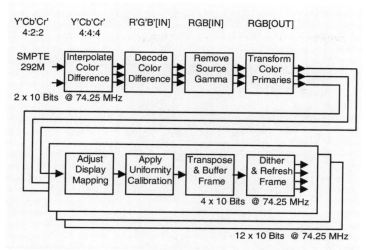

Figure 9.49 Video processing architecture for a Scanned Linear GLV system. (*After* [23]. *Courtesy of Silicon Light Machines.*)

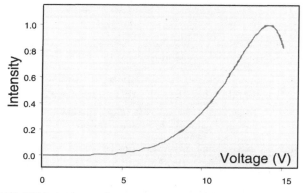

Figure 9.50 GLV electro-optic response. (*After* [23]. *Courtesy of Silicon Light Machines.*)

The electro-optic response of a GLV pixel is quite uniform because it is based on a simple and repeatable system of electrostatic (attractive) and mechanical spring (restorative) forces. Because of this mechanical simplicity, the GLV response is highly predictable and can be mathematically calculated from relatively simple models.

The SMPTE 292M input is *row-centric* (i.e., the video data is presented sequentially by row). Because the scanned linear GLV system as currently implemented scans left to right by column, a frame buffer is used to store data by rows and transpose it into column data for display. Because higher refresh rates produce better image quality, the frame buffer accepts progressive data at the source rate and sends it out at a faster rate for display. The frame

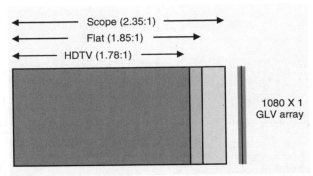

Figure 9.51 The GLV horizontally scanned linear system supports variable aspect ratios. (*After* [23]. *Courtesy of Silicon Light Machines.*)

buffer in the prototype system typically reads data in at 24 or 30 f/s and refreshes the display up to four times the input rate.

Scanning horizontally has several benefits:

- It requires a smaller and less expensive linear GLV array (1080 pixels vs. 1920 pixels, a 44 percent pixel count reduction).

- The smaller modulator allows additional system cost savings, such as smaller recombination and projection optics, smaller look-up tables, and so on.

- A horizontal scan enables electronic support for variable aspect ratios (Figure 9.51). For example, a horizontal scan system can change from 4:3 to 16:9 for HDTV or to *cinemascope* (2.35) for electronic cinema, without requiring anamorphic lenses or complex scaling algorithms that tend to degrade image quality.

By refreshing the display 3 or 4 times per frame, it is possible to achieve 1.6 or 2 additional effective bits of grayscale through dithering [23]. The prototype system uses conventional drivers, similar to those that might be used as LCD column drivers. Through *temporal dithering*, the system exploits the inherent speed and the scanned line approach of the GLV device to achieve 10-bit gray scale using simple, low cost 8-bit drivers.

9.4.3 Discrete Element Display Systems

The essential characteristic defining discrete-element display devices is division of the viewing surface (volume for three-dimensional displays) into separate segments that are individually controlled to generate an image. Each element, therefore, embodies a dedicated controlling switch or valve as opposed to raster display devices, which employ one (or a small number of) control device(s) for activation of all display elements.

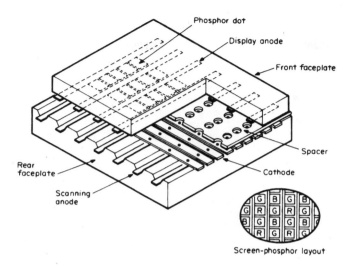

Figure 9.52 DC plasma display panel for color video. (*After* [10].)

Luminescent Panels

The prevalent technology of large screen luminescent panels employs plasma excitation of phosphors overlying each display cell. The plasma is formed in low-pressure gas that may be excited by either ac or dc fields. Figure 9.52 depicts a color dc plasma panel. Such devices have been constructed in a variety of sizes and cell structures. Color rendition requires red/green/blue trios that are excited preferentially or proportionally, as in a color CRT, which triples the number of actual cells (three cells comprise one picture element).

Plasma display panels have been produced with an optical resolution exceeding 100 pixels/in and a picture resolution of over 4,000,000 pixels. Other distinguishing features include:

- High contrast ratio (1000:1 possible)
- Wide viewing angle (160° or more vertical and horizontal)
- Memory mode capability
- Eight bit gray scale (256)
- Multiplexing capability (produced at 2,048 lines and above)
- Full-color capability

Because of their favorable size and weight considerations, PDP systems are well suited to large screen (40-in diagonal) full-color emissive display applications.

Both ac and dc PDPs have been developed to produce full-color, flat-panel large area dot-matrix devices. PDP systems operate in a memory mode. Memory operation is achieved

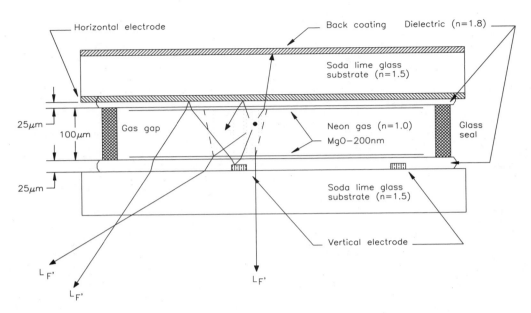

Figure 9.53 Cross section of an AC-PDP cell capacitor. (*After* [26].)

through the use of a thick-film current-limiting series capacitor at each cell site. The internal cell memory capability eliminates the need for a refresh scan. The cell memory holds an image until it is erased. This eliminates flicker because each cell operates at a duty cycle of one. The series cell capacitor in an AC-PDP is fabricated using a thick-film screen-printed dielectric glass, as illustrated in Figure 9.53. The gap separation between the two substrates is typically 4 mils. The surface of the thick-film dielectric is coated with a thin-film dielectric material such as magnesium oxide.

Flood Beam CRT

Large displays (20 m diagonal and larger) have been employed for mass audiences in stadiums to provide special coverage of sporting events. These consist of flood beam CRT arrays in which each device fulfills the function of a single phosphor dot in a delta-gun shadow-mask CRT. Thus, each display element consists of a trio of flood beam tubes, one red, one green, one blue, each with a 1 to 6-in (2.5 to 15 cm) diameter. A fully NTSC-capable display would require in excess of 400,000 such tubes (147,000 of each color) to be individually addressed. Practical implementations have employed less than 40,000, thus requiring substantially reduced drive complexity.

Matrix Addressing

Provision for individual control of each element of a discrete element display implies a proliferation of electrical connections and, therefore, rapidly expanding cost and complexity. These factors are minimized in discrete-element displays through *matrixing* tech-

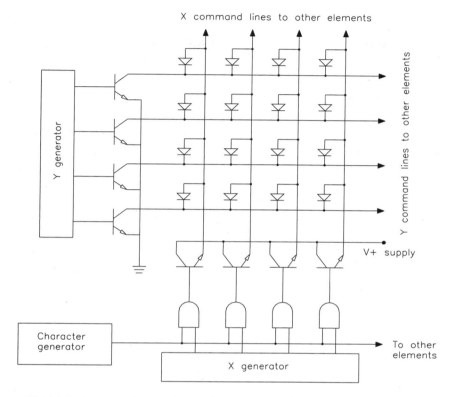

Figure 9.54 Matrix addressing for flat-panel display. (*After* [10].)

niques wherein a display element must receive two or more gating signals before it is activated. Geometric arrangements of the matrixing connectors guarantee that only the appropriate elements are activated at a given time.

Matrix addressing requires that the picture elements of a panel be arranged in rows and columns, as illustrated in Figure 9.54 for a simplified 4 × 4 matrix. Each element in a row or column is connected to all other elements in that row or column by active devices (typically transistors). Thus, it is possible to select the drive signals so that when only a row or column is driven, no elements in that row or column will be activated. However, when both the row and column containing a selected element are driven, the sum of the two signals will activate a given element. The resulting reduction in the number of connections required, compared to a discrete-element drive, is significant because the number of lines is reduced from one-per-element to the sum of the number of rows and columns in the panel.

For active matrix addressing to work properly, it is necessary to have a sharp discontinuity in the transfer characteristic of the device at each pixel at the intersection of the row and column address lines. In some cases, the characteristics of the display device itself can be used to provide this discontinuity. The breakdown potential of a gas discharge can be used

348 Video Display Engineering

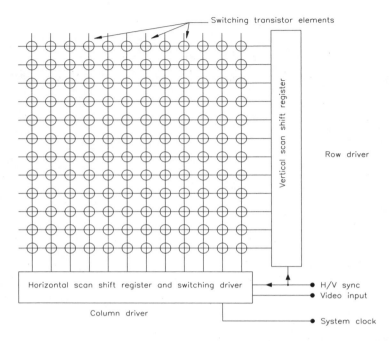

Figure 9.55 Principle of active matrix addressing for a large screen display. (*After* [27].)

(for plasma displays) or alternatively, liquid crystals can be made to have a reasonably sharp transfer curve.

Stadium applications require color video displays that can be viewed in sunlight. In such cases, the high brightness required is achieved with a matrix system. A bright source at each pixel is controlled by its own drive circuit, as illustrated in Figure 9.55. Light sources are typically either color incandescent light bulbs or electron-excited fluorescent phosphors (discussed in the previous section). The switching matrix must control the intensity of each color in each pixel of the display. For video, the system must have a continuous gray scale. Because of the cost and complexity of providing individual control of each pixel light source, such displays usually have somewhat lower resolution than those using light valves.

A number of stadium systems have been produced, including:

- *Astrovision* (Panasonic). This system uses incandescent light sources to produce high light output. Optical efficiency, however, is low.

- *Diamond Vision* (Mitsubishi). This system uses electron-excited phosphors in a switched matrix to generate light. Efficiency is relatively high. Response time is faster than incandescent lamp-based systems.

- *Jumbotron* (Sony). Like Diamond Vision, this system uses electron-excited phosphors in a switched matrix. Efficiency is high and response time is fast (relative to incandescent).

Active matrix addressing can also be applied to light valve display systems. The nonlinear element is provided by a transistor or diode at the junctions of the matrix. In recent years, thin film transistor technology has been used to allow large matrices to be constructed on a transparent substrate. Liquid crystals are generally used with the active matrix, although a deformable light modulator can also be used.

Active matrix-addressed liquid crystal panels are commercially available, configured either as direct-view devices or as light valve projectors. The direct view panels use back-lighting. Color is obtained by placing colored filters over three sub-pixels in a cluster.

Active matrix LC displays (AMLCD) are commercially available from a number of sources, and several companies have produced HDTV AMLCD projectors. In these systems, three active matrix drivers are used, one for each primary color. The three colored images are superimposed using dichroic mirrors to project the image with a single projection lens.

Large screen projectors can be converted into flat panel displays by substituting a fiber optic panel for the screen. In this configuration, the image is projected onto a small input port adjacent to the panel. The light is carried to the large display surface by optical fibers. Black spacers are placed between the fibers on the display surface in much the same fashion as the black matrix shadow-mask of the CRT. The screen, consequently, appears black during no-signal conditions, producing good contrast under high ambient light levels.

9.4.4 Projectors for Cinema Applications

Screen brightness is a critical element in providing an acceptable HDTV large screen display. Without adequate brightness, the impact on the audience is reduced. Theaters have historically utilized front projection systems for 35 mm film. This arrangement provides for efficient and flexible theater seating. Large screen HDTV is likely to maintain the same arrangement. Typical motion picture theater projection system specifications are as follows:

- Screen width 30 ft
- Aspect ratio 2.35:1
- Contrast ratio 300:1
- Screen luminance 15 ft·L
- Center-to-edge brightness uniformity 85 percent

These specifications meet the expectations of motion picture viewers. It follows that for HDTV projectors to be competitive in theatrical display applications, they must meet similar specifications.

9.4.5 Operational Considerations

Accurate convergence is critical to display resolution. Figure 9.56 shows the degradation in resolution resulting from convergence error. It is necessary to keep convergence errors to less than half the distance between the scanning lines to hold resolution loss below 3 dB. Errors in convergence also result in color contours in a displayed image. Estimates have

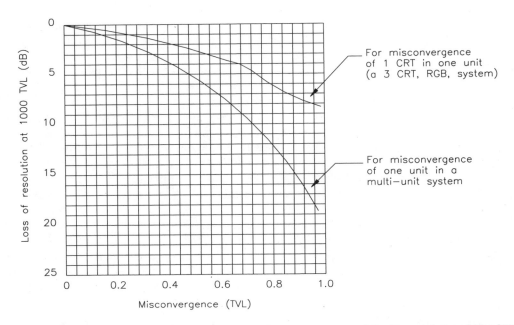

Figure 9.56 The relationship between misconvergence and display resolution. (*After* [28].)

Table 9.6 Appearance of Scan Line Judder in a Projection Display (*After* [28].)

Condition of Adjacent Lines	S/N (p-p)
Clearly overlapped	Less than 69 dB
Just before overlapped	75 dB
No judder	Greater than 86 dB

put the detectable threshold of color contours at 0.75–0.5 minutes of arc. This figure also indicates that convergence error must be held under 0.5 scanning lines.

Raster stability influences the short-term stability of the display. The signal-to-noise ratio (S/N) and raster stability relationships for deflection circuits are shown in Table 9.6. A S/N equivalent of 1/5 scanning line is necessary to obtain sufficient raster stability. In HDTV applications, this translates to approximately 80 dB. Other important factors are high speed and improved efficiency of the deflection circuit.

Many manufacturers have incorporated automatic convergence systems into their products. These usually take the form of a CCD camera sensor that scans various portions of the screen as test patterns are displayed.

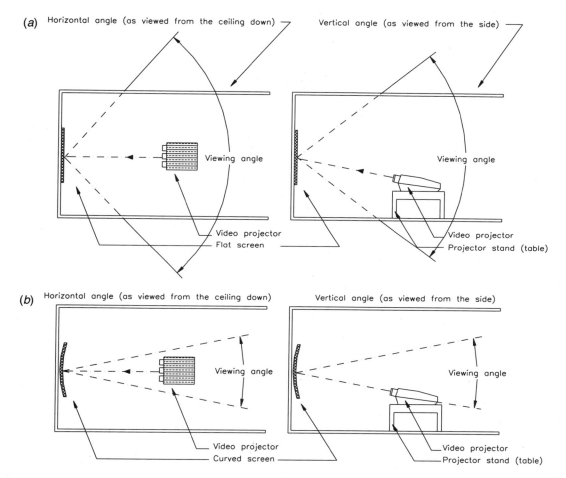

Figure 9.57 Relationship between screen type and viewing angle (*a*) conventional (flat) screen, (*b*) high gain (curved) screen.

Screen Considerations

Because most screens reflect both ambient light in the room and the incident light from the projector (in a front projection system), lighting levels are an important consideration. A rear screen system can reduce this problem. Ambient light on the screen's projector side reflects harmlessly back towards the projector, away from the audience. Light on the viewing side of the screen may fall on the screen but the human eye has experience in filtering out this kind of reflection. The result is that a rear screen projector provides an increased contrast ratio (all other considerations being equal); it can, therefore, be used in lighter areas. Special screen coatings can also help scatter ambient light away from the viewers, enhancing the rear screen image. Coatings can also increase the apparent brightness of

352 Video Display Engineering

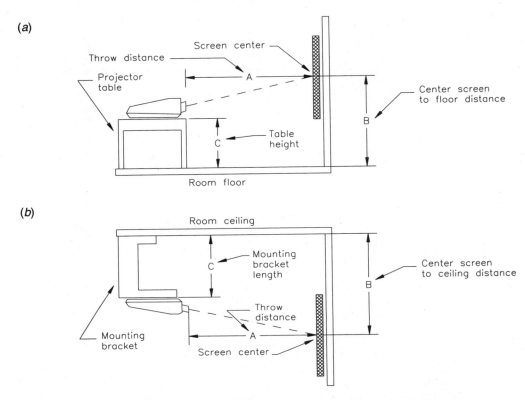

Figure 9.58 Mounting options for a projection system: (*a*) table top, (*b*) ceiling mount.

screens operated from the front. The biggest gains, however, are made by using a screen that is curved slightly. Such screens can increase the apparent brightness of an image. Curved screens, however, also narrow the acceptable field of view. High gain screens provide a typical viewing angle of ±50° from the center line of the screen. Conventional (low gain) screens can provide viewing of up to ±90° from the center line. Figure 9.57 illustrates the tradeoffs involved. Low gain (flat) screens are best suited for computer graphics applications. High gain screens are best limited to video and low resolution graphics applications.

Projectors normally work best if the are exactly perpendicular to the screen.

Mounting Considerations

The two basic mounting options for a projection system are table top and ceiling. Table top mounting is the simplest. There are several disadvantages to this approach, however. The first is the need to decorate the room around the projector (for a front projection system). Also, people may get in the way of the projector's beam, and the projectors themselves can be distracting because they may be bright, noisy, and/or give off heat.

Ceiling mounting goes a long way toward eliminating these problems, but if the ceiling is not sufficiently high, the projector can be obtrusive. Dangling projectors can be retracted using special hoists available from several sources. The projector must also be designed to operate from the ceiling. Several makes provide jumpers to change the scan pattern for the desired configuration.

Figure 9.58 illustrates the two mounting options. Projection display manufacturers typically provide the optimum mounting dimensions (A, B, and C in the figure), which vary from one model to the next.

9.5 References:

1. Blaha, Richard J.: "Large Screen Display Technology Assessment for Military Applications," *Large-Screen Projection Displays II*, William P. Bleha, Jr., (ed.), Proc. SPIE 1255, SPIE, Bellingham, Wash., pp. 80–92, 1990.
2. Fink, D. G. (ed.): *Television Engineering Handbook*, McGraw-Hill, New York, N.Y., 1957.
3. Luxenberg, H., and R. Kuehn: *Display Systems Engineering*, McGraw-Hill, New York, N.Y., 1968.
4. McKechnie, S: Philips Laboratories (NA) report, 1981, unpublished.
5. Kikuchi, M., et al.: "A New Coolant-Sealed CRT for Projection Color TV," *IEEE Trans.*, vol. CE-27, no. 3, pp. 478–485, August 1981.
6. Van, M. W., and J. Von Esdonk: "A High Luminance High-Resolution Cathode-Ray Tube for Special Purposes," *IEEE Trans Electron Dev.*, IEEE, New York, N.Y., ED-30, pg. 193, 1983.
7. Cheng, J. B., and Q. H. Wang: "Studies on YAG Phosphor Screen for HDTV Projector," *Proc SPIE 2892*, SPIE, Bellingham, Wash., pg. 36, 1996.
8. Wang, Q. H., J. B. Cheng, and Z. L. Lin: "A New YAG Phosphor Screen for Projection CRT," *Electron Lett.*, vol. 34, no. 14, pg. 1420, 1998.
9. Wang, Qionghua, Jianbo Cheng, Zulun Lin, and Gang Yang: "A High-Luminance and High-Resolution CRT for Projection HDTV Display," *Journal of the SID*, Society for Information Display, San Jose, Calif., vol. 7, no. 3, pg 183-186, 1999.
10. Benson, K. B., and D. G. Fink: *HDTV: Advanced Television for the 1990s*, McGraw-Hill, New York, N.Y., 1990.
11. Pease, Richard W.: "An Overview of Technology for Large Wall Screen Projection Using Lasers as a Light Source," *Large Screen Projection Displays II*, William P. Bleha, Jr., (ed.), Proc. SPIE 1255, SPIE, Bellingham, Wash., pp. 93–103, 1990.
12. Takeuchi, Kazuhiko, et. al.: "A 750-TV-Line Resolution Projector using 1.5 Megapixel a-Si TFT LC Modules," *SID 91 Digest*, Society for Information Display, San Jose, Calif., pp. 415–418, 1991.
13. Tomioka, M., and Y. Hayshi: "Liquid Crystal Projection Display for HDTV," *Proceedings of the International Television Symposium*, ITS, Montreux, Switzerland, 1991.
14. Tsuruta, Masahiko, and Neil Neubert: "An Advanced High Resolution, High Brightness LCD Color Video Projector," *SMPTE Journal*, SMPTE, White Plains, N.Y., pp. 399–403, June 1992.

15. Fritz, Victor J.: "Full-Color Liquid Crystal Light Valve Projector for Shipboard Use," *Large Screen Projection Displays II*, William P. Bleha, Jr., (ed.), Proc. SPIE 1255, SPIE, Bellingham, Wash., pp. 59–68, 1990.
16. Phillips, Thomas E., et. al.: "1280 × 1024 Video Rate Laser-Addressed Liquid Crystal Light Valve Color Projection Display," *Optical Engineering*, Society of Photo-Optical Instrumentation Engineers, vol. 31, no. 11, pp. 2300–2312, November 1992.
17. Bleha, W. P.: "Image Light Amplifier (ILA) Technology for Large-Screen Projection," *SMPTE Journal*, SMPTE, White Plains, N.Y., pp. 710–717, October 1997.
18. Hornbeck, Larry J.: "Digital Light Processing for High-Brightness, High-Resolution Applications," *Projection Displays III*, Electronic Imaging '97 Conference, SPIE, Bellingham, Wash., February 1997.
19. Florence, J., and L. Yoder: "Display System Architectures for Digital Micromirror Device (DMD) Based Projectors," *Proc. SPIE*, SPIE, Bellingham, Wash., vol. 2650, Projection Displays II, pp. 193–208, 1996.
20. Gove, R. J., V. Markandey, S. Marshall, D. Doherty, G. Sextro, and M. DuVal: "High-Definition Display System Based on Digital Micromirror Device," *International Workshop on HDTV* (HDTV '94), International Institute for Communications, Turin, Italy (October 1994).
21. Sextro, G., I. Ballew, and J. Lwai: "High-Definition Projection System Using DMD Display Technology," *SID 95 Digest*, Society for Information Display, pp. 70–73, 1995.
22. Younse, J. M.: "Projection Display Systems Based on the Digital Micromirror Device (DMD)," SPIE Conference on Microelectronic Structures and Microelectromechanical Devices for Optical Processing and Multimedia Applications, Austin, Tex., *SPIE Proceedings*, SPIE, Bellingham, Wash., vol. 2641, pp. 64–75, Oct. 24, 1995.
23. Corrigan, R. W., B. R. Lang, D.A. LeHoty, and P.A. Alioshin: "An Alternative Architecture for High Performance Display," Silicon Light Machines, Sunnyvale, Calif., 1999. Presented at the 141st SMPTE Technical Conference (paper 141-25).
24. Amm, D. T., and R. W. Corrigan: "Optical Performance of the Grating Light Valve Technology," Projection Displays V Symposium, *SPIE Proceedings*, SPIE, San Jose, Calif., vol. EI 3634-10, February 1999.
25. Amm, D. T., and R.W. Corrigan: "Grating Light Valve Technology: Update and Novel Applications," SID Symposium—Anaheim, SID, San Jose, Calif., May 1998.
26. Wedding, Donald K., Sr.: "Large Area Full Color ac Plasma Display Monitor," *Large Screen Projection Displays II*, William P. Bleha, Jr., (ed.), Proc. SPIE 1255, SPIE, Bellingham, Wash., pp. 29–35, 1990.
27. Glenn, William E.: "Large Screen Displays for Consumer and Theater Use," *Large Screen Projection Displays II*, William P. Bleha, Jr., (ed.), Proc. SPIE 1255, SPIE, Bellingham, Wash., pp. 36–43, 1990.
28. Mitsuhashi, Tetsuo: "HDTV and Large Screen Display," *Large-Screen Projection Displays II*, William P. Bleha, Jr., (ed.), Proc. SPIE 1255, SPIE, Bellingham, Wash., pp. 2–12, 1990.

9.6 Bibliography

Ashizaki, S., Y. Suzuki, K. Mitsuda, and H. Omae: "Direct-View and Projection CRTs for HDTV," *IEEE Transactions on Consumer Electronics*, IEEE, New York, N.Y., 1988.

Bates, W., P. Gelinas, and P. Recuay: "Light Valve Projection System for HDTV," *Proceedings of the ITS*, International Television Symposium, Montreux, Switzerland, 1991.

Bates, W., P. Gelinas, and P. Recuay: "Light Valve Projection System for HDTV," *Proceedings of the International Television Symposium*, ITS, Montreux, Switzerland, 1991.

Bauman, E.: "The Fischer Large-Screen Projection System," *SMPTE Journal*, SMPTE, White Plains, N.Y., vol. 60, pg. 351, 1953.

Bleha, William P., Jr., (ed.): *Large-Screen Projection Displays II*, Proc. SPIE 1255, SPIE, Bellingham, Wash., 1990.

Cheng, Jia-Shyong, et. al.: "The Optimum Design of LCD Parameters in Projection and Direct View Applications," *Display Technologies*, Shu-Hsia Chen and Shin-Tson Wu (eds.), Proc. SPIE 1815, SPIE, Bellingham, Wash., pp. 69–80, 1992.

Corrigan, R. W., D. T. Amm, P. A. Alioshin, B. Staker, D .A. LeHoty, K. Gross, and B. R. Lang: "Calibration of a Scanned Linear Grating Light Valve TM Projection System," SID Symposium—San Jose, SID, San Jose, Calif., May 1999.

Fink, D. G.: *Color Television Standards*, McGraw-Hill, New York, N.Y., 1986.

Fink, D. G., et al.: "The Future of High-Definition Television," *SMPTE Journal*, SMPTE, White Plains, N.Y., vol. 89, February/March 1980.

Fujio, T., J. Ishida, T. Komoto, and T. Nishizawa: "High-Definition Television Systems—Signal Standards and Transmission," *SMPTE Journal*, SMPTE, White Plains, N.Y., vol. 89, August 1980.

Gerhard-Multhaupt, R.: "Light Valve Technologies for HDTV Projection Displays: A Summary," *Proceedings of the ITS*, International Television Symposium, Montreux, Switzerland, 1991.

Glenn, William. E.: "Principles of Simultaneous Color Projection Using Fluid Deformation," *SMPTE Journal*, SMPTE, White Plains, N.Y., vol. 79, pg. 788, 1970.

Good, W.: "Recent Advances in the Single-Gun Color Television Light-Valve Projector," *Soc. Photo-Optical Instrumentation Engrs.*, vol. 59, 1975.

Good, W.: "Projection Television," *IEEE Trans.*, vol. CE-21, no. 3, pp. 206–212, August 1975.

Gretag AG: "What You May Want to Know about the Technique of Eidophor," Regensdorf, Switzerland.

Grinberg, J. et al.: "Photoactivated Birefringent Liquid-Crystal Light Valve for Color Symbology Display," *IEEE Trans. Electron Devices*, vol. ED-22, no. 9, pp. 775–783, September 1975.

Hardy, A. C., and F. H. Perrin: *The Principles of Optics*, McGraw-Hill, New York, N.Y., 1932.

Howe, R., and B. Welham: "Developments in Plastic Optics for Projection Television Systems," *IEEE Trans.*, vol. CE-26, no. 1, pp. 44–53, February 1980.

Hubel, David H.: *Eye, Brain and Vision*, Scientific American Library, New York, N.Y., 1988.

Itah, N., et al.: "New Color Video Projection System with Glass Optics and Three Primary Color Tubes for Consumer Use," *IEEE Trans. Consumer Electronics*, vol. CE-25, no. 4, pp. 497–503, August 1979.

Judd, D. B.: "The 1931 C.I.E. Standard Observer and Coordinate System for Colorimetry," *Journal of the Optical Society of America*, vol. 23, 1933.

Kingslake, Rudolf (ed.): *Applied Optics and Optical Engineering*, vol. 1, Chap. 6, Academic, New York, N.Y., 1965.

"Kodak Filters for Scientific and Technical Uses," Eastman Kodak Co., Rochester, N.Y.

Kurahashi, K., et al.: "An Outdoor Screen Color Display System," *SID Int. Symp. Digest 7*, Technical Papers, vol. XII, Society for Information Display, San Jose, Calif., pp. 132–133, April 1981.

Lakatos, A. I., and R. F. Bergen: "Projection Display Using an Amorphous-Se-Type Ruticon Light Valve," *IEEE Trans. Electron Devices*, vol. ED-24, no. 7, pp. 930–934, July 1977.

Maseo, Imai, et. al.: "High-Brightness Liquid Crystal Light Valve Projector Using a New Polarization Converter," *Large Screen Projection Displays II*, William P. Bleha, Jr., (ed.), Proc. SPIE 1255, SPIE, Bellingham, Wash., pp. 52–58, 1990.

Morizono, M.: "Technological Trends in High Resolution Displays Including HDTV," *SID International Symposium Digest*, paper 3.1, Society for Information Display, San Jose, Calif., May 1990.

Nasibov, A., et al.: "Electron-Beam Tube with a Laser Screen," *Sov. J. Quant. Electron.*, vol. 4, no. 3, pp. 296–300, September 1974.

Pfahnl, A.: "Aging of Electronic Phosphors in Cathode Ray Tubes," *Advances in Electron Tube Techniques*, Pergamon, New York, N.Y., pp. 204–208.

Pitts, K., and N. Hurst: "How Much Do People Prefer Widescreen (16 × 9) to Standard NTSC (4 × 3)?," *IEEE Transactions on Consumer Electronics*, IEEE, New York, N.Y., August 1989.

Pointer, R. M.: "The Gamut of Real Surface Colors," *Color Res. App.*, vol. 5, 1945.

Poorter, T., and F. W. deVrijer: "The Projection of Color Television Pictures," *SMPTE Journal*, SMPTE, White Plains, N.Y., vol. 68, pg. 141, 1959.

Robertson, A.: "Projection Television—1 Review of Practice," *Wireless World*, vol. 82, no. 1489, pp. 47–52, September 1976.

Schiecke, K.: "Projection Television: Correcting Distortions," *IEEE Spectrum*, IEEE, New York, N.Y., vol. 18, no. 11, pp. 40–45, November 1981.

Sears, F. W.: *Principles of Physics, III*, Optics, Addison-Wesley, Cambridge, Mass., 1946.

Sherr, S.: *Fundamentals of Display System Design*, Wiley-Interscience, New York, N.Y., 1970.

Taneda, T., et al.: "A 1125-Scanning Line Laser Color TV Display," *SID 1973 Symp. Digest Technical Papers*, Society for Information Display, San Jose, Calif., vol. IV, pp. 86–87, May 1973.

Wang, S., et al.: "Spectral and Spatial Distribution of X-Rays from Color Television Receivers," *Proc. Conf. Detection and Measurement of X-radiation from Color Television Receivers*, Washington, D.C., pp. 53–72, March 28–29, 1968.

Whitaker, Jerry C., and K. Blair Benson (eds.): *Standard Handbook of Video and Television Engineering*, 3rd ed., McGraw-Hill, New York, N.Y., 2000.

Williams, Charles S., and Becklund, Orville A.: *Optics: A Short Course for Engineers and Scientists*, Wiley Interscience, New York, N.Y., 1972.

"X-Radiation Measurement Procedures for Projection Tubes," TEPAC Publication 102, Electronic Industries Association, Washington, D. C.

Yamamoto, Y., Y. Nagaoka, Y. Nakajima, and T. Murao: "Super-compact Projection Lenses for Projection Television," *IEEE Transactions on Consumer Electronics*, IEEE, New York, N.Y., August 1986.

Chapter 10
Flat Panel Displays

10.1 Introduction

A wide variety of flat panel devices are available in the marketplace. For each basic technology, several variations typically exist. Because of the focus on this book is displays for video applications, this chapter will concentrate on those flat panel display technologies that are applicable for video purposes, specifically:

- Liquid crystal display (LCD), including *active-matrix liquid crystal display* (AMLCD) and *plasma-addressed liquid crystal* (PALC)
- Plasma display panel (PDP)
- Field emission display (FED)

It is fair to point out that many of the display technologies that do not now support full-color video may do so in the near future. Flat panel development is racing along at an unprecedented pace, driven largely by the demand of consumers for display devices that are more compact than CRTs (for a given screen size), lighter, and more power-efficient.

It is commonplace to regard flat panel displays and solid-state displays as essentially the same product category. While this is—generally speaking—true, it should be remembered that some flat panel displays have as much in common with vacuum devices as they do with solid-state devices (the FED is a convenient example).

10.2 Liquid Crystal Displays

With the exception of the *light-emitting diode* (LED), the liquid crystal display (LCD) is the best known and most highly developed flat panel display device. Applications range from simple monochrome displays for cellular telephones to full-color large-screen video systems.

10.2.1 Principles of Operation

Materials classified as *liquid crystals* usually exist in a liquid form at high temperatures and in a solid form at low temperatures. In between these two extremes they exhibit the

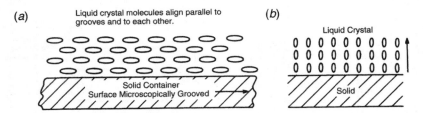

Figure 10.1 The liquid-crystal/grooved interface: (*a*) no field applied, (*b*) an electric field greater than the critical value applied. (*After* [2].)

characteristics of both states [1]. There are a number of different types of liquid crystals, although *nematic crystals* are by far the most common found in display devices.

The essential feature of a liquid crystal is the long rod-like molecule physical shape. In a nematic crystal, the molecules align in a predictable manner according to certain rules (Figure 10.1). If a surface—such as a display device panel—is microscopically groved, the interface molecules will be aligned by the groves and intermolecule forces will maintain the orientation across the liquid crystal, as illustrated in Figure 10.1.*a*. The molecules will align in an electric field, but beyond a certain *critical value*, the field will be sufficient to overcome the alignment, as shown in Figure 10.1*b*. In the practical application of liquid crystal materials, the transition is not so abrupt; groove alignment persists at the interface itself (Figure 10.2). The process of alignment in the electric field results from the *anisotropic dielectric constant* characteristics of the liquid crystal material.

The principle of the *twisted nematic cell* is a bit more complicated. As illustrated in Figure 10.3, confining plates, typically spaced 10 μm apart, are grooved orthogonally, forcing the molecular orientation to spiral through 90°. Two polarizers and a mirror are added, as shown in the figure; incident ambient light is polarized and enters the liquid-crystal cell with the plane of polarization parallel to the molecular orientation. As the light traverses the cell, the plane of polarization is rotated by the twist in the liquid crystal so that it reaches the opposite face with a polarization 90° to the original direction, but now parallel to the direction of the second polarizer, through which it can now pass. The light is next reflected from the mirror and passed back through the cell, reversing the prior sequence.

When an electric field of a potential greater than the critical field is applied between the transparent electrodes, the 90° twist in the crystal is disrupted; the molecules align parallel to the field so that the rotation of the light's plane of polarization cannot be sustained. The *crossed polarizers*, therefore, effectively block reflection of the incident light from the backing mirror. The surface thus appears to be dark. The contrast ratio can be further improved through the use of a so-called *super twisted nematic crystal*, where the molecular orientation is rotated through 270° rather than 90°.

The LCD device thus described uses ambient light as the illumination source in conjunction with the mirror element. The LCD can also be oriented to serve a transmissive rather than reflective function.

The transmission LCD functions in a manner similar to the reflective LCD except that the mirror is eliminated, replaced by a powered backlight source. Although the energy-efficient advantage of the passive device is lost in this arrangement, monochromatic backlighting

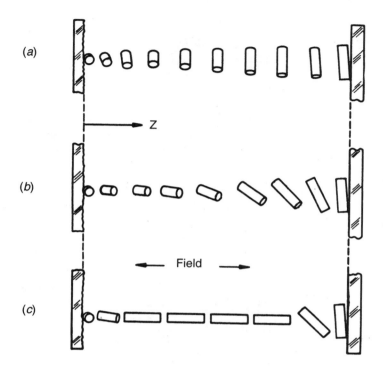

Figure 10.2 The orientation of the liquid-crystal axis in a cell: (*a*) no field applied, (*b*) a field of approximately twice the critical field applied, (*c*) a field several times the critical field applied. (*After* [3].)

provides one simple means of constructing displays with varying (albeit limited) background colors.

A more sophisticated method of achieving color in an LCD is through the use of *cholesteric crystals* in the light-controlling element. The three primary types of liquid crystals—nematic, cholesteric, and *smectic*—are distinguished by the different types of molecular ordering that they display. In the cholesteric crystal, the direction of molecular alignment rotates in each successive parallel plane, illustrated in Figure 10.4. As shown, the spatial period of the rotation is known as the *pitch*. Braggs reflections occur when the wavelength of the incident light meets the condition $\lambda = p/n$, where p is the pitch and n is an integer. The liquid-crystal can thus appear to be colored in incident white light, although it is important to point out that in practice the color is strongly depending upon the temperature of the cell.

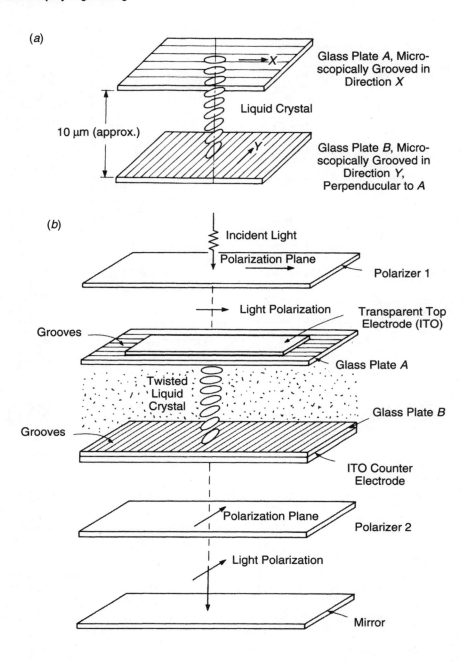

Figure 10.3 LCD architecture: (*a*) twisted nematic cell, (*b*) LCD element. (*After* [2].)

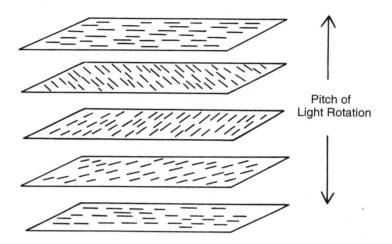

Figure 10.4 Cholesteric ordering: a large number of planes of nematic ordering are formed where the directors rotate as light moves along a direction perpendicular to the planes. (*After* [4].)

Display Addressing

Addressing techniques for liquid-crystal devices can be classified into three basic types [5]:

- Direct (static) addressing
- Passive matrix addressing
- Active matrix addressing.

Direct addressing is used for displays of low information content, such as numeric, bar graph, and other fixed pattern displays. For high information content applications, this approach is impractical because of the huge number of interconnects required.

Passive matrix addressing is the simplest and least expensive approach for medium-to-high information content displays. Each liquid-crystal cell is addressed through a row and column matrix, as shown in Figure 10.5. Although simple and intuitive, this approach quickly gets out of hand in a display of large dimensions. Furthermore, issues relating to cell turn-on and turn-off times come in to play, making the timing of the multiplexing system critical in many practical applications.

Active matrix addressing removes the limitations inherent in the multiplexing approach by incorporating a nonlinear control element in series with each pixel. This provides a 100 percent duty cycle for the pixel using the charge stored at the pixel during the row addressing time. The active matrix pixel array, with a control element at each pixel, is illustrated in Figure 10.6.

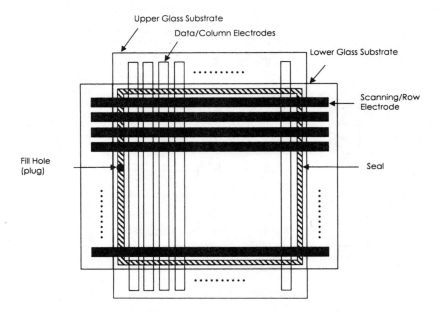

Figure 10.5 Passive matrix addressing. (*From* [5]. *Used with permission.*)

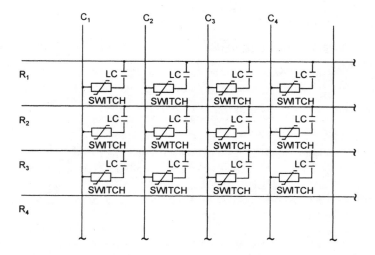

Figure 10.6 Active matrix addressing scheme. (*From* [5]. *Used with permission.*)

The active-matrix LCD (AMLCD) typically uses a three terminal *thin film transistor* (TFT) as the active matrix switching element [5]. This provides superior image quality as a result of the complete isolation of each pixel from its neighboring pixel. A common active

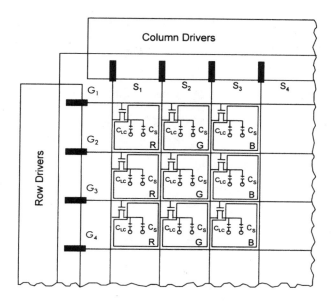

Figure 10.7 Addressing topology using three-terminal TFT devices. (*From* [5]. *Used with permission.*)

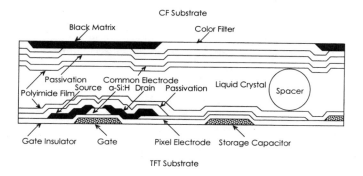

Figure 10.8 Cross-section of an AMLCD display panel. (*From* [5]. *Used with permission.*)

matrix array employing three terminal TFT devices is shown in Figure 10.7. A storage capacitor (C_s in the figure) is usually incorporated at each pixel to reduce the pixel voltage offset, improve display uniformity, and broaden the acceptable operating temperature range.

Figure 10.8 shows a cross-section of a TFT-LCD pixel illustrating the various layers of the TFT, specifically:

- The pixel electrode
- Storage capacitor

- Liquid-crystal cell with polyimide alignment layers, spacers, color filters, and black matrix

10.2.2 AMLCD Performance Parameters

The performance of an AMLCD can be characterized by the following parameters [5]:

- Viewing angle
- Contrast ratio
- Gray-level luminance stability
- Chromaticity stability
- Color gamut
- Response time
- Reflectance
- Power consumption
- Luminance
- Imaging "sticking"
- Crosstalk among pixels

It is intuitive that certain performance aspects are more important in some applications than in others, i.e. portable applications require the lowest possible power consumption, while at the other end of the applications scale, video requires high resolution, high luminance, and wide color gamut (among other things).

One parameter that has seen considerable improvement in recent years is the viewing angle. This specification, the range of horizontal and vertical viewing angles (the *viewing envelope*), has improved the point that AMLCD displays are practical for instrumentation applications such as airborne avionics.

One of the techniques used to accomplish improved viewing angles is *in-plane switching* (IPS) [5]. In the IPS scheme, illustrated in Figure 10.9, the bottom TFT substrate contains both the pixel electrode and the common electrodes in an interdigitated configuration. The top color filter substrate does not contain any electrode. The liquid crystal is aligned homogeneously, and the input polarizer axis is parallel to the liquid-crystal director orientation. The output polarizer is a 90° cross-polarizer configuration. For the field-off condition, the input polarizer light travels unaltered through the cell, and then is blocked by the crossed exit polarizer. This provides the normally black state. The field-on state causes in-plane rotation of the liquid-crystal modules (to align with the field), and the resulting optical retardation causes rotation of the polarization axis of the input light, which then emerges through the output polarizer.

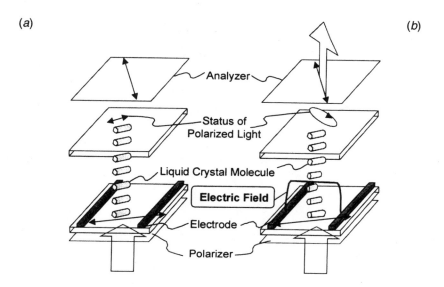

Figure 10.9 The principle of in-plane switching for the AMLCD: (*a*) off state, (*b*) on state. (*From* [5]. *Used with permission.*)

AMLCD Applications

High-performance AMLCDs have successfully replaced CRTs in demanding applications ranging from airline cockpit displays to standard- and high-definition video. Current AMLCD work is focusing on direct view devices with sizes of 25 to 30 in diagonal.

AMLCDs have become the dominant technology for portable display applications, such as laptop computers because they offer a unique combination of low-voltage drive, low-power consumption, low cost, and acceptable image quality.

10.2.3 Plasma-Address Liquid Crystal Display

Building upon the inherent simplicity and scalability of the AMLCD addressing scheme, *plasma-addressed liquid crystal display* (PALC) technology has shown considerable progress within the past few years. PALC is an active-matrix liquid crystal display that uses an alternative matrix technique that eliminates the silicon-based integrated circuit processing used with TFT designs [6]. PALC instead relies on the electrical properties of plasma to provide large-sized AMLCDs suitable for high-definition applications. Screen sizes of 42-in are practical at this writing.

Although structurally different, the PALC and TFT active substrates are functionally equivalent. Figure 10.10 illustrates additional similarities.

The color filter plate contains transparent conductors, patterned in a manner identical to those used in passively addressed LCDs, and color filters identical to those used in passive

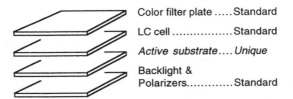

Figure 10.10 Simplified illustration of the PALC and AMLCD stack. Compared to conventional LCDs, only the active substrate is unique. (*From* [6]. *Used with permission.*)

and active matrix devices. These color filters provide the excellent color rendition common to the AMLCD family of devices. In addition, color filters—in conjunction with the polarizers—are responsible for the good contrast performance of AMLCDs.

The liquid crystal cell used in the PALC device is typically a standard TN cell. The use of nematic liquid crystals results in continuous grayscale performance, i.e. the number of displayed gray levels can be as large as desired; it is limited only by the number of bits supported by the drive circuitry.

Backlighting requirements for the AMLCD separates the image generation process from light production, which allows each parameter to be optimized independently.

Application Considerations

Progressively updating a 1000 line display at a 60 Hz rate requires a line-addressing rate of approximately 16 µs. The plasma therefore must capture data and then decay to a non-active state within the 16 µs time-line. In a PALC display, the decay time of the plasma is strongly influenced by gas decomposition within the channels. Available compositions exhibit decay rates as short as 2.5 µs, well within the requirements for HD imaging [7].

The resolution performance of a large PALC display is, at this writing, rather modest by AMLCD standards [6]. For example, a full-color pixel pitch of 0.5 mm is required for a 42-in. HD display.

The measured performance lifetime of common PALC panels has exceeded 30,000 hrs. The PALC utilizes a dc plasma to generate the charge carriers necessary to address the liquid crystal pixels. Because plasma light generation is not required, power levels are quite low. Furthermore, pixel addressing requires only a few microseconds, and so the plasma duty cycle is likewise short.

Another important consideration in panel life is the distribution of power over the device. PALC dissipation is independent of the displayed image, eliminating the problem of image burn-in.

Many of the advancements made in AMLCD viewing angles are applicable to PALC displays. One approach has extended the horizontal and vertical viewing angle to 140° [8].

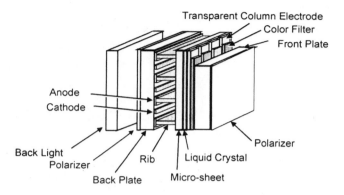

Figure 10.11 Physical structure of the PALC. (*From* [9]. *Used with permission.*)

HD Display Device

The demands of high-definition imaging place considerable requirements on a display device. Despite the challenges, a 42-in. HD PALC has been developed into a practical product [9]. The display incorporates 960 × 1920 color pixels. Twice the vertical resolution was realized while maintaining the VGA-pitch plasma structure. The key feature of this technology is the precise control of local plasma conductivity.

To make a fine-pitch PALC with conventional technologies, two fine-pitch substrates are prepared: one is a substrate with fine-pitch color filter and transparent column electrode elements (ITO), and the other is a fine-pitch plasma substrate. (See Figure 10.11.) For the color filter and ITO substrate, the structure for a PALC is basically identical to a passive matrix LCD. The issue, therefore, lies in making a fine pattern on a large panel.

For the plasma substrate, however, a fine pitch structure reduces the aperture ratio, which results in lower brightness.

In this device, twice the vertical resolution is achieved by precisely controlling the spread of the plasma switch. (See Figure 10.12.) The plasma structure itself is the same as for a VGA panel. As shown in the figure, an electrode that acts as the cathode during regular operation is formed at the center of the channel. The other electrode, which acts as the anode during regular operation, is formed beneath the rib. During regular operation, one channel that forms between two ribs corresponds to one scan line. In the high-definition device, one channel is divided into two scan lines and through manipulation of the addressing and driving systems, the resolution is effectively doubled.

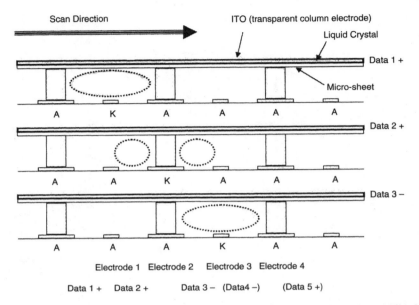

Figure 10.12 Addressing scheme for a high-definition PALC device. (*After* [9]. *Used with permission.*)

10.3 Plasma Displays

Plasma display panel (PDP) devices have emerged as one of the dominant flat-panel display technologies [10].[1] The major market success has come in portable instrumentation and laptop computers. PDP devices can be grouped into three fundamental technologies:

- **dc displays**. Subdivisions include those devices with memory and those using dc refresh.
- **ac displays**. Subdivisions include those devices with memory and those using ac refresh.
- **ac-dc hybrids**.

In its basic design, the plasma display panel has the following attributes:

- Large nonlinearity resulting from the electrical characteristics of the gas discharge used in the device; that is, below a certain threshold voltage, the gas discharge will emit no light. This sharp nonlinearity allows the plasma display to be multiplexed virtually without limit, which means that very large plasma displays are practical.

---

1 Portions of this section were adapted from: Weber, Larry F.: "Plasma Displays," in *The Electrical Engineering Handbook*, Richard C. Dorf (ed.), CRC Press, Boca Raton, Fla., 1786-1798, 1993. Used with permission.

- Memory-based display cells are practical. Stored directly in the glass plasma panel, memory is a desirable feature for a flat panel display because it facilitates high brightness, eliminates flicker, and can simplify the interface scanning system. For a memory-based display cell, the duty cycle is unity; the cells that should be illuminated are always on.

- Discharge switching is practical. In a PDP, the gas discharge itself can perform high-voltage switching operations that allow certain silicon-based switching devices to be eliminated, and others to be reduced in number. Because the circuits that drive a plasma panel constitute a large portion of the final cost of a PDP product, reducing the number of required drivers can have a dramatic effect on product price and, therefore, market penetration.

- Long operating life. Lifetime projections for ac-based PDPs range up to 350,000 hrs. Somewhat lower numbers are assigned for dc-based devices. Whether these very long lifetimes can ever be realized (350,000 hrs ≈ 40 years), it is safe to say that plasma displays will last considerably longer than other emissive devices.

- Performance parameters, including brightness, luminance efficiency, and viewing angle are good.

10.3.1 Gas Discharge Characteristics

Figure 10.13 illustrates the important reactions that occur in a gas discharge [10]. The reactions in the gas volume include the following:

- Ionization (I)

- Excitation (E)

- Metastable generation (M)

- Penning ionization (P)

The three surface reactions that occur at the cathode cause electrons to be ejected from the cathode by a bombarding neon ion, a neon metastable atom, or a high-energy photon. The most important volume reaction is ionization (I), which can cause the generation of an avalanche in the gas volume, as shown in the figure. This avalanche is started by an electron near the cathode and, as it grows toward the anode, the avalanche generates a large number of election-ion pairs. The number of electron-ion pairs increases with increasing applied voltage across the gas. Ions, photos, or metastable atoms that are transported to the cathode can then eject electrons with a cathode surface-dependent probability, and these ejected electrons then initiate further avalanches. These mechanisms act as a positive feedback system that becomes unstable when the loop gain is greater than unity. The onset of the unstable condition is defined as the *gas firing voltage*. Above this firing voltage, the discharge current will continue to grow without bounds if the initial avalanche is primed with at least a single electron.

Figure 10.14 shows the voltage-current characteristics for a typical gas discharge found in a plasma display. Note the strong nonlinearity at the firing voltage. This is a major attribute of gas discharge that allows matrix addressing. When the discharge current has sufficient magnitude, space charge distortion sets in and the characteristic achieves a *negative re-

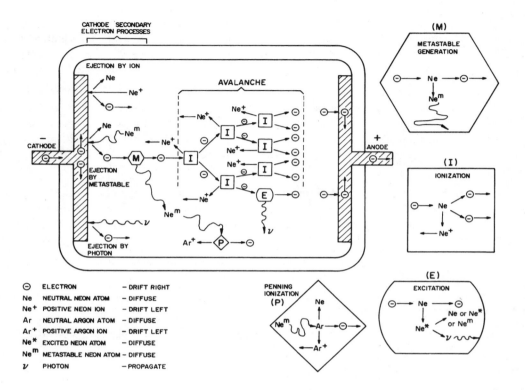

Figure 10.13 Gas discharge reactions model for a plasma display panel. (*From* [10]. *Used with permission.*)

sistance region. Most plasma displays operate near the junction of the normal and the abnormal glow regions of the characteristic.

Figure 10.15 shows the characteristics of the glow discharge commonly found in a plasma display. The light comes from two luminous regions:

- The *negative glow region*
- Positive column region

These regions are caused by the space charge distribution of the electrons and ions that distort the electric field and voltage distribution. Figure 10.16 shows the wavelength distribution found in plasma displays. It is the classic line spectrum of neon gas.

To avoid device failure resulting from the arc regions shown in Figure 10.14, some form of current limiting must be provided. The two most common basic approaches are illustrated in Figure 10.17. As shown, for the dc plasma display, a series resistor (usually external to the device) is used for current limiting. For the ac plasma display, current limiting is accom-

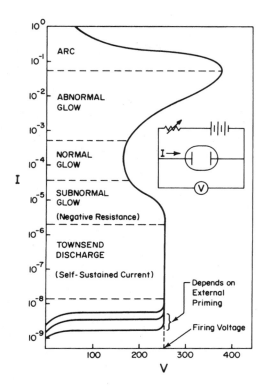

Figure 10.14 Typical voltage-current characteristics of a gas discharge. (*From* [10]. *Used with permission.*)

plished using an internal glass dielectric that couples the electrodes capacitively to the gas discharge.

Memory capability for an ac plasma display can be implemented with relative simplicity. During fabrication, a dielectric layer is constructed with a specified impedance and a current-limiting capacitor is placed in series with each pixel. When a voltage pulse is applied to the ac panel, the discharge deposits a charge on the wall that reduces the voltage across the gas. After a short time, the discharge will extinguish and the light output will end until the applied voltage reverses polarity and a new discharge pulse occurs. This charge allows the ac plasma display to operate in a memory mode, which greatly increases the brightness of large area displays.

10.3.2 Memory-Type ac Plasma Display

The basic ac PDP topology is show in Figure 10.18. This type of device is made by depositing thin-film electrodes on the front and back substrates and then covering those electrodes with a thin dielectric glass [10]. As mentioned in the previous section, the dielectric glass serves as a capacitor that limits current discharge. This dielectric also is used to store the charge, which gives this type of panel its inherent memory. The two substrates are then

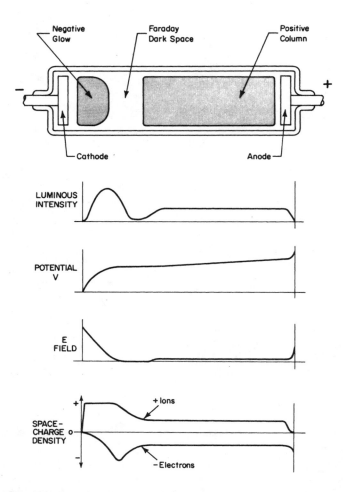

Figure 10.15 The luminous regions of a gas discharge. (*From* [10]. *Used with permission.*)

sealed around the perimeter and filled with neon gas. There are no barrier ribs between the discharge areas; pixels are isolated simply through the action of electric fields.

The inner surface of the dielectric, which is coated with magnesium oxide, is in contact with the neon gas. This MgO layer facilitates low operating voltages and long life.

The ac PDP requires an ac signal, referred to as the *sustaining voltage*, applied during operation. When a pixel is discharging, charge collects on the dielectric glass walls and influences the voltage across the gas. The component of voltage resulting from this charge is known as the *wall voltage*. When a pixel is in the on state, the wall voltage changes for each polarity reversal of the sustain voltage. This change in wall voltage coincides with a pulse of light because of the gas discharge. When the pixel is off, there are no light pulses, and the wall voltage remains zero.

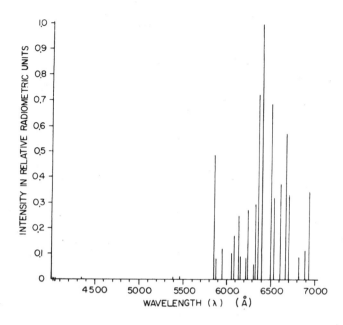

Figure 10.16 Light wavelength distribution for a neon gas plasma display. (*From* [10]. *Used with permission.*)

Pixel addressing is achieved through a partial discharge introduced by an address pulse, timed between the sustain voltage pulses. A write pulse causes the wall voltage to transit from zero volts to the final equilibrium wall voltage. In the same manner, an erase pulse causes the wall voltage to return to zero.

10.3.3 Hybrid ac-dc Plasma Display

As show in Figure 10.19, the hybrid PDP is basically a dc plasma panel with the normal anode and cathode dc electrodes and standard barrier ribs to isolate the discharges. The distinction of this panel is the *trigger electrode* that is buried under the dc cathodes. This ac electrode is not used for display purposes but rather to create *priming particles* for the dc discharge.

The addressing sequence is initiated with a negative pulse applied to the ac trigger electrode. This creates the priming particles that eventually are used to ignite the normal dc discharge that is used for the display. These priming particles have two beneficial effects:

- They allow the use of lower voltages to drive the normal dc address electrodes
- They permit a shorter time delay between the start of the address pulse and the ignition of the discharge

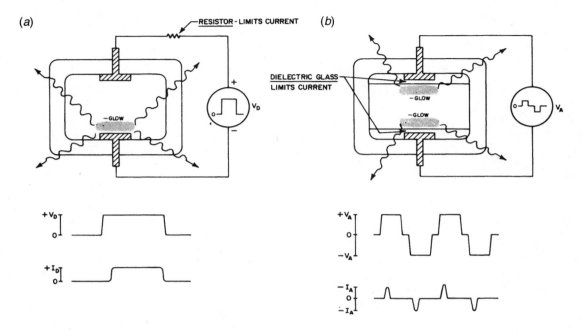

Figure 10.17 Common current-limiting techniques for PDP devices: (*a*) dc display, (*b*) ac display. (*From* [10]. *Used with permission.*)

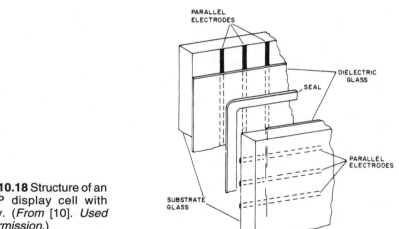

Figure 10.18 Structure of an ac PDP display cell with memory. (*From* [10]. *Used with permission.*)

10.3.4 Color Plasma Displays

Full-color operation is possible with a PDP by introducing *vacuum ultraviolet* (VUV) excitation phosphors into the discharge cell [11]. (See Figure 10.20.) Because the plasma

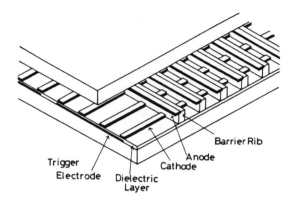

Figure 10.19 Structure of a hybrid ac-dc plasma panel. (*Courtesy of Sony. From* [10]. *Used with permission.*)

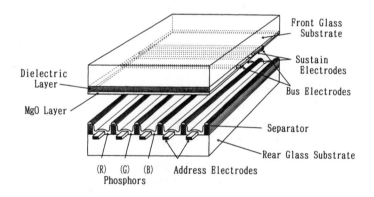

Figure 10.20 Full-color ac memory plasma display cell. (*Courtesy of Fujitsu. From* [10]. *Used with permission.*)

display controls electron motions, a response time of less than 1 µs is possible, allowing the device to be used for displaying HDTV images with 8-bit (256) intensity levels.

The primary color red, green, and blue phosphors are deposited onto walls of each discharge cell. In order to retain the same resolution of an equivalent monochrome display, the discharge *cell pitch* (*sub-pixel pitch*) must be reduced by a factor of three. The deposition process on the bottom and side walls of the discharge cells is somewhat complicated. Practical methods include:

- Thick-film printing. This well-proved technique requires careful adjustments of the binder contents and printing conditions.

- Photo-sensitive phosphor paste. In this method, a layer having special optical properties is first deposited on the rib surface to which the phosphor is to be deposited. Through irradiation with UV light, the layer exhibits tackiness, which causes the phosphor powder to adhere to the layer.

- Lamination. Phosphor in the form of thin layers is laminated along the rib surface and then etched using conventional photo processing techniques.

Display devices for video applications require at least 8-bit gray levels (256). In neon-red monochrome displays, gray scale expression is achieved by intensity modulation of the discharge current. This technique, however, is not applicable to color plasma devices whose output luminance has a strong saturation characteristic with respect to the discharge current.

Gray scales for color PDPs, rather, are obtained by using a *pulse-number-modulation* scheme. One video field is divided into eight sub-fields, each of which consists of an *address period* and a *display period*. The lengths of the display periods are arranged according to the binary sequence, 1:2:4:8:16:32:64:128. The display periods are filled with trains of constant width and constant period pulses. With the 8 sub-field technique, gray levels of 2^8 can be achieved by combining appropriate sub-fields.

The technique of gray scale expression using sub-fields is quite adequate for still images, however, disturbances of gray scales are often observed when displaying moving images [12]. When a gradation pattern is stationary, there is no disturbance, but as soon as the pattern starts moving and the eyes follow it, disturbances are perceived. When the eyes stop following the image, the disturbances immediately disappear. Given this pattern movement, color disturbances occur as well.

Various methods of reducing these motional artifacts have been proposed [11], with varying degrees of success. One method, referred to as *distributed-address and sustain* (DAS), has shown promise [13].

In the DAS approach, the address period and the display period are not separated. In this manner, the problem of reduced light-emitting time in conventional approaches is eliminated. DAS is implemented using revised driver timing and modified driver waveforms.

10.3.5 Performance Issues for Video Applications

Although the performance of PDP devices is acceptable for home video (consumer) use, several issues remain [11]. Prime among these is image quality. While high luminance is necessary for attractive images, it is more important to achieve high contrast under high ambient light environments, particularly in consumer applications. Contrast performance can be improved through the use of one or more of the following techniques:

- Increasing emission intensity from the phosphor

- Minimizing background lights by reducing emission from control discharges or by hiding auxiliary discharge emissions

- Reducing ambient light reflections at the front glass and at the phosphor surface through the use of a neutral density filter or a circular polarizing filter

Of course, any filter placed in front of an image-producing surface also reduces the output luminance, requiring higher luminance and luminous efficiency. In this regard there is, naturally, the necessity for compromise among operating parameters.

Reflected light can also be reduced with color filters (a red-transmitting filter in front of each red cell, and so on). With ideal color filters, the contrast ratio can be improved by a factor of three without a reduction in output luminance.

For video applications, certain color gamut issues—principally involving the red phosphor—also remain to be solved for PDP devices [11].

Despite these challenges, PDPs hold great promise and are expected to claim a portion of the small- to medium-sized projection display market. The cost economies of the display elements used in building PDPs (standardization of glass substrate, use of non-rare-earth phosphors, and combined filters) permit large displays to be constructed at prices competitive with projection systems. On the negative side insofar as cost is concerned, however, is the large number of high voltage circuit drivers required. The cost of the drive electronics in a plasma display usually exceeds that of the panel itself.

Work continues to extend PDP performance, largely concentrating on the following parameters:

- Large screen sizes, in the range of 50 to 60 in.
- Improved contrast ratio of as high as 100:1 in normal ambient light
- Increased peak white luminance to as high as 700 cd/m^2 to compete with CRT devices
- Improved luminous efficiency to 5 lm/W or higher

At this writing the technologies necessary to achieve these goals were believed to exist. Implementation remained a challenge, however.

10.3.6 Advanced XGA Device

Among the products on the cutting edge of PDP device performance is a high-luminance 50-in. diagonal color display (Pioneer, [14]). Key performance parameters include:

- Luminous efficiency of 1.0 lm/W in a 50-in. diagonal XGA/wide panel display
- Sub-pixel pitch of 0.286 mm
- 768(V) × 1260(H) pixel configuration

The display cell, utilizing "T"-shaped electrodes, is shown in Figure 10.21*a*. This configuration is contrasted with a more conventional rectangular cell, shown in *b*.

10.4 Field Emission Displays

A field emission display (FED) is—in essence—an offspring of the CRT. As shown in Figure 10.22, the FED is a vacuum tube display device in which electrons from millions of individual (and very small) cathodes travel to a multicolor phosphor viewing screen to create an image [15]. A matrix of emitters makes up the displayed raster.

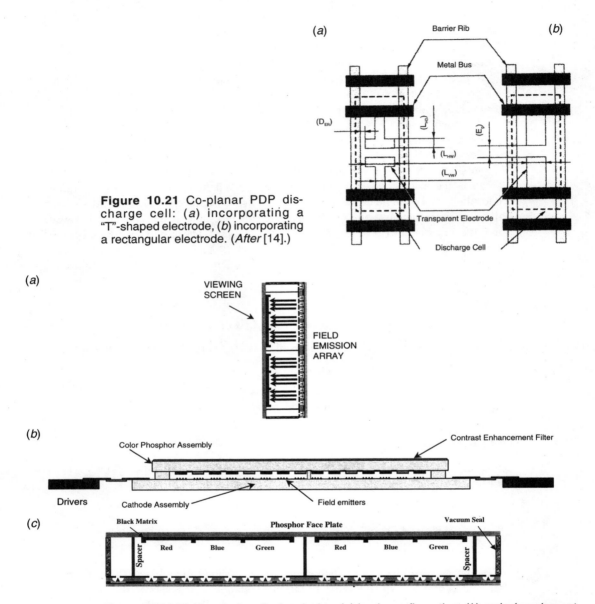

Figure 10.21 Co-planar PDP discharge cell: (*a*) incorporating a "T"-shaped electrode, (*b*) incorporating a rectangular electrode. (*After* [14].)

Figure 10.22 Field emission display device: (*a*) basic configuration, (*b*) emission element, (*c*) emission element detail. (*From* [15]. *Used with permission*.)

Owing to its CRT heritage, the FED enjoys the benefits of the CRT over other display technologies, while foregoing the physical size penalties (depth) of the CRT.

Figure 10.23 A planar carbon emitter. (*From* [15]. *Used with permission.*)

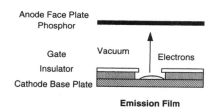

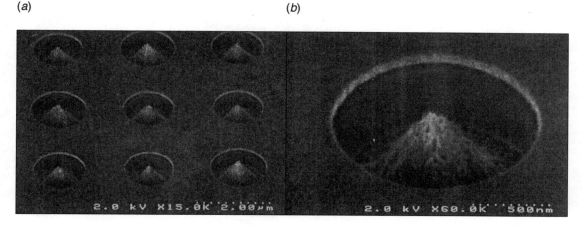

Figure 10.24 Electron microphotograph of a Spindt array: (*a*) group of nine devices, (*b*) closeup of a single emitter. (*From* [15]. *Used with permission.*)

10.4.1 Operational Elements

As with the CRT, the ultimate performance of the display is determined to a large extent by the emissive surfaces—*microtips* for the FED. Figure 10.23 shows a simple planar carbon emitter. Advantages of this approach include simplicity of the structure and the resulting low manufacturing costs.

A variety of other emitter shapes are possible, ranging from cylindrical to pyramid. The shape chosen for a given device involves a trade-off between maximum emitter efficiency, long term performance, and manufacturing constraints. A microphotograph of a high-performance emitter is shown in Figure 10.24. Emission for this configuration—a *Spindt array*—strongly depends on the tip height and sharpness, and the position of the tip within the well.

Whatever form the emitter takes, preserving the life of the cathode is an important design criteria. Conditions that can lead to cathode destruction include the following:

- Emitter damage resulting from uncontrolled emission current.

- Undesirable vacuum levels, which can lead to impact ionization of the residual gases, followed by ion bombardment damage to the emitter.

- Electron-stimulated desorption from surfaces and device degradation resulting from the gas-emitter surface interactions.

- Electron beam induced desorption of gas from the phosphor and subsequent absorption on emitter tips.

Addressing of the FED is based on a matrix where rows are addressed sequentially in time. During each row address time, all columns are addressed simultaneously. The row pattern repeats at the frame rate and is content independent.

Gray scale can be achieved by pulse width modulation of column voltages, amplitude modulation of column voltages, or a combination of the two. Among the challenges of gray scale performance are that emission uniformity is difficult to control across the dimensions of a display screen. Using voltage modulation to control gray scale can, therefore, be problematic [16]. Pulse modulation, on the other hand, provides uniform gray levels compared to voltage modulation, but requires resistive layers in the FED panel to get an averaging effect and enhanced emission current uniformity.

As with most other vacuum devices, getters are used to maintain a good vacuum in the package. Vacuum levels in range of 10^{-6} to 10^{-7} Torr are required for long life.

10.5 Fixed Pixel Pattern

The flat panel devices discussed in this section are on the cutting edge of display technology. All, however, share one common drawback: a fixed pixel pattern. It is quite practical—even commonplace—for CRT-based displays to offer multisync operation, where the scanning parameters of the display conform to the input signal. With a fixed pattern device, however, the input signal must conform to the display. In many applications this issue is not important, but in others it can be a serious limiting factor.

It is important also to point out that several of the projection devices discussed in the previous chapter share this limitation.

10.6 References

1. Morris, James E.: "Liquid-Crystal Displays," in *The Electrical Engineering Handbook*, Richard C. Dorf (ed.), CRC Press, Boca Raton, Fla., 1993.

2. Allison, J.: *Electronic Engineering Semiconductors and Devices*, 2nd ed., McGraw-Hill, London, pg. 308–309, 1990.

3. Baur, G.: *The Physics and Chemistry of Liquid Crystal Devices*, G. J. Sprokel (ed.), Plenum, New York, N.Y., pg. 62, 1980.

4. Wilson, J., and J. F. B. Hawkes: *Optoelectronics: An Introduction*, Prentice-Hall, London, pg. 145, 1989.

5. Sarma, Kalluri R.: "Active-Matrix LCDs," in *Seminar Lecture Notes*, Society for Information Display, San Jose, Calif., vol. 1, pp. M3/3–M3/45, 1999.

6. Buzak, Thomas S.: "Recent Advances in PALC Technology," in *Proceedings of the 18th International Display Research Conference*, Society for Information Display, San Jose, Calif., Asia Display '98, pp. 273–276, 1998.

7. Ilcisen, K. J., et. al.: *Eurodisplay '96*, pg. 595, 1996.

8. Burgmans, A., et. al.: *Information Display*, pg. 14, April/May 1998.
9. Hayashi, M., N. Yamada, and B. Sastra: "Development of a 42-in. High-Definition Plasma-Addressed LCD," in *SID International Symposium Digest of Technical Papers*, Society for Information Display, San Jose, Calif., pp.280–284, 1999.
10. Weber, Larry F.: "Plasma Displays," in *The Electrical Engineering Handbook*, Richard C. Dorf (ed.), CRC Press, Boca Raton, Fla., 1786–1798, 1993.
11. Mikoshiba, Shigeo: *Color Plasma Displays*, Society for Information Display, San Jose, Calif., pp M-4/3–M-4/68, 1999.
12. Masuda, T., et. al: *Conference Record—International Display Resolution Conference*, pg. 357, 1994.
13. Lim, G. S., et. al.: "New Driving Method for Improvement of Picture Quality in 40-in. AC PDP," in *Asia Display '98—Proceedings of the 18th International Display Research Conference*, Society for Information Display, San Jose, Calif., pp. 591–594, 1998.
14. Nishio, T., and K. Amemiya: "High-Luminance and High-Definition 50-in.-Diagonal Co-Planar Color PDPs with T-Shaped Electrodes," in *SID International Symposium Digest of Technical Papers*, Society for Information Display, San Jose, Calif., pp.268–272, 1999.
15. Dworsky, Lawrence N., and Babu R. Chalamala: *Field-Emission Displays*, Society for Information Display, San Jose, Calif., pp. F-1/3–F1/66, 1999.
16. Na, Young-Sun, et. al.: "A New Data Driver Circuit for Field Emission Display," in *Asia Display '98—Proceedings of the 18th International Display Research Conference*, Society for Information Display, San Jose, Calif., pp. 137–140, 1998.

List of Figures and Tables

Figure 1.1 Applications for high-definition imaging in business and industrial facilities. 17

Figure 1.2 Viewing angle vs. screen distance for a conventional video image. 18

Figure 1.3 Viewing angle vs. screen distance for an HDTV image. 19

Figure 1.4 Illustration of the differences in screen capture capabilities of convention and HDTV images. 20

Figure 1.5 The effects of listener positioning on center image shift 24

Figure 1.6 Optimum speaker system placement for HDTV. 26

Figure 1.7 Lighting methods for optimum viewing conditions: (*a*) position of lighting components, (*b*) detail of light assembly. 29

Figure 1.8 Filter cutoff characteristics for yellow-minus-yellow lighting system. 31

Figure 1.9 Information display capability of various common technologies 36

Figure 1.10 Flat panel LCD display for desktop computer applications. 38

Figure 1.11 Overview of the MD-11 "glass cockpit." 39

Figure 1.12 Navigation display from the MD-11 aircraft. 40

Figure 2.1 The electromagnetic spectrum. 46

Figure 2.3 Spectral distribution of solar radiant power density at sea level, showing the ozone, oxygen, and carbon dioxide absorption bands. 47

Figure 2.2 The radiating characteristics of tungsten: (trace *A*) radiant flux from 1 cm^2 of a blackbody at 3000 K, (trace *B*) radiant flux from 1 cm^2 of tungsten at 3000 K, (trace *C*) radiant flux from 2.27 cm^2 of tungsten at 3000 K (equal to curve *A* in the visible region). 47

Figure 2.4 Power distribution of a monochrome video picture tube light source. 48

Figure 2.5 The photopic luminosity function. 53

Figure 2.6 Scotopic luminosity function (trace *A*) as compared with photopic luminosity function (trace *B*). 53

Figure 2.7 Weber's fraction $\Delta B/B$ as a function of luminance *B* for a dark-field surround. 55

Figure 2.8 Test chart for high-definition television applications produced by a signal waveform generator. The electronically-produced pattern is used to check resolution, geometry, bandwidth, and color reproduction. 57

Figure 2.9 Critical frequencies as they relate to retinal illumination and luminance (1 ft·L $\cong$ cd/m^2; 1 troland = retinal illuminance per square millimeter pupil area from the surface with luminance of 1 cd/m^2). 59

Figure 2.10 Solid angle ω subtended by surface *S* with its normal at angle θ from the line of propagation. 63

Figure 2.11 Light-transfer characteristics of typical video camera tubes. 66

Figure 2.12 The measurement of diffuse transmittance. 67

Figure 2.13 The measurement of reflectance. 68

Figure 3.1 Spectral sensitivities of the three types of cones in the human retina. The curves have been normalized so that each is unity at its peak. 72

Figure 3.2 Tristimulus color matching instruments: (a) conventional colorimeter, (b) addition of a primary color to perform the match. 74

Figure 3.3 Color-matching functions of the CIE standard observer based on matching stimuli of wavelengths 700.0, 546.1, and 435.8 nm, with units adjusted to be equal for a match to an equienergy stimulus. 76

Figure 3.4 A geometrical model of perceptual color space for reflecting objects. 79

Figure 3.5 The color triangle, showing the use of trilinear coordinates. The amounts of the three primaries needed to match a given color are proportional to r, g, and b. 81

Figure 3.6 A chromaticity diagram. The amounts of the three primaries needed to match a given color are proportional to r, g, and b ($= 1 - r - g$). 82

Figure 3.7 The center of gravity law in the chromaticity diagram. The additive mixture of color stimuli represented by C_1 and C_2 lies at C, whose location on the straight line C_1C_2 is given by $d_1T_1 = d_2T_2$, where T_1 and T_2 are the total tristimulus values of the component stimuli. 84

Figure 3.8 The spectrum locus and alychne of the CIE *1931 Standard Observer* plotted in a chromaticity diagram based on matching stimuli of wavelengths 700.0, 546.1, and 435.8 nm. The locations of the CIE primary stimuli X, Y, and Z are shown. 85

Figure 3.9 The CIE 1931 color-matching functions. 87

Figure 3.10 The CIE 1931 chromaticity diagram showing spectrum locus and wavelengths in nanometers. 92

Figure 3.11 The CIE 1931 chromaticity diagram: (a) regions corresponding to various color names derived from observations of self-luminous areas against a dark background, (b) equivalent wavelengths of major divisions, (c) illustration combining all elements of the chromaticity diagram into a single drawing. 93

Figure 3.12 The relative spectral power distributions of CIE standard illuminants A, B, C, and D_{65}. 95

Figure 3.13 The color triangle defined by a standard test of color television receiver phosphors compared with the maximum real color gamut on a u', v' chromaticity diagram. 96

Figure 3.14 The combination of vectors. 97

Figure 3.15 The relationship between color space and the CIE chromaticity diagram. 98

Figure 3.16 A drawing of the 1931 CIE color standard illustrating all three dimensions, x, y, and Y. 99

Figure 3.17 The 1976 CIE UCS diagram. The u', v' chromaticity coordinates for any real color are located within the bounds the horse-shoe-shaped spectrum locus and the line of purples that joins the spectrum ends. 100

Figure 3.18 The triangle representing the range of chromaticities generally achievable using the additive mixture of typical red, green, and blue phosphors on a CRT display. 101

Figure 3.19 The CIELUV color space illustrating the relationship between the opponent color axes and the axis representing 1976 CIE metric lightness. 102

Figure 3.20 The CIELUV object-color solid showing constant lightness planes, $L^* = 10.0$, 20.0, 40.0, 60.0, 80.0, and 90.0. 104

Figure 3.21 The RGB color model cube. 106

Figure 3.22 The CMY color model primaries and their mixtures. 108

Figure 3.23 The single-hexcone HSV color model. The $V=1$ plane contains the RGB model $R=1$, $G=1$, and $B=1$ planes in the regions illustrated. 108

Figure 4.1 The interlace scanning pattern (raster) of the television image. 113

Figure 4.2 The NTSC color television waveform: (a) principle components, (b) detail of picture elements. 115

Figure 4.3 Detail of sync and color subcarrier pulse widths for the NTSC system. 116

Figure 4.4 Vectorscope representation of vector and chroma amplitude relationships in the NTSC system for a color bars signal. 118

Figure 4.5 Comb filtering: (a) circuit introducing a one scan-line delay, (b) the luminance and chrominance passband. 120

Figure 4.6 Comb filtering: (a) circuit introducing a two scan-line delay; (b) the luminance passband. 121

Figure 4.7 Block diagram of a simplified color television system. 122

Figure 4.8 CIE 1931 chromaticity diagram showing three sets of phosphors used in color television displays. 125

Figure 4.9 Geometry of the field of view occupied by a television image. 126

Figure 4.10 Comparison of the aspect ratios of television and motion pictures. 128

Figure 5.1 Visual acuity (the ability to resolve details of an image) as a function of the luminosity to which the eye is adapted. 136

Figure 5.2 Visual acuity of the human eye as a function of luminosity and contrast (experimental data). 137

Figure 5.3 Contrast sensitivity as a function of background luminosity. 138

Figure 5.4 The effects of surround luminance on contrast sensitivity of the human eye. 139

Figure 5.5 Video signal spectra: (a) camera scanning spot, shown with a guassian distribution, passing over a luminance boundary on the scanning line, (b) corresponding camera output signal resulting from convolution of the spot and luminance distributions. 140

Figure 5.6 Scanning patterns of interest in analyzing conventional video signals: (a), (b), (c) flat fields useful for determining color purity and transfer gradient (gamma); (d) horizontal half-field pattern for measuring low-frequency performance; (e) vertical half field for examining high-frequency transient performance; (f) display of oblique bars; (g) in monochrome, a tonal wedge for determining contrast and luminance transfer characteristics; in color, a display used for hue measurements and adjustments; (h) wedge for measuring horizontal resolution; (i) wedge for measuring vertical resolution. 142

Figure 5.7 An array of image patterns corresponding to indicated values of m and n. 144

Figure 5.8 The typical spectrum of a video signal, showing the harmonics of the line-scanning frequency surrounded by clusters of components separated at intervals equal to the field-scanning frequency. 145

Figure 5.9 The CIE 1931 chromaticity diagram illustrating use of the center of gravity law (i.e., $T_r d_1 = T_g d_2$, $T_{c1} = T_r + T_g$, $T_{c1} d_3 = T_b d_4$). 146

Figure 5.10 Ellipses of equally perceptible color differences. 150

Figure 5.11 Multiburst video test waveform: (*a*, left) picture display, (*b*, right) multiburst signal as viewed on a waveform monitor (1-*H*). 153

Figure 5.12 Conventional sweep frequency test waveform: (*a*, left) picture display, (*b*, right) waveform monitor display, with markers (1-*H*). 154

Figure 5.13 Single horizontal frequency test signal from a zone plate generator: (*a*, left) picture display, (*b*, right) waveform monitor display (1-*H*). 154

Figure 5.14 Horizontal frequency sweep test signal from a zone plate generator: (*a*, left) picture display, (*b*, right) waveform monitor display (1-*H*). 155

Figure 5.15 Single vertical frequency test signal: (*a*, left) picture display, (*b*, right) magnified vertical rate waveform, showing the effects of scan sampling. 155

Figure 5.16 Vertical frequency sweep picture display. 156

Figure 5.17 Combined horizontal and vertical frequency sweep picture display. 156

Figure 5.18 Combined horizontal and vertical frequency sweeps, selected line waveform display (1-*H*). This figure shows the maintenance of horizontal structure in the presence of vertical sweep. 157

Figure 5.19 The best-known zone plate pattern, combined horizontal and vertical frequency sweeps with zero frequency in the center screen. 157

Figure 5.20 Vertical frequency sweep picture display. 158

Figure 5.21 The same vertical sweep as Figure 5.20 except that appropriate pattern motion has been added to "freeze" the beat pattern in the center screen for photography or other analysis. 159

Figure 5.22 A hyperbolic variation of the two-axis zone plate frequency sweep. 159

Figure 5.23 A two-axis frequency sweep in which the range of frequencies is swept several times in each axis. Complex patterns such as this may be created for specific test requirements. 159

Figure 6.1 A generalized schematic of a cathode ray tube using electrostatic deflection. 164

Figure 6.2 Particle size distribution of four representative phosphors. 167

Figure 6.3 Energy transitions of electrons leading to luminescence. 168

Figure 6.4 Light output of aluminized and non-aluminized phosphor screens. 170

Figure 6.5 Linearity characteristics of five common phosphors. Conditions: E_{a2} = 25 kV. (1 ft·L = 3,426 cd/m^2) 172

Figure 6.6 The loss in phosphor efficiency as the screen heats at high current operation (rare-earth green). 173

Figure 6.7 Typical spectral energy distribution of color primaries at equal current density. 173

Figure 6.8 The Kelly chart of color designations for lights showing NTSC and current commercial primary phosphors and locus of dyes, paints, and pigments. 174

Figure 6.9 Typical phosphor rise time decay curve. 175

Figure 6.10 Contrast ratio as a function of ambient illumination for a given CRT brightness. 176

Figure 6.11 Modulation transfer function (MTF) characteristics of high-resolution phosphor screens prepared by various methods. 178

Figure 6.12 The brightness output of high-resolution phosphor screens prepared by various methods. Brightness is expressed as a function of current density at constant phosphor voltage. 179

Figure 6.13 Triode electron gun structure. 180

Figure 6.14 The electron trajectory from an electron gun using the parameters specified in the text. 180

Figure 6.15 Basic structure of the tetrode electron gun. 181

Figure 6.16 Generalized schematic of a CRT grid structure and accelerating electrode in a device using electrostatic deflection. 182

Figure 6.17 The basic principles of electron optics. 183

Figure 6.18 A unipotential lens. 184

Figure 6.19 Electron beam shape. 184

Figure 6.20 Generalized schematic of a CRT gun structure using electromagnetic focus and deflection. 185

Figure 6.21 Generalized schematic of a CRT with electrostatic focus and deflection. An *Einzel* focusing lens is depicted. 186

Figure 6.22 Illustration of chromatic aberration. 187

Figure 6.23 Illustration of coma. 187

Figure 6.24 Illustration of astigmatism. 188

Figure 6.25 Pincushion and barrel distortion. 188

Figure 6.26 Spherical aberration. 189

Figure 6.27 Magnetic-focusing elements on the neck of a CRT. 190

Figure 6.28 Spiral distortion in magnetic-lens images. 191

Figure 6.29 Idealized cathode with a spherical field. 192

Figure 6.30 Monochrome electron gun configuration and the corresponding field and beam plots. 193

Figure 6.31 Variation in CRT spot size as a function of beam current. 194

Figure 6.32 The interdependence of beam current, line width, and brightness in a CRT. 195

Figure 6.33 Construction of a typical storage CRT. 196

Figure 6.34 A typical bistable storage CRT: (*a*) basic construction, (*b*) detail of screen storage system. 197

Figure 6.35 Wide aspect ratio resolution chart produced by an electronic signal generator. 203

Figure 7.1 Electrostatic deflection CRT. 215

Figure 7.2 The elements of an electrostatic deflection CRT: (*a*) overall tube geometry, (*b*) detail of deflection region. 216

Figure 7.3 Trajectory of an electron beam. 217

Figure 7.4 Beam spread nomograph. 219

Figure 7.5 Principle quantities of magnetic deflection. 222

Figure 7.6 Linearity distortion resulting from CRT screen curvature. 223

Figure 7.7 The mechanics of flat-face CRT linearity distortion. 224

Figure 7.8 Basic linearity correction circuit for a CRT. 230

Figure 7.9 A plot of E_o versus E_i where $E_o = KI^3$. 231

Figure 7.10 Block diagram of X/Y linearity correction circuit. 231

Figure 7.11 Basic design of a dynamic focus circuit. 232

Figure 7.12 Plot of a dynamic focusing correction signal. 233

Figure 7.13 Pincushion distortion: (*a*) overall effect on the displayed image, (*b*) corresponding composite correction signal, (*c*) horizontal pincushion distortion component, (*d*) horizontal correction signal, (*e*) vertical pincushion distortion component, (*f*) vertical correction signal. 234

Figure 7.14 Block diagram of a pincushion correction circuit. 235

Figure 8.1 Basic concept of a shadow-mask color CRT: (*a*) overall mechanical configuration, (*b*) delta gun arrangement on tube base, (*c*) shadow-mask geometry. 238

Figure 8.2 Basic concept of the Trinitron color CRT: (*a*) overall mechanical configuration, (*b*) in-line gun arrangement on tube base, (*c*) mask geometry. 240

Figure 8.3 Operating principles of the Chromatron CRT: (*a*) overall mechanical configuration, (*b*) no deflection voltage applied to the grid and electrons strike the green phosphor strips, (*c*) deflection voltage applied to the grid with a polarity such that electrons are deflected to the blue phosphor strips (the opposite polarity produces red). 241

Figure 8.4 Mechanical layout of the beam penetration color CRT. 242

Figure 8.5 Shadow-mask CRT using in-line guns and round mask holes: (*a*) overall tube geometry, (*b*) detail of phosphor dot layout. 243

Figure 8.6 Shadow-mask CRT using in-line guns and vertical stripe mask holes. 244

Figure 8.7 Delta-gun, round-hole mask negative guard band tri-dot screen. The taper on the mask holes is shown in the detail drawing only. 245

Figure 8.8 The relationship between beam shift at local doming and the effective radius of the faceplate with various mask materials. 250

Figure 8.9 Comparison of small area mask doming for a conventional CRT and a flat-tension-mask CRT. 251

Figure 8.10 Mechanical structure of a taut shadow-mask CRT. 252

Figure 8.11 Relationship between screen size and CRT weight. 254

Figure 8.12 Principal dimensions of a 38V-in (965 mm) CRT: (*a*) front view, (*b*) side view. 257

Figure 8.13 Increase in spot size as a function of beam current for a large-screen CRT. 258

Figure 8.14 Beam current versus luminance characteristics for a large-screen CRT. 259

Figure 8.15 Simplified mechanical structure of a bipotential color electron gun. 260

Figure 8.16 Bipotential electron gun configuration, shown with the corresponding field and beam plots. 261

Figure 8.17 Tripotential electron gun configuration, shown with the corresponding field and beam plots. 262

Figure 8.18 Electrode arrangement of a UniBi gun. 263

Figure 8.19 Electrode arrangement of a BiUni gun. 263

Figure 8.20 Electrode arrangement of the Trinitron gun. 264

Figure 8.21 Spot-size comparison of high-resolution guns versus commercial television guns: (*a*) 13-in vertical display, (*b*) 19-in vertical display. 266

Figure 8.22 Astigmatism errors in a color CRT. 267

Figure 8.23 Astigmatism and coma errors. 267

Figure 8.24 Misconvergence in the four corners of the raster in a color CRT. 268

Figure 8.25 Field configurations suitable for correcting misconvergence in a color CRT. 268

Figure 8.26 Four-pole field used for misconvergence correction in a color CRT. 269

Figure 8.27 Self-converging deflection field in an in-line gun device: (*a*) horizontal deflection field, (*b*) vertical deflection field. 270

Figure 8.28 Simplified block diagram of a CRT magnetic deflection amplifier. 270

Fiigure 8.29 The effect of beam parallax. 271

Figure 8.30 The effect of beam tilt. 272

Figure 8.31 Color dot displacement: (*a*) horizontal deflection, (*b*) vertical deflection. 273

Figure 8.32 Color dot displacement: (*a*) horizontal shift of vertical green bars, (*b*) vertical shift of horizontal blue bars. 274

Figure 8.33 Parabolic correction waveforms: (*a*) first harmonic only, (*b*) first and second harmonic signals. 274

Figure 8.34 The image field showing beam crossover. 275

Figure 8.35 Self-converging in-line system: (*a*) vertical, (*b*) horizontal. 275

Figure 8.36 Design of a flat CRT (Sony): (*a*) front view, (*b*) side view. 276

Figure 8.37 Channel multiplying flat CRT (Philips): (*a*) exploded front view, (*b*) side view. 277

Figure 8.38 Beam guide flat CRT (RCA): (*a*) front view, (*b*) vertical beam deflection system, (*c*) horizontal beam deflection system, 278

Figure 8.39 Matrix drive and deflection flat CRT (Matsushita). 280

Figure 8.40 Horizontal address vertical deflection flat CRT: (*a*) exploded view, (*b*) detail of electron path under no-deflection condition, (*c*) detail of electron path under deflection condition. 281

Figure 9.1 Basic grading of full-color display technologies vs. display area for military applications. 290

Figure 9.2 Refraction and reflection at air-glass surface: (*a*) beam incident upon glass from air, (*b*) beam incident upon air from glass. 291

Figure 9.3 The result when the angle of refraction exceeds the critical angle and the ray is totally reflected. 292

Figure 9.4 Various forms of simple converging and diverging lenses. 294

Figure 9.5 Lens effects: (*a*) parallel rays incident upon the lens pass through the second focal point, (*b*) rays passing through the first focal point incident upon the lens emerge parallel. 294

Figure 9.6 Magnification of a simple lens. 295

Figure 9.7 Focal points and principal planes for a thick lens. 296

Figure 9.8 Surfaces of best focus, illustrating lens astigmatism. 298

Figure 9.9 Lens system treated as a single thick lens. 299
Figure 9.10 Arrangement of a dichroic mirror beam-splitting system. 300
Figure 9.11 Transmission characteristics of typical dichroic mirrors. 300
Figure 9.12 Response curve of a spectral trim filter (yellow). 301
Figure 9.13 Response curve of a neutral density filter (visible spectrum). 302
Figure 9.14 Principal elements of a video projection system. 305
Figure 9.15 Projection system screen characteristics: (*a*) front projection, (*b*) rear projection. 307
Figure 9.16 High-contrast rear-projection system. 308
Figure 9.17 Optical projection configurations: (*a*) single tube, single lens rear projection system; (*b*) crossed-mirror trinescope rear projection; (*c*) three tube, three lens rear projection system; (*d*) folded optics, front projection with three tubes in-line and a dichroic mirror; (*e*) folded optics, rear projection. 309
Figure 9.18 Three tube in-line array. The optical axis of the outboard (red and blue) tubes intersect the axis of the green tube at screen center. Red and blue rasters show trapezoidal (keystone) distortion and result in mis-convergence when superimposed on the green raster, thus requiring electrical correction of scanning geometry and focus. 311
Figure 9.19 Mechanical construction of a liquid-cooled projection CRT. 314
Figure 9.20 Projection CRT with integral Schmidt optics. 314
Figure 9.21 Mechanical configuration of the Eidophor projector optical system. 318
Figure 9.22 Functional operation of the General Electric single-gun light valve system. 319
Figure 9.23 Functional operation of the two channel HDTV light valve system. 320
Figure 9.24 Color characteristics of the GE HDTV light valve system. 321
Figure 9.25 Block diagram of a laser-scanning projection display. 322
Figure 9.26 Configuration of a color laser projector. 323
Figure 9.27 Rotating polygon line scanner for a laser projector. 324
Figure 9.28 Laser-screen projection CRT. 325
Figure 9.29 Structure of a liquid-cooled LC panel. 326
Figure 9.30 LC panel temperature dependence on lamp power in a liquid crystal projector. 327
Figure 9.31 LC projection system for HDTV applications. 327
Figure 9.32 Optical System for an NTSC LCD projector. 328
Figure 9.33 Layout of a homeotropic liquid crystal light valve. 329
Figure 9.34 Block diagram of an LCLV system. 330
Figure 9.35 Functional optical system diagram of an LCLV projector. 331
Figure 9.36 Layout of the fluid filled prism assembly for an LCLV projector. 332
Figure 9.37 Block diagram of a color laser-addressed light valve display system. 333
Figure 9.38 Cross-sectional diagram of the Image Light Amplifier light valve. 334
Figure 9.39 Optical channel of the basic ILA projector system. 335
Figure 9.40 Full-color optical system for the ILA projector (Series 300/400). 336

Figure 9.41 Color gamut of the ILA projector (Series 300/400) compared with SMPTE 240M. 337

Figure 9.42 A pair of DMD pixels with one mirror shown at −10° and the other at +10°. 338

Figure 9.43 The basics of DMD optical switching. 339

Figure 9.44 A scanning electron microscope view of the DMD yoke and spring tips (the mirror has been removed). 339

Figure 9.45 Optical system of the DLP 3-chip implementation. 340

Figure 9.46 DMD display resolution as a function of chip diagonal. 341

Figure 9.47 A typical GLV pixel showing alternate ribbons being addressed. 341

Figure 9.48 The diffractive (bright) state of a GLV pixel. 342

Figure 9.49 Video processing architecture for a Scanned Linear GLV system. 343

Figure 9.50 GLV electro-optic response. 343

Figure 9.51 The GLV horizontally scanned linear system supports variable aspect ratios. 344

Figure 9.52 DC plasma display panel for color video. 345

Figure 9.53 Cross section of an AC-PDP cell capacitor. 346

Figure 9.54 Matrix addressing for flat-panel display. 347

Figure 9.55 Principle of active matrix addressing for a large screen display. 348

Figure 9.56 The relationship between misconvergence and display resolution. 350

Figure 9.57 Relationship between screen type and viewing angle (*a*) conventional (flat) screen, (*b*) high gain (curved) screen. 351

Figure 9.58 Mounting options for a projection system: (*a*) table top, (*b*) ceiling mount. 352

Figure 10.1 The liquid-crystal/grooved interface: (*a*) no field applied, (*b*) an electric field greater than the critical value applied. 360

Figure 10.2 The orientation of the liquid-crystal axis in a cell: (*a*) no field applied, (*b*) a field of approximately twice the critical field applied, (*c*) a field several times the critical field applied. 361

Figure 10.3 A twisted nematic cell. 362

Figure 10.4 Cholesteric ordering: a large number of planes of nematic ordering are formed where the directors rotate as light moves along a direction perpendicular to the planes. 363

Figure 10.5 Passive matrix addressing. 364

Figure 10.6 Active matrix addressing scheme. 364

Figure 10.7 Addressing topology using three-terminal TFT devices. 365

Figure 10.8 Cross-section of an AMLCD display panel. 365

Figure 10.9 The principle of in-plane switching for the AMLCD: (*a*) off state, (*b*) on state. 367

Figure 10.10 Simplified illustration of the PALC and AMLCD stack. Compared to conventional LCDs, only the active substrate is unique. 368

Figure 10.11 Physical structure of the PALC. 369

Figure 10.12 Addressing scheme for a high-definition PALC device. 370

Figure 10.13 Gas discharge reactions model for a plasma display panel. 372

Figure 10.14 Typical voltage-current characteristics of a gas discharge. 373

Figure 10.15 The luminous regions of a gas discharge. 374

Figure 10.16 Light wavelength distribution for a neon gas plasma display. 375

Figure 10.17 Common current-limiting techniques for PDP devices: (*a*) dc display, (*b*) ac display. 376

Figure 10.18 Structure of an ac PDP display cell with memory. 376

Figure 10.19 Structure of a hybrid ac-dc plasma panel. 377

Figure 10.20 Full-color ac memory plasma display cell. 377

Figure 10.21 Co-planar PDP discharge cell: (*a*) incorporating a "T"-shaped electrode, (*b*) incorporating a rectangular electrode. 380

Figure 10.22 Field emission display device: (*a*) basic configuration, (*b*) emission element, (*c*) emission element detail. 380

Figure 10.23 A planar carbon emitter. 381

Figure 10.24 Electron microphotograph of a Spindt array: (*a*) group of nine devices, (*b*) closeup of a single emitter. 381

Tables

Table 1.1 Relative Merits of HDTV Display Technologies 12

Table 1.2 Display Requirements for Cinema Projection of HDTV Images 21

Table 1.3 Ultimate Requirements for Cinema Projection of HDTV Images 21

Table 1.4 Electrical to Optical Efficiency of Various Electronic Projection Display Technologies 22

Table 1.5 Comparison of Common Radar-Room Lighting Systems 32

Table 1.6 Content Descriptors for Common Computer and Video Systems 33

Table 1.7 Comparison of Technology Problems Facing High Information Content Display Systems 35

Table 2.1 Psychophysical and Psychological Characteristics of Color 49

Table 2.2 Relative Luminosity Values for Photopic and Scotopic Vision 50

Table 2.3 Conversion Factors for Luminance and Retinal Illuminance Units 61

Table 2.4 Typical Luminance Values 62

Table 2.5 Conversion Factors for Illuminance Units 62

Table 3.1 The Perceptual Terms and Their Psychophysical Correlates 86

Table 3.2 CIE Colorimetric Data (1931 Standard Observer) 88

Table 4.1 Video and Sync Levels in IRE Units 116

Table 5.1 Relative Flicker Threshold for Various Luminances 139

Table 6.1 Monochrome CRT Electrode Potentials 166

Table 6.2 Characteristics of Standardized Phosphor Types 169

Table 6.3 Typical Characteristics of Common Phosphors 171

Table 7.1 Comparison of Common Electromagnetic and Electrostatic Deflection CRTs 214

Table 8.1 Color CRT Diagonal Dimension vs. Weight 253
Table 8.2 Comparative Resolution of Shadow-mask Designs 255
Table 8.3 Specifications for a 38V-in High-Resolution Trinitron CRT 256
Table 9.1 Performance Requirements for Large Screen Military Display Applications 289
Table 9.2 Performance Levels of Video and Theater Displays 306
Table 9.3 Screen Gain Characteristics for Various Materials 308
Table 9.4 Performance of Various Types of Projection Lenses 312
Table 9.5 Operating Characteristics of Basic Light Valve Projection Systems 317
Table 9.6 Appearance of Scan Line Judder in a Projection Display 350

List of References

Aiken, J. A.: "A Thin Cathode Ray Tube," *Proc. IRE*, vol. 45, pg. 1599, 1957.

Aiken, W. R.: "A Thin Cathode Ray Tube," *Proc. IRE*, vol. 45, no. 12, pp. 1599–1604, December 1957.

Allen, E. S.: *Six-Place Tables*, 7th ed., McGraw-Hill, New York, N.Y., 1947.

"Alignment of NTSC Color Picture Monitors," SMPTE Recommended Practice RP 167-1995, SMPTE, White Plains, N.Y., 1995.

Allison, J.: *Electronic Engineering Semiconductors and Devices*, 2nd ed., McGraw-Hill, London, pg. 308–309, 1990.

Anstey, G., and M. J. Dore: "Automatic Measurement of Cathode Ray Tube MTFs," Royal Signals and Radar Establishment, 1980.

Amm, D. T., and R.W. Corrigan: "Grating Light Valve Technology: Update and Novel Applications," SID Symposium—Anaheim, SID, San Jose, Calif., May 1998.

Amm, D. T., and R. W. Corrigan: "Optical Performance of the Grating Light Valve Technology," Projection Displays V Symposium, *SPIE Proceedings*, SPIE, San Jose, Calif., vol. EI 3634-10, February 1999.

Ashizaki, S., Y. Suzuki, K. Mitsuda, and H. Omae: "Direct-View and Projection CRTs for HDTV," *IEEE Transactions on Consumer Electronics*, vol. 34, no. 1, pp. 91–98, February 1988.

Baldwin, M. W., Jr.: "Subjective Sharpness of Additive Color Pictures," *Proc. IRE*, vol. 39, pp. 1173–1176, October 1951.

Baldwin, M. W., Jr.: "The Subjective Sharpness of Simulated Television Images," *Proc. IRE*, vol. 28, p. 458, October 1940.

Baldwin, M., Jr.: "The Subjective Sharpness of Simulated Television Images," *Proceedings of the IRE*, vol. 28, July 1940.

Barbin, R., and R. Hughes: "New Color Picture Tube System for Portable TV Receivers," *IEEE Trans. Broadcast TV Receivers*, vol. BTR-18, no. 3, pp. 193–200, August 1972.

Barkow, W. H., and J. Gross: "The RCA Large Screen 110° Precision In-line System," ST-5015, RCA Entertainment, Lancaster, Penn.

Barkow, W. H., and J. Gross: "The RCA Large Screen 110° Precision In-line System," ST-5015, RCA Entertainment, Lancaster, Penn.

Barten, P. J. G.: "Spot Size and Current Density Distribution of CRTs," *Proceedings of the Society for Information Display*, Society for Information Display, San Jose, Calif., vol. 25, no. 3, pp. 155–159, 1984.

Barten, Peter G. J.: "Physical Model for the Contrast Sensitivity of the Human Eye," *Human Vision, Visual Processing, and Digital Display III*, Bernice E. Rogowitz ed., Proc. SPIE 1666, SPIE, Bellingham, Wash., pp. 57–72, 1992.

Bartleson, C. J., and E. J. Breneman: "Brightness Reproduction in the Photographic Process," *Photog. Sci. Eng.*, vol. 11, pp. 254–262, 1967.

Bates, W., P. Gelinas, and P. Recuay: "Light Valve Projection System for HDTV," *Proceedings of the ITS*, International Television Symposium, Montreux, Switzerland, 1991.

Battison, John:, "Making History," *Broadcast Engineering*, Intertec Publishing, Overland Park, Kan., June 1986.

Bauman, E.: "The Fischer Large-Screen Projection System," *SMPTE Journal*, SMPTE, White Plains, N.Y., vol. 60, pg. 351, 1953.

Baur, G.: *The Physics and Chemistry of Liquid Crystal Devices*, G. J. Sprokel (ed.), Plenum, New York, N.Y., pg. 62, 1980.

Bedell, R. J.: "Modulation Transfer Function of Very High Resolution Miniature Cathode Ray Tubes," *IEEE Transactions on Electron Devices*, vol. ED-22, no. 9, pp. 793–796, September 1975.

Belton, J.: "The Development of the CinemaScope by Twentieth Century Fox," *SMPTE Journal*, SMPTE, White Plains, N.Y., vol. 97, September 1988.

Bender, Walter, and Alan Blount: "The Role of Colorimetry and Context in Color Displays," *Human Vision, Visual Processing, and Digital Display III*, Bernice E. Rogowitz ed., Proc. SPIE 1666, SPIE, Bellingham, Wash., pp. 343–348, 1992.

Benson, K. B., and D. G. Fink: *HDTV: Advanced Television for the 1990s*, McGraw-Hill, New York, N.Y., 1990.

Benson, K. B., and J. C. Whitaker (eds.): *Television Engineering Handbook*, revised ed., McGraw-Hill, New York, N.Y., 1991.

Benson, K. B., and J. C. Whitaker: *Television and Audio Handbook For Engineers and Technicians*, McGraw-Hill, New York, N.Y., 1989.

Benson, K. B., and J. C. Whitaker: *Television Engineering Handbook*, revised ed., McGraw-Hill, New York, N.Y., 1991.

Benson, K. B.: "Report on Sources of Variability in Color Reproduction as Viewed on the Home Television Receiver," *IEEE Trans. BTR*, vol. 19, pp. 269–275, 1973.

Bingley, F. J.: "Colorimetry in Color Television—Pt. I," *Proc. IRE*, vol. 41, pp. 838–851, 1953.

Bingley, F. J.: "Colorimetry in Color Television—Pts. II and III," *Proc. IRE*, vol. 42, pp. 48–57, 1954.

Bingley, F. J.: "The Application of Projective Geometry to the Theory of Color Mixture," *Proc. IRE*, vol. 36, pp. 709–723, 1948.

Birks, J. B.: "Electrophoretic Deposition of Insulating Materials," *Progr. Dielectrics*, vol. 1, 1959.

Blacker, A., et al.: "A New Form of Extended Field Lens for Use in Color Television Picture Tube Guns," *IEEE Trans. Consumer Electronics*, pp. 238–246, August 1966.

Blaha, R.: "Degaussing Circuits for Color TV Receivers," *IEEE Trans. Broadcast TV Receivers*, vol. BTR-18, no. 1, pp. 7–10, February 1972.

Blaha, Richard J.: "Large Screen Display Technology Assessment for Military Applications," *Large-Screen Projection Displays II*, William P. Bleha, Jr., (ed.), Proc. SPIE 1255, SPIE, Bellingham, Wash., pp. 80–92, 1990.

Bleha, W. P.: "Image Light Amplifier (ILA) Technology for Large-Screen Projection," *SMPTE Journal*, SMPTE, White Plains, N.Y., pp. 710–717, October 1997.

Bleha, William P., Jr., (ed.): *Large-Screen Projection Displays II*, Proc. SPIE 1255, SPIE, Bellingham, Wash., 1990.

Bock, Wolfgang: "Some European Capabilities in Satellite Cinema Exhibition," in *Large-Screen Projection Displays II*, William P. Bleha, Jr., (ed.), Proc. SPIE 1255, SPIE, Bellingham, Wash., pp. 13–20, 1990.

Boers, J.: "Computer Simulation of Space Charge Flows," Rome Air Development Command RADC-TR-68-175, University of Michigan, 1968.

Boynton, R. M.: *Human Color Vision*, Holt, New York, N.Y., 1979.

Boynton, R.M.: *Human Color Vision*, Holt, New York, N.Y., p. 404, 1979.

Brodeur, R., K. R. Field, and D. H. McRae: "Measurement of Color Rendition in Color Television," in M. Pearson (ed.), *Proc. ISCC Conf. Optimum Reproduction of Color*, Williamsburg, Va., 1971, Graphic Arts Research Center, Rochester, N.Y., 1971.

Burgmans, A., et. al.: *Information Display*, pg. 14, April/May 1998.

Buzak, Thomas S.: "Recent Advances in PALC Technology," in *Proceedings of the 18th International Display Research Conference*, Society for Information Display, San Jose, Calif., Asia Display '98, pp. 273-276, 1998.

Carpenter, C.: et al., "An Analysis of Focusing and Deflection in the Post-Deflection Focus Color Kinescope," *IRE Trans. Electron Devices*, vol. 2, pp. 1–7, 1955.

Castellano, Joseph A.: *Display Systems*, Academic Press, New York, N.Y., 1992.

Casteloano, Joseph A.: *Handbook of Display Technology*, Academic, New York, N.Y., 1992.

Cathode Ray Tube Displays, MIT Radiation Laboratory Series, vol. 22, McGraw-Hill, New York, N.Y., 1953.

Cerulli, N. F.: "Method of Electrophoretic Deposition of Luminescent Materials and Product Resulting Therefrom," U.S. Patent 2,851,408, September 9, 1958.

Chang, I.: "Recent Advances in Display Technologies," *Proc. SID*, Society for Information Display, San Jose, Calif., vol. 21, no. 2, pg. 45, 1980.

Chen, H., and R. Hughes: "A High Performance Color CRT Gun with an Asymmetrical Beam Forming Region," *IEEE Trans. Consumer Electronics*, vol. CE-26, pp. 459–465, August, 1980.

Chen, K. C., W. Y. Ho and C. H. Tseng: "Invar Mask for Color Cathode Ray Tubes, in *Display Technologies*, Shu-Hsia Chen and Shin-Tson Wu (eds.), Proc. SPIE 1815, SPIE, Bellingham, Wash., pp.42–48, 1992.

Cheng, J. B., and Q. H. Wang: "Studies on YAG Phosphor Screen for HDTV Projector," *Proc SPIE 2892*, SPIE, Bellingham, Wash., pg. 36, 1996.

Cheng, Jia-Shyong, et. al.: "The Optimum Design of LCD Parameters in Projection and Direct View Applications," *Display Technologies*, Shu-Hsia Chen and Shin-Tson Wu (eds.), Proc. SPIE 1815, SPIE, Bellingham, Wash., pp. 69–80, 1992.

Clapp, R., et al.: "A New Beam Indexing Color Television Display System," *Proc. IRE*, vol. 44, no. 9, pp. 1108–1114, September 1956.

Cloz, R., et al.: "Mechanism of Thin Film electroluminescence," *Conference Record*, SID Proceedings, Society for Information Display, San Jose, Calif., vol. 20, no. 3, 1979.

Cohen, C.: "Sony's Pocket TV Slims Down CRT Technology," *Electronics*, pg. 81, February 10, 1982.

"Colorimetry," Publication no. 15, Commission Internationale de l'Eclairage, Paris, 1971.

Committee on Colorimetry, Optical Society of America: *The Science of Color*, New York, N.Y., 1953.

Corrigan, R. W., B. R. Lang, D.A. LeHoty, and P.A. Alioshin: "An Alternative Architecture for High Performance Display," Silicon Light Machines, Sunnyvale, Calif., 1999. Presented at the 141st SMPTE Technical Conference (paper 141-25).

Corrigan, R. W., D. T. Amm, P. A. Alioshin, B. Staker, D .A. LeHoty, K. Gross, and B. R. Lang: "Calibration of a Scanned Linear Grating Light Valve TM Projection System," SID Symposium—San Jose, SID, San Jose, Calif., May 1999.

Credelle, T. L., et al.: "Cathodoluminescent Flat Panel TV Using Electron Beam Guides," *SID Int. Symp. Digest*, Society for Information Display, San Jose, Calif., pg. 26, 1980.

Credelle, T. L., et al.: *Japan Display '83*, pg. 26, 1983.

Credelle, T. L.: "Modular Flat Display Device with Beam Convergence," U.S. Patent 4,131,823.

"Critical Viewing Conditions for Evaluation of Color Television Pictures," SMPTE Recommended Practice RP 166-1995, SMPTE, White Plains, N.Y., 1995.

Crost, Munsey E.: "Display Devices and the Human Observer," *Proc. Interlab. Sem. Component Technol. Pt. 1, R&D Tech. Rep.* ECOM-2865. U.S. Army Electronics Command, Fort Monmouth, N.J., August, 1967.

"CRT Control Grid Having Orthogonal Openings on Opposite Sides," U.S. Patent 4,242,613, Dec. 30, 1980.

Curie, D., and G. F. J. Garlick: *Luminescence in Crystals*, Wiley, New York, N.Y., 1963.

Daly, Scott: "The Visible Differences Predictor: An Algorithm for the Assessment of Image Fidelity, *Human Vision, Visual Processing, and Digital Display III*, Bernice E. Rogowitz ed., Proc. SPIE 1666, SPIE, Bellingham, Wash., pp. 2–15, 1992.

Dasgupta, B. B.: "Recent Advances in Deflection Yoke Design," *SID International Symposium Digest of Technical Papers*, Society for Information Display, San Jose, Calif., pp. 248–252, May 1999.

Davis, C., and D. Say: "High Performance Guns for Color TV—A Comparison of Recent Designs," *IEEE Trans. Consumer Electronics*, vol. CE-25, August 1979.

Davson, H.: *Physiology of the Eye*, 4th ed., Academic, New York, N.Y., 1980.

DeMarsh, L. E.: "Color Rendition in Television," *IEEE Trans. CE*, vol. 23, pp. 149–157, 1977.

DeMarsh, L. E.: "Colorimetric Standards in US Color Television," *J. SMPTE*, vol. 83, pp. 1–5, 1974.

Diakides, N. A.: "Phosphors," *Proc. Soc. Photo-Opt. Instrum. Eng.*, vol 42, August 1973.

Dickenson, W.: "Monochrome Picture Tubes—Status Report," *IEEE Trans. Broadcast TV Receivers*," vol. BTR-13, no. 3, pp. 46–48, 1967.

Donofrio, R.: "Color in Color TV—A Phosphor Approach," *Color Engineering*, pp. 11–14, February 1971.

Donofrio, R.: "Image Sharpness of a Color Picture Tube by Modulation Transfer Techniques," *IEEE Tran. Broadcast Television Receivers*, vol. BTR-18, no. 1, p. 16, February 1972.

Donofrio, R.: "Image Sharpness of a Color Picture Tube by MTF Techniques," *IEEE Trans. Broadcast TV Receivers*, vol. BTR-18, no. 1, pp. 1–6, February 1972.

Donofrio, R.: "Low Current Density Aging," *Proc. Electrochemical Society*, May 12, 1981.

"Dr. Vladimir K. Zworykin: 1889-1982," *Electronic Servicing and Technology*, Intertec Publishing, Overland Park, Kan., October 1982.

Dressler, R.: "The PDF Chromatron—A Single or Multi-Gun CRT," *Proc. IRE*, vol. 41, no. 7, July 1953.

Dwight, H. B.: *Tables of Integrals and Other Mathematical Data*, 4th ed., Macmillan, New York, N.Y., 1961.

Dworsky, Lawrence N., and Babu R. Chalamala: *Field-Emission Displays*, Society for Information Display, San Jose, Calif., pp.F-1/3-F1/66, 1999.

Eccles, D. A., and Y. Zhang: "Digital-Television Signal Processing and Display Technology," SID 99 Digest, Society for Information Display, San Jose, Calif., pp. 108–111, 1999.

EG&G Gamma Scientific Inc.: Digital Spatial Scanning System Software, SPATL-C11/MTF Ver. 5.29, San Diego, Calif.

Ekstrand, R.: "A Flesh-Tone Correction Circuit," *IEEE Trans. BTR*, vol. 17, pp. 182–189, 1971.

"Electron Gun with Astigmatic Flare-Reducing Beam Forming Region," U.S. Patent 4,234,814, Nov. 18, 1980.

Epstein, D. W.: "Colorimetric Analysis of RCA Color Television System," *RCA Review*, vol. 14, pp. 227–258, 1953.

Eshbach, O. W.: *Handbook of Engineering Fundamentals*, 2d ed., Wiley, New York, N.Y., 1936.

Evans, R. M., W. T. Hanson, Jr., and W. L. Brewer: *Principles of Color Photography*, Wiley, New York, N.Y., 1953.

Fink, D. G. (ed.): *Television Engineering Handbook*, McGraw-Hill, New York, N.Y., 1957.

Fink, D. G., et. al.: "The Future of High Definition Television," *SMPTE Journal*, vol. 89, SMPTE, White Plains, N.Y., February/March 1980.

Fink, D. G.: "Perspectives on Television: The Role Played by the Two NTSCs in Preparing Television Service for the American Public," *Proceedings of the IEEE*, vol. 64, IEEE, New York, N.Y., September 1976.

Fink, D. G.: *Color Television Standards*, McGraw-Hill, New York, N.Y., 1986.

Fink, D. G.: *Television Engineering*, 2nd ed., McGraw-Hill, New York, N.Y., 1952.

Fink, Donald G., and Donald Christiansen (eds.): *Electronics Engineers' Handbook*, 2d ed., McGraw-Hill, New York, N.Y., 1982.

Fink, Donald G., and H. Wayne Beaty (eds.): *Standard Handbook for Electrical Engineers*, 11th ed., McGraw-Hill, New York, N.Y., 1978.

Fink, Donald, (ed.): *Television Engineering Handbook*, McGraw-Hill, New York, N.Y., 1957.

Fink, Donald, and Donald Christiansen (eds.): *Electronics Engineers Handbook*, 3rd ed., McGraw-Hill, New York, N.Y., 1989.

Fiore, J., and S. Kaplan: "A Second Generation Color Tube Providing More than Twice the Brightness and Improved Contrast," Spring Conf. Broadcast and Television Receivers, IEEE, June 1969.

Flechsig, W.: "CRT for the Production of Multicolored Pictures on a Luminescent Screen," French Patent 866,065, 1939.

Fleschig, W.: German Patent No. 736,575, July 2, 1938.

Florence, J., and L. Yoder: "Display System Architectures for Digital Micromirror Device (DMD) Based Projectors," *Proc. SPIE*, SPIE, Bellingham, Wash., vol. 2650, Projection Displays II, pp. 193–208, 1996.

Foley, James D., et al.: *Computer Graphics: Principles and Practice*, 2nd ed., Addison-Wesley, Reading, Mass., pp. 584–592, 1991.

Fowle, F. E.: *Smithsonian Physical Tables*, 9th ed., Smithsonian Institution, Washington, D.C., 1954.

Foyle, David C.: "Proposed Evaluation Framework for Assessing Operation Performance with Multisensor Displays," *Human Vision, Visual Processing, and Digital Display III*, Bernice E. Rogowitz ed., Proc. SPIE 1666, SPIE, Bellingham, Wash., pp. 514–525, 1992.

Fritz, Victor J.: "Full-Color Liquid Crystal Light Valve Projector for Shipboard Use," *Large Screen Projection Displays II*, William P. Bleha, Jr., (ed.), Proc. SPIE 1255, SPIE, Bellingham, Wash., pp. 59–68, 1990.

Fujio, T., J. Ishida, T. Komoto and T. Nishizawa: "High-Definition Television Systems-Signal Standards and Transmission," *SMPTE Journal*, SMPTE, White Plains, N.Y., vol. 89, August 1980.

Fujio, T., J. Ishida, T. Komoto, and T. Nishizawa: "High-Definition Television Systems—Signal Standards and Transmission," *SMPTE Journal*, SMPTE, White Plains, N.Y., vol. 89, August 1980.

Fyler, N. F., et al.: "The CBS Colortron," *Proc. IRE*, vol 42, pp. 326–334, January 1954.

Gallaro, A. V., and R. A. Hedler: "Process for Forming a Color CRT Screen Structure Having Optical Filter Therein," U.S. Patent 3,884,694, 1973.

Geer, C. W.: U.S. Patent No. 2,480,848, July 11, 1944.

Gerhard-Multhaupt, R.: "Light Valve Technologies for HDTV Projection Displays: A Summary," *Proceedings of the ITS*, International Television Symposium, Montreux, Switzerland, 1991.

Gerritson, J.: "Soft Flash Picture Tubes," *IEEE Trans. Consumer Electronics*, vol. CE-24, no. 4, pp. 560–561, November 1978.

Glenn, William E.: "Display Requirements for the High-Definition Electronic Cinema," *SID 91 Digest*, Society for Information Display, San Jose, Calif., pp. 144–145, 1991.

Glenn, William E.: "Large Screen Displays for Consumer and Theater Use," *Large Screen Projection Displays II*, William P. Bleha, Jr., (ed.), Proc. SPIE 1255, SPIE, Bellingham, Wash., pp. 36–43, 1990.

Glenn, William. E.: "Principles of Simultaneous Color Projection Using Fluid Deformation," *SMPTE Journal*, SMPTE, White Plains, N.Y., vol. 79, pg. 788, 1970.

Godfrey, R., et al.: "Development of the Permachrome Color Picture Tube," *IEEE Trans. Broadcast TV Receivers*, vol. BTR-14, no. 1, 1968.

Goede, Walter F: "Electronic Information Display Perspective," *SID Seminar Lecture Notes*, Society for Information Display, San Jose, Calif., vol. 1, pp. M-1/3–M1/49, May 17, 1999.

Goldberg, P.: *Luminescence of Inorganic Solids*, Academic, New York, N.Y., 1966.

Gomery, Douglas: "Theater Television: A History," *SMPTE Journal*, SMPTE, White Plains, N.Y., February 1989.

Good, W.: "Projection Television," *IEEE Trans.*, vol. CE-21, no. 3, pp. 206–212, August 1975.

Good, W.: "Recent Advances in the Single-Gun Color Television Light-Valve Projector," *Soc. Photo-Optical Instrumentation Engrs.*, vol. 59, 1975.

Gove, R. J., V. Markandey, S. Marshall, D. Doherty, G. Sextro, and M. DuVal: "High-Definition Display System Based on Digital Micromirror Device," *International Workshop on HDTV* (HDTV '94), International Institute for Communications, Turin, Italy (October 1994).

Gow, J., and R. Door: "Compatible Color Picture Presentation with the Single-Gun Tri Color Chromatron," *Proc. IRE*, vol. 42, no. 1, pp. 308–314, January 1954.

Green, Marc: "Temporal Sampling Requirements for Stereoscopic Displays, *Human Vision, Visual Processing, and Digital Display III*, Bernice E. Rogowitz ed., Proc. SPIE 1666, SPIE, Bellingham, Wash., pp. 101–111, 1992.

Gretag AG: "What You May Want to Know about the Technique of Eidophor," Regensdorf, Switzerland.

Grinberg, J. et al.: "Photoactivated Birefringent Liquid-Crystal Light Valve for Color Symbology Display," *IEEE Trans. Electron Devices*, vol. ED-22, no. 9, pp. 775–783, September 1975.

Grogan, Timothy A.: "Image Evaluation with a Contour-Based Perceptual Model," *Human Vision, Visual Processing, and Digital Display III*, Bernice E. Rogowitz ed., Proc. SPIE 1666, SPIE, Bellingham, Wash., pp. 188–197, 1992.

Guild, J.: "The Colorimetric Properties of the Spectrum," *Phil. Trans. Roy. Soc. A.*, vol. 230, pp. 149–187, 1931.

Hamasaki, Kimio: "How to Handle Sound with Large Screen," *Proceedings of the ITS, International Television Symposium*, Montreux, Switzerland, 1991.

Hardy, A. C., and F. H. Perrin: *The Principles of Optics*, McGraw-Hill, New York, N.Y., 1932.

Harwood, L. A.: "A Chrominance Demodulator IC with Dynamic Flesh Correction," *IEEE Trans. CE*, vol. 22, pp. 111–118, 1976.

Hasker, J.: "Astigmatic Electron Gun for the Beam Indexing Color TV Display," *IEEE Trans. Electron Devices*, vol. ED-18, no. 9, pg. 703, September 1971.

Hayashi, M., N. Yamada, and B. Sastra: "Development of a 42-in. High-Definition Plasma-Addressed LCD," in *SID International Symposium Digest of Technical Papers*, Society for Information Display, San Jose, Calif., pp.280–284, 1999.

Hecht, S., S. Shiaer, and E. L. Smith: "Intermittent Light Stimulation and the Duplicity Theory of Vision," Cold Spring Harbor Symposia on Quantitative Biology, vol. 3, pg. 241, 1935.

Hecht, S.:"The Visual Discrimination of Intensity and the Weber-Fechner Law," *J. Gen Physiol.*, vol. 7, pg. 241, 1924.

Henney, Keith, (ed.): *Radio Engineering Handbook*, 5th ed., McGraw-Hill, New York, N.Y., 1959.

Henney, Keith, and Beverly Dudley: *Handbook of Photography*, McGraw-Hill, New York, N.Y., 1939.

Herman, S.: "The Design of Television Color Rendition," *J. SMPTE*, SMPTE, White Plains, N.Y., vol. 84, pp. 267–273, 1975.

Herold, E.: "A History of Color TV Displays," *Proc. IEEE*, vol. 64, no. 9, pp. 1331–1337, September 1976.

Hockenbrock, Richard: "New Technology Advances for Brighter Color CRT Displays," in *Display System Optics II*, Harry M. Assenheim (ed.), Proc. SPIE 1117, SPIE, Bellingham, Wash., pp. 219–226, 1989.

Hornbeck, Larry J.: "Digital Light Processing for High-Brightness, High-Resolution Applications," *Projection Displays III*, Electronic Imaging '97 Conference, SPIE, Bellingham, Wash., February 1997.

Hoskoshi, K., et al.: "A New Approach to a High Performance Electron Gun Design for Color Picture Tubes," 1980 IEEE Chicago Spring Conf. Consumer Electronics.

Howe, R., and B. Welham: "Developments in Plastic Optics for Projection Television Systems," *IEEE Trans.*, vol. CE-26, no. 1, pp. 44–53, February 1980.

Hu, C., Y. Yu and K. Wang: "Antiglare/Antistatic Coatings for Cathode Ray Tube Based on Polymer System, in *Display Technologies*, Shu-Hsia Chen and Shin-Tson Wu (eds)., Proc. SPIE 1815, SPIE, Bellingham, Wash., pp.42–48, 1992.

Hubel, David H.: *Eye, Brain and Vision*, Scientific American Library, New York, N.Y., 1988.

Hudson, R. G.: *The Engineer's Manual*, Wiley, New York, N.Y., 1939.

Hunt, R. W. G.: *The Reproduction of Colour*, 3d ed., Fountain Press, England, 1975.

Hunter, R.: *The Measurement of Appearance*, Wiley, New York, N.Y., 1975.

Hutter, Rudolph G. E., "The Deflection of Electron Beams," in *Advances in Image Pickup and Display*, B. Kazan (ed.), vol. 1, pp. 212–215, Academic, New York, N.Y., 1974.

IEEE Standard Dictionary of Electrical and Electronics Terms, 2nd ed., Wiley, New York, N.Y., 1977.

IES Lighting Handbook, Illuminating Engineering Society of North America, New York, N.Y., 1981.

Ilcisen, K. J., et. al.: *Eurodisplay '96*, pg. 595, 1996.

Infante, C.: "On the Resolution of Raster-Scanned CRT Displays," *Proceedings of the Society for Information Display*, Society for Information Display, San Jose, Calif., vol. 26, no. 1, pp. 23–36, 1985.

Itah, N., et al.: "New Color Video Projection System with Glass Optics and Three Primary Color Tubes for Consumer Use," *IEEE Trans. Consumer Electronics*, vol. CE-25, no. 4, pp. 497–503, August 1979.

Jenkins. A. J.: "Modulation Transfer Function (MTF) Measurements on Phosphor Screens," *Assessment of Imaging Systems: Visible and Infrared* (Sira), SPIE, Bellingham, Wash., vol. 274, pp. 154–158, 1981.

Johnson, A.: "Color Tubes for Data Display—A System Study," Philips ECG, Electronic Tube Division.

Jordan, Edward C. (ed.): *Reference Data for Engineers: Radio, Electronics, Computer, and Communications*, 7th ed., Howard W. Sams, Indianapolis, IN, 1985.

Judd, D. B., and G. Wyszencki: *Color in Business, Science, and Industry*,. 3rd ed., Wiley, New York, N.Y., pp. 44-45, 1975.

Judd, D. B.: "The 1931 C.I.E. Standard Observer and Coordinate System for Colorimetry," *Journal of the Optical Society of America*, vol. 23, 1933.

Judd, D., and G. Wyszecki: *Color in Business, Science and Industry*, Wiley, New York, N.Y., 1975.

Kallman, H. P., and G. M. Spurch: *Luminescence of Organic and Inorganic Materials*, Wiley, New York, N.Y., 1962.

Kaplan, Sam H.: "The History of Color Picture Tubes and Some Future Projections," *SMPTE Journal*, SMPTE, White Plains, N.Y., pp. 396–400, May 1990.

Kaufman, J. E. (ed.): *IES Lighting Handbook-1981 Reference Volume*, Illuminating Engineering Society of North America, New York, N.Y., 1981.

Keller, Thomas B.: "Proposal for Advanced HDTV Audio," *1991 HDTV World Conference Proceedings*, National Association of Broadcasters, Washington, D.C., April 1991.

Kelly, K. L.: "Color Designation of Lights," *Journal of the Optical Society of America*, vol. 33, 1943.

Kelly, R. D., A. V. Bedbord and M. Trainer: "Scanning Sequence and Repetition of Television Images," *Proceedings of the IRE*, vol. 24, April 1936.

Kikuchi, M., et al.: "A New Coolant-Sealed CRT for Projection Color TV," *IEEE Trans.*, vol. CE-27, no. 3, pp. 478–485, August 1981.

Kingslake, R. (ed.): *Applied Optics and Optical Engineering*, vol. 1, Academic, New York, N.Y., 1965.

Kingslake, Rudolf (ed.): *Applied Optics and Optical Engineering*, vol. 1, Chap. 6, Academic, New York, N.Y., 1965.

Kobari, Y., et al.: "A Novel Arc Suppression Technique for CRTs," *IEEE Trans. Consumer Electronics*, vol. CE-26, no. 3, pp. 446–450, August 1980.

"Kodak Filters for Scientific and Technical Uses," Eastman Kodak Co., Rochester, N.Y.

Koller, L. R.: "Thin Film Phosphors," Electrochemical Society Meeting, Washington D.C., May 13, 1957.

Kucherrov, G. V., et. al.: "Application of the Modulation Transfer Function Method to the Analysis of Cathode-Ray Tubes," *Radio Engineering and Electronics Physics*, vol. 19, pp. 150–152, February 1974.

Kurahashi, K., et al.: "An Outdoor Screen Color Display System," *SID Int. Symp. Digest 7*, Technical Papers, vol. XII, Society for Information Display, San Jose, Calif., pp. 132–133, April 1981.

Lagadec, Roger: "Audio for Television: Digital Sound in Production and Transmission," *Proceedings of the ITS, International Television Symposium*, Montreux, Switzerland, 1991.

Lakatos, A. I., and R. F. Bergen: "Projection Display Using an Amorphous-Se-Type Ruticon Light Valve," *IEEE Trans. Electron Devices*, vol. ED-24, no. 7, pp. 930–934, July 1977.

Langford-Smith, F., (ed.): *Radiotron Designer's Handbook*, 4th ed., Radio Corporation of America, Harrison, N.J., 1953.

Langmuir, D.: "Limitations of Cathode Ray Tubes," *Proc. IRE*, vol. 25, pp. 977–991, 1937.

Law, H. B.: "A Three-Gun Shadow Mask Color Kinescope, *Proc. IRE*, vol. 39, pp. 1186–1194, October 1951.

Lee, Marshall M.: *Winning with People: The First 40 Years of Tektronix*, Tektronix, Beaverton, Ore., 1986.

Lehmann, W.: "Method of Forming a Uniform Layer of Luminescent Material on a Surface," U.S. Patent 2,798,821, July 1957.

Leverenz, H. W.: *An introduction to Luminescence of Solids*, Dover, New York, N.Y., 1968.

Lim, G. S., et. al.: "New Driving Method for Improvement of Picture Quality in 40-in. AC PDP," in *Asia Display '98—Proceedings of the 18th International Display Research Conference*, Society for Information Display, San Jose, Calif., pp. 591-594, 1998.

Lincoln, Donald: "TV in the Bay Area as Viewed from KPIX," *Broadcast Engineering*, Intertec Publishing, Overland Park, Kan., May 1979.

Lucchesi, B., and M. Carpenter: "Pictures of Deflected Electron Spots from a Computer," *IEEE Trans. Consumer Electronics*, vol. CE-25, no. 4, pp. 468–474, 1979.

Luxenberg, H. R., and R. L. Kuehn (eds.): *Display Systems Engineering*, McGraw-Hill, New York, N.Y. 1968.

M. Weston: "The Zone Plate: Its Principles and Applications," *EBU Review—Technical*, no. 195, October 1982.

MacAdam, D. L.: "Visual Sensitivities to Color Differences in Daylight," *J. Opt. Soc. Am.*, vol. 32, pp. 247–274, 1942.

Maeda, M.: *Japan Display '83*, pg. 2, 1971.

Mannos, J. L., and R. W. Tracy: "Cathode-Ray Tube (CRT) Softcopy Image Display Evaluation," *Advances in Display Technology*, SPIE, Bellingham, Wash., vol. 199, pp. 146–150, 1979.

Martin, Russel A., Albert J. Ahumanda, Jr., and James O. Larimer: "Color Matrix Display Simulation Based Upon Luminance and Chromatic Contrast Sensitivity of Early Vision," *Human Vision, Visual Processing, and Digital Display III*, Bernice E. Rogowitz ed., Proc. SPIE 1666, SPIE, Bellingham, Wash., pp. 336–342, 1992.

Maseo, Imai, et. al.: "High-Brightness Liquid Crystal Light Valve Projector Using a New Polarization Converter," *Large Screen Projection Displays II*, William P. Bleha, Jr., (ed.), Proc. SPIE 1255, SPIE, Bellingham, Wash., pp. 52–58, 1990.

Masterson, W., and R. Barbin: "Designing Out the Problems of Wide-Angle Color TV Tube," *Electronics*, pp. 60–63, April 26, 1971.

Masuda, T., et. al: *Conference Record—International Display Resolution Conference*, pg. 357, 1994.

McCroskey, Donald: "Setting Standards for the Future," *Broadcast Engineering*, Intertec Publishing, Overland Park, Kan., May 1989.

McKechnie, S: Philips Laboratories (NA) report, 1981, unpublished.

Mears, N., "Method and Apparatus for Producing Perforated Metal Webs," U.S. Patent 2,762,149, 1956.

Mertz, P., and F. Gray: "A Theory of Scanning and Its Relation to the Characteristics of the Transmitted Signal in Telephotography and Television," *Bell System Tech. J.*, vol. 13, pp. 464–515, July 1934.

Mertz, P.: "Television-The Scanning Process," *Proc. IRE*, vol. 29, pp. 529–537, October 1941.

Middlebrook, B., and M. Day: "Measure CRT Spot Size to Pack More Information into High-Speed Graphic Displays: You Can do it with the Vernier Line Method," *Electronic Design*, vol. 15, pp. 58–60, July 19, 1975.

Mikoshiba, Shigeo: *Color Plasma Displays*, Society for Information Display, San Jose, Calif., pp M-4/3-M-4/68, 1999.

Miller, Howard: "Options in Advanced Television Broadcasting in North America," *Proceedings of the ITS, International Television Symposium*, Montreux, Switzerland, 1991.

Mitsuhashi, Tetsuo: "HDTV and Large Screen Display, in *Large-Screen Projection Displays II*, William P. Bleha, Jr., ed., Proc. SPIE 1255, SPIE, Bellingham, Wash., pp. 2—12, 1990.

Mokhoff, N.: "A Step Toward Perfect Resolution," *IEEE Spectrum*, IEEE, New York, N.Y., vol. 18, no. 7, pp. 56–58, July 1981.

Moon, P.: *The Scientific Basis of Illuminating Engineering*, Dover, New York, N.Y., 1961.

Morell, A. M., et al.: "Color Television Picture Tubes," in *Advances in Image Pickup and Display*, vol. 1, B. Kazan (ed.), pg. 136, Academic, New York,N.Y., 1974.

Morgan, M. J., and R. J. Watt: "The Modulation Transfer Function of a Display Oscilloscope: Measurements and Comments," *Vision Research*, vol. 22, pp. 1083–1085, Great Britain, 1981.

Morizono, M.: "Technological Trends in High Resolution Displays Including HDTV," *SID International Symposium Digest*, Society for Information Display, San Jose, Calif., paper 3.1, May 1990.

Morrell, A., et al.: *Color Television Picture Tubes*, Academic, New York, N.Y., 1974.

Morrell, A.: "Color Picture Tube Design Trends," *Proc. SID*, Society for Information Display, San Jose, Calif., vol. 22, no. 1, pp. 3–9, 1981.

Morris, James E.: "Liquid-Crystal Displays," in *The Electrical Engineering Handbook*, Richard C. Dorf (ed.), CRC Press, Boca Raton, Fla., 1993.

Moss, H.: *Narrow Angle Electron Guns and Cathode Ray Tubes*, Academic, New York, N.Y., 1968.

Munsell Book of Color. Munsell Color Co., 2441 No. Calvert Street, Baltimore, MD 21218.

Na, Young-Sun, et. al.: "A New Data Driver Circuit for Field Emission Display," in *Asia Display '98—Proceedings of the 18th International Display Research Conference*, Society for Information Display, San Jose, Calif., pp. 137-140, 1998.

Naiman, Avi C., and Walter Makous: "Spatial Non-linearities of Grayscale CRT pixels," *Human Vision, Visual Processing, and Digital Display III*, Bernice E. Rogowitz ed., Proc. SPIE 1666, SPIE, Bellingham, Wash., pp. 41–56, 1992.

Nasibov, A., et al.: "Electron-Beam Tube with a Laser Screen," *Sov. J. Quant. Electron.*, vol. 4, no. 3, pp. 296–300, September 1974.

Neal, C. B.: "Television Colorimetry for Receiver Engineers," *IEEE Trans. BTR*, vol. 19, pp. 149–162, 1973.

Newhall, S. M., D. Nickerson, and D. B. Judd: "Final Report of the OSA Subcommittee on the Spacing of the Munsell Colors," *Journal of the Optical Society of America*, vol. 33, pp. 385–418, 1943.

Nickerson, D.: "History of the Munsell Color System, Company and Foundation, I," *Color Res. Appl.* vol. 1, pp. 7–10, 1976.

Nickerson, D.: "History of the Munsell Color System, Company and Foundation, II: Its Scientific Application," *Color Res. Appl.* vol. 1, pp. 69–77, 1976.

Nickerson, D.: "History of the Munsell Color System, Company and Foundation, III," *Color Res. Appl.* vol. 1, pp. 121–130, 1976.

Nishio, T., and K. Amemiya: "High-Luminance and High-Definition 50-in.-Diagonal Co-Planar Color PDPs with T-Shaped Electrodes," in *SID International Symposium Digest of Technical Papers*, Society for Information Display, San Jose, Calif., pp.268-272, 1999.

Nix, L.: "Spot Growth Reduction in Bright, Wide Deflection Angle CRTs," *SID Proc.*, Society for Information Display, San Jose, Calif., vol. 21, no. 4, pg. 315, 1980.

Novick, S. B.: "Tone Reproduction from Colour Telecine Systems," *Br. Kin. Sound TV*, vol. 51, pp. 342–347, 1969.

Oess, F.: "CRT Considerations for Raster Dot Alpha Numeric Presentations," *Proc. SID*, Society for Information Display, San Jose, Calif., vol. 20, no. 2, pp. 81–88, second quarter, 1979.

Ohkoshi, A., et al.: "A New 30V" Beam Index Color Cathode Ray Tube," *IEEE Trans. Consumer Electronics*, vol. CE-27, p. 433, August 1981.

"Optical Characteristics of Cathode Ray Tube Screens," JEDEC Publication 16B, Electron Tube Council, Washington, D.C., 1971.

"Optical Characteristics of Cathode Ray Tube Screens," TEPAC Publication 116, Electronic Industries Association, Washington, D. C., 1980.

Pakswer, S., and P. J. Intiso: *Journal of the Electrochemical Society*, vol. 99, 1952.

Palac, K.: Method for Manufacturing a Color CRT Using Mask and Screen Masters, U.S. Patent 3,989,524, 1976.

Pearson, M. (ed.): *Proc. ISCC Conf. on Optimum Reproduction of Color*, Williamsburg, Va., 1971, Graphic Arts Research Center, Rochester, N.Y., 1971.

Pease, Richard W.: "An Overview of Technology for Large Wall Screen Projection Using Lasers as a Light Source," *Large Screen Projection Displays II*, William P. Bleha, Jr., (ed.), Proc. SPIE 1255, SPIE, Bellingham, Wash., pp. 93–103, 1990.

Pender, H., and K. McIlwain (eds.), *Electrical Engineers Handbook*, Wiley, New York, N.Y., 1950.

Pfahnl, A.: "Aging of Electronic Phosphors in Cathode Ray Tubes," *Advances in Electron Tube Techniques*, Pergamon, New York, N.Y., pp. 204–208, 1961.

Phillips, Thomas E., et. al.: "1280 × 1024 Video Rate Laser-Addressed Liquid Crystal Light Valve Color Projection Display," *Optical Engineering*, Society of Photo-Optical Instrumentation Engineers, vol. 31, no. 11, pp. 2300–2312, November 1992.

Pitts, K. and N. Hurst: "How Much Do People Prefer Widescreen (16 × 9) to Standard NTSC (4 × 3)?," *IEEE Transactions on Consumer Electronics*, IEEE, New York, N.Y., August 1989.

Pointer, M. R.: "The Gamut of Real Surface Colours," *Color Res. Appl.*, vol. 5, pp. 145–155, 1980.

Pointer, R. M.: "The Gamut of Real Surface Colors, *Color Res. App.*, vol. 5, 1945.

Polysak, S. L.: *The Retina*, University of Chicago Press, Chicago, Ill., 1941.

Poole, H. H.: *Fundamentals of Display Systems*, Spartan, Washington, D.C., 1966.

Poorter, T., and F. W. deVrijer: "The Projection of Color Television Pictures," *SMPTE Journal*, SMPTE, White Plains, N.Y., vol. 68, pg. 141, 1959.

Popodi, A. E., "Linearity Correction for Magnetically Deflected Cathode Ray Tubes," *Elect. Design News*, vol. 9, no. 1, January 1964.

Pritchard, D. H.: "US Color Television Fundamentals—A Review," *IEEE Trans. CE*, vol. 23, pp. 467–478, 1977.

Quinn, S. F., and C. A. Siocos: "PLUGE Method of Adjusting Picture Monitors in Television Studios—A Technical Note," *SMPTE Journal*, SMPTE, White Plains, N.Y., vol. 76, pg. 925, September 1967.

Rash, C. E., and R. W. Verona: "Temporal Aspects of Electro-Optical Imaging Systems," *Imaging Sensors and Displays*, SPIE, Bellingham, Wash., vol. 765, pp. 22–25, 1987.

Ratliff, Earl: "A Survey of Display Technologies for Military Aircraft Cockpit Applications, in *High-Resolution Displays and Projection Systems*, Elliott Schlam and Marko Slusarczuk (eds.), Proc. SPIE, SPIE, Bellingham, Wash., vol 1664, pp.66–89, 1992.

"Recommended Practice for Measurement of X-Radiation from Direct View TV Picture Tubes," Publication TEP 164, Electronics Industries Association, Washington, D.C., 1981.

Reese, Greg: "Enhancing Images with Intensity-Dependent Spread Functions," *Human Vision, Visual Processing, and Digital Display III*, Bernice E. Rogowitz ed., Proc. SPIE 1666, SPIE, Bellingham, Wash., pp. 253–261, 1992.

Reinhart, William F.: "Gray-scale Requirements for Anti-aliasing of Stereoscopic Graphic Imagery," *Human Vision, Visual Processing, and Digital Display III*, Bernice E. Rogowitz ed., Proc. SPIE 1666, SPIE, Bellingham, Wash., pp. 90–100, 1992.

Richards, C. J.: *Electronic Display and Data Systems: Constructional Practice*, McGraw-Hill, London, pg. 98, 1973.

Richmond, J. C.: "Image Quality of Photoelectric Imaging Systems and its Evaluation," *Proceedings of Symposium on Photo-Electronic Image Devices*, 6th ed., National Bureau of Standards, Washington, D. C., pp. 519–538, 1976.

Robbins, J., and D. Mackey: "Moire Pattern in Color TV," *IEEE Trans. Consumer Electronics*, vol. CE-28, no. 1, pp. 44–55, February 1982.

Robertson, A. R.: "Colour Differences," *Die Farbe*, vol. 29, pp. 273–296, 1981.

Robertson, A.: "Projection Television—1 Review of Practice," *Wireless World*, vol. 82, no. 1489, pp. 47–52, September 1976.

Robinder R., D. Bates and P. Green: "A High Brightness Shadow Mask Color CRT for Cockpit Displays," *SID Digest*, Society for Information Display, San Jose, Calif., vol. 14, pp. 72–73, 1983.

Rublack, W.: "In-Line Plural Beam CRT with an Aspherical Mask," U.S. Patent 3,435,668, 1969.

Rychlewski, T., and R. Vogel: "Phosphor Persistence in Color Television Screens," *Electrochemical Technology*, vol. 4, no. 1-2, pp. 9–12, January–February 1966.

Sadowski, M.: *RCA Review*, vol 95, 1957.

Sakamoto, Y.: and E. Miyazaki, *Japan Display '83*, pg. 30, 1983.

Sarma, Kalluri R.: "Active-Matrix LCDs," in *Seminar Lecture Notes*, Society for Information Display, San Jose, Calif., vol. 1, pp. M3/3-M3/45, 1999.

Say, D.: "Picture Tube Spot Analysis Using Direct Photography," *IEEE Trans. Consumer Electronics*, vol. CE-23, pp. 32–37, February 1977.

Say, D.: "The High Voltage Bi-Potential Approach to Enhanced Color Tube Performance," *IEEE Trans. Consumer Electronics*, vol. CE-24, no. 1, 1978.

Say, D.: "The High Voltage Bi-Potential Approach to Enhanced Color Tube Performance," *IEEE Trans. Consumer Electronics*, vol. CE-24, no. 1, 1978.

Schade, O. H.: "Electro-optical Characteristics of Television Systems," *RCA Review*, vol. 9, pp. 5–37, 245–286, 490–530, 653–686, 1948.

Schiecke, K.: "Projection Television: Correcting Distortions," *IEEE Spectrum*, IEEE, New York, N.Y., vol. 18, no. 11, pp. 40–45, November 1981.

Schow, Edison: "A Review of Television Systems and the Systems for Recording Television," *Sound and Video Contractor*, Intertec Publishing, Overland Park, Kan., May 1989.

Schwartz, J., and M. Fogelson: "Recent Developments in Arc Suppression for Picture Tubes," *IEEE Trans. Consumer Electronics*, vol. CE-25, no. 1, pp. 82–90, February 1979.

Schwartz, J.: "Electron Beam Cathodoluminescent Panel Display," U.S. Patent 4,137,486.

Sears, F. W.: *Principles of Physics, III*, Optics, Addison-Wesley, Cambridge, Mass., 1946.

"Setting Chromaticity and Luminance of White for Color Television Monitors Using Shadow Mask Picture Tubes," SMPTE Recommended Practice 71-1977 SMPTE, White Plains, N.Y., 1977.

Sextro, G., I. Ballew, and J. Lwai: "High-Definition Projection System Using DMD Display Technology," *SID 95 Digest*, Society for Information Display, pp. 70–73, 1995.

Sherr, S., *Electronic Displays*, Wiley, New York, N.Y. 1979.

Sherr, S.: *Fundamentals of Display System Design*, Wiley-Interscience, New York, N.Y., 1970.

Silverman, Alan: "The Future of the Electronic Cinema, New Math, Old Economics, and Restaurant Cuisine, *SID 91 Digest*, Society for Information Display, San Jose, Calif., 1991.

Sinclair, Clive, "Small Flat Cathode Ray Tube," *SID Digest*, Society for Information Display, San Jose, Calif., pp. 138–139, 1981.

Skolnik, M. Il, (ed.):*Radar Handbook*, McGraw-Hill, New York, N.Y., pp. 6–8, 1970.

Slamin, Brendan: "Sound for High-Definition Television," *Proceedings of the ITS, International Television Symposium*, Montreux, Switzerland, 1991.

Slobodin, David E.: "ARPA High-Definition Systems Program," in *Advanced Flat Panel Display Technologies*, Peter S. Freidman (ed.), Proc. SPIE 2174, SPIE, Bellingham, Wash. pp. 2–3, 1994.

Smith, A. R.: "Color Gamut Transform Pairs," *SIGGRAPH 78*, 12–19, 1978.

Smith, P. F., and W. R. Longley: *Mathematical Tables and Formulas*, Wiley, New York, N.Y., 1929.

Smith, V. C., and J. Pokorny: "Spectral Sensitivity of the Foveal Cone Pigments Between 400 and 500 nm," *Vision Res.* vol. 15, pp. 161–171, 1975.

Spangenberg, K. R., *Vacuum Tubes*, McGraw-Hill, New York, N.Y., 1948.

Sproson, W. N.: *Colour Science in Television and Display Systems*, Adam Hilger, Bristol, England, 1983.

Standards and Definitions Committee, Society for Information Display, San Jose, Calif.

Stanley, T.: "Flat Cathode Ray Tube," U.S. Patent 4,031,427.

"Suitable Sound Systems to Accompany High Definition and Enhanced Television Systems," Report 1072, Recommendations and Reports to the CCIR, 1986, Broadcast Service-Sound, International Telecommunications Union, Geneva, 1986.

Sullivan, James R., and Lawrence A. Ray: "Secondary Quantization of Gray-Level Images for Minimum Visual Distortion," *Human Vision, Visual Processing, and Digital Display III*, Bernice E. Rogowitz ed., Proc. SPIE 1666, SPIE, Bellingham, Wash., pp. 27–40, 1992.

Summers, Christopher J.: "The Phosphor Technology Center of Excellence: Research, Education, Industrial Interactions," in *Advanced Flat Panel Display Technologies*, Peter S. Freidman (ed.), Proc. SPIE 2174, SPIE, Bellingham, Wash., pp. 9–15, 1994.

Swartz, J.: "Beam Index Tube Technology," *SID Proceedings*, Society for Information Display, San Jose, Calif., vol. 20, no. 2, p. 45, 1979.

Takeuchi, Kazuhiko, et. al.: "A 750-TV-Line Resolution Projector using 1.5 Megapixel a-Si TFT LC Modules," *SID 91 Digest*, Society for Information Display, San Jose, Calif., pp. 415–418, 1991.

Taneda, T., et al.: "A 1125-Scanning Line Laser Color TV Display," *SID 1973 Symp. Digest Technical Papers*, Society for Information Display, San Jose, Calif., vol. IV, pp. 86–87, May 1973.

Tannas, Lawrence E., Jr.: *Flat Panel Displays and CRTs*," Van Nostrand Reinhold, New York, N.Y., pg. 18, 1985.

Tektronix application note #21W-7165: "Colorimetry and Television Camera Color Measurement," Tektronix, Beaverton, Ore., 1992.

"Television Pioneering," *Broadcast Engineering*, Intertec Publishing, Overland Park Kan., May 1979.

Terman, F. E.: *Radio Engineers' Handbook*, McGraw-Hill, New York, N.Y., 1943.

"The Role of Film in TV Programming," *Broadcast Engineering*, Intertec Publishing, Overland Park, Kan., May 1979.

Thomas, G. A.: "An Improved Zone Plate Test Signal Generator," *Proceedings, Eleventh International Broadcasting Conference*, Brighton, UK, pp. 358–361, 1986.

Thomas, Woodlief, Jr. (ed.): *SPSE Handbook for Photographic Science and Engineering*, Wiley, New York, N.Y., 1973.

Toet, A.: "Hierarchical Image Fusion," *Machine Vision and Applications*, SPIE, Bellingham, Wash., vol. 3, pp. 111, 1990.

Tomioka, M., and Y. Hayshi: "Liquid Crystal Projection Display for HDTV," *Proceedings of the International Television Symposium*, ITS, Montreux, Switzerland, 1991.

Tong, Hua-Sou: "HDTV Display—A CRT Approach," in *Display Technologies*, Shu-Hsia Chen and Shin-Tson Wu (eds.), Proc. SPIE, SPIE, Bellingham, Wash., pp. 2–4, 1992.

Torick, Emil L.: "HDTV: High Definition Video—Low Definition Audio?," *1991 HDTV World Conference Proceedings*, National Association of Broadcasters, Washington, D.C., April 1991.

Torick, Emil L.: "HDTV: High Definition Video—Low Definition Audio?," *1991 HDTV World Conference Proceedings*, National Association of Broadcasters, Washington, D.C., April 1991.

True, R.: "Space Charge Limited Beam Forming Systems Analyzed by the Method of Self-Consistent Fields with Solution of Poisson's Equation on a Deformable Relaxation Mesh," Ph.D. thesis, University of Connecticut, Storrs, 1968.

Tsuruta, Masahiko, and Neil Neubert: "An Advanced High Resolution, High Brightness LCD Color Video Projector," *SMPTE Journal*, SMPTE, White Plains, N.Y., pp. 399–403, June 1992.

Uba, T., K. Omae, R. Ashiya, and K. Saita: "16:9 Aspect Ratio 38V-High Resolution Trinitron for HDTV," *IEEE Transactions on Consumer Electronics*, IEEE, New York, N.Y., February 1988.

Ulbrich, E., and Walters C.: "Controls and Displays for Douglas Aircraft for the 1990s," in *High-Resolution Displays and Projection Systems*, Elliott Schlam and Marko Slusarczuk (eds.), Proc. SPIE, SPIE, Bellingham, Wash., vol. 1664, pp. 96–106, 1992.

Underwriters Laboratory Report UL492.8, January 25, 1974.

van Raalte, John A.: "CRT Technologies for HDTV Applications," *1991 HDTV World Conference Proceedings*, National Association of Broadcasters, Washington, D.C., April 1991.

Van, M. W., and J. Von Esdonk: "A High Luminance High-Resolution Cathode-Ray Tube for Special Purposes," *IEEE Trans Electron Dev.*, IEEE, New York, N.Y., ED-30, pg. 193, 1983.

Verona, R. W., H. L. Task, V. C. Arnold, and J. H. Brindle: "A Direct Measure of CRT Image Quality," U. S. Army Aeromedical Research Laboratory, USAARL Report No. 79-14, 1979.

Verona, Robert W.: "Comparison of CRT Display Measurement Techniques, in *Helmet-Mounted Displays III*, Thomas M. Lippert (ed.), Proc. SPIE 1695,SPIE, Bellingham, Wash., pp. 117–127, 1992.

Vogel, R.: "Contrast Measurements in Color T.V. Tubes," IEEE Conf. Chicago, 1970.

Wang, Q. H., J. B. Cheng, and Z. L. Lin: "A New YAG Phosphor Screen for Projection CRT," *Electron Lett.*, vol. 34, no. 14, pg. 1420, 1998.

Wang, Qionghua, Jianbo Cheng, Zulun Lin, and Gang Yang: "A High-Luminance and High-Resolution CRT for Projection HDTV Display," *Journal of the SID*, Society for Information Display, San Jose, Calif., vol. 7, no. 3, pg 183-186, 1999.

Wang, S., et al.: "Spectral and Spatial Distribution of X-Rays from Color Television Receivers," *Proc. Conf. Detection and Measurement of X-radiation from Color Television Receivers*, Washington, D.C., pp. 53–72, March 28–29, 1968.

Weber, Larry F.: "Plasma Displays," in *The Electrical Engineering Handbook*, Richard C. Dorf (ed.), CRC :Press, Boca Raton, Fla., 1786-1798, 1993.

Wedding, Donald K., Sr.: "Large Area Full Color ac Plasma Display Monitor," *Large Screen Projection Displays II*, William P. Bleha, Jr., (ed.), Proc. SPIE 1255, SPIE, Bellingham, Wash., pp. 29–35, 1990.

Wentworth, J. W.: *Color Television Engineering*, McGraw-Hill, New York, N.Y., 1955.

Whitaker, Jerry C., and K. Blair Benson (eds): *Standard Handbook of Video and Television Engineering*, McGraw-Hill, New York, N.Y., 1999.

Williams, Charles S., and Becklund, Orville A.: *Optics: A Short Course for Engineers and Scientists*, Wiley Interscience, New York, N.Y., 1972.

Wilson, J., and J. F. B. Hawkes: *Optoelectronics: An Introduction*, Prentice-Hall, London, pg. 145, 1989.

Wintringham, W. T.: "Color Television and Colorimetry," *Proc. IRE*, vol. 39, pp. 1135–1172, 1951.

Wolfe, Richard M.: "HDTV for the Cinema: Friend or Foe?," *SID 91 Digest*, Society for Information Display, San Jose, Calif., 1991.

Woodhead, A., et al.: *1982 SID Digest*, Society for Information Display, San Jose, Calif., pg. 206, 1982.

Wright, W. D.: "A Redetermination of the Trichromatic Coefficients of the Spectral Colours," *Trans. Opt. Soc.*, vol. 30, pp. 141–164, 1928–1929.

Wright, W. D.: *Researches on Normal and Defective Colour Vision*, Mosby, St. Louis, Mo., 1947.

Wright, W. D.: *The Measurement of Colour*, 4th ed., Adam Hilger, London, England, 1969.

Wyszecki, G., and W. S. Stiles: *Color Science*, 2nd ed., Wiley, New York, N.Y., 1982.

Wyszecki, G., and W. Stiles: *Color Science*, Wiley, New York, N.Y., 1967.

Wyszecki, G.: "Proposal for a New Color-Difference Formula," *J. Opt. Soc. Am.*, vol. 53, pp. 1318–1319, 1963.

"X-Radiation Measurement Procedures for Projection Tubes," TEPAC Publication 102, Electronic Industries Association, Washington, D. C.

Yamamoto, Y., Y. Nagaoka, Y. Nakajima, and T. Murao: "Super-compact Projection Lenses for Projection Television," *IEEE Transactions on Consumer Electronics*, IEEE, New York, N.Y., August 1986.

Yoshida, S., et al.: "25-V Inch 114-Degree Trinitron Color Picture Tube and Associated New Development," *Trans. BTR*, pp. 193-200, August 1974.

Yoshida, S., et al.: "A Wide Deflection Angle (114°) Trinitron Color Picture Tube," *IEEE Trans. Electron Devices*, vol. 19, no. 4, pp. 231–238, 1973.

Yoshida. S.: et al., "The Trinitron—A New Color Tube," *IEEE Trans. Consumer Electronics*, vol. CE-28, no. 1, pp. 56–64, February 1982.

Younse, J. M.: "Projection Display Systems Based on the Digital Micromirror Device (DMD)," SPIE Conference on Microelectronic Structures and Microelectromechanical Devices for Optical Processing and Multimedia Applications, Austin, Tex., *SPIE Proceedings*, SPIE, Bellingham, Wash., vol. 2641, pp. 64–75, Oct. 24, 1995.

Zworykin, V. K., and G. Morton: *Television*, 2d ed., Wiley, New York, N.Y., 1954.

Subject Index

A

ac plasma display panel 34
ac thin film display 35
accelerating electrode 181
accelerating lens 181
accelerator 185
achromatic colors 78
active matrix addressing 363
active matrix pixel array 363
active-matrix LCD 364
active-matrix liquid crystal display 359
address period 378
advanced systems 13
advanced-definition television 13
afterglow 175
Allen B. DuMont 4
alphanumeric and graphics 34
aluminized 176
aluminizing 169
alychne 84
ambient illumination 307
anamorphic lens 299
angle of incidence 290
anisotropic dielectric constant 360
aperture correction 207
aperture grille 239, 282
aspect ratio 12 - 13, 207
aspheric corrector lens 313
astigmatism 185, 267, 298
Astrovision 348
auto-sync 316

B

balance point 146
band emitter 173
bar edge measurement 205
barrel distortion 187, 298
barrel-type electron gun 265
beam collimation 263
beam crossover 191
beam guide CRT 279

beam parallax 271
beam penetration color CRT 241
beam tilt 271
beam-limited process 177
bipotential lens 261
bistable storage tube 197
BiUni electron gun 263
black matrix 246
black matrix process 176
black matrix/black surround 177
blackbody 60
blanking 112
Braggs reflections 361
Brewster's law 304
brightness 48, 54, 78, 239
brightness shading 208
broad blue band systen 28
broadband polarizer 331

C

Callier Q coefficient 67
carry-over smear 130
cathode ray tube 2, 11, 33, 163
cathodoluminescence 167
cathodoluminescent emission 169
cell pitch 377
center of deflection 244
center of gravity law 83
channel electron multiplier 277
channel multiplying CRT 276
cholesteric crystals 361
chroma 79
chromatic aberration 186, 298
chromaticity coordinates 81
chromaticity diagrams 81
chromaticity space 98
Chromatron 240
chrominance amplitude 207
chrominance phase 207
chrominance signal 117
CIE 86

CIE 1931 Standard Observer 88
CIE 1964 Supplementary Standard Observer 89
CIE colorimetric system 71
CIE primaries 98
CIELUV color space 102 - 103
circles of least confusion 298
CMY model 107
coefficient of thermal expansion 250
collection efficiency 312
collector mesh 196
color CRT 237
 tube geometry 243
color display measurement 145
color fringing 299
color gamut 129, 379
color matching 208
color purity 28
color rendition 148
color spaces 98
color stimulus 75
color targets 104
color temperature 149
color triangle 80
color vector quantities 96
color video displays 26
colorimeter 73
colorimetric color reproduction 148
colorimetry 71
color-matching functions 76, 90
color-mixing functions 76
color-reproducing system 95
coma 186, 297
comparison field 52
composite video 114
composite video waveform 113
conjugate points 296
constant luminance principal 7, 117
contrast 13
contrast ratio 151, 176, 321, 379
contrast sensitivity 137
contrast vision 137
control grid 181
convergence 316
corresponding color reproduction 148
coulomb aging 170

critical angle 291, 301
critical frequency 51, 59
critical fusion frequency 59
cross color 119
cross luminance 119
crossed polarizer 360
crosshatch signal 207
CRT bulb 166
CRT measurement 199
CRT projection 288
curvature of field 186, 267, 298
cylindrical properties 299

D

D6500 95
darkfield performance 321
data projector 288
dc plasma display panel 34
deflection amplifier 269
deflection correction circuitry 228
deflection defocusing 225
deflection plane 244
deflection region 163
deflection system 265
deflection voltage 218
deflection yoke 225, 239
degaussing 239
delta-gun configuration 239
depth perception 128
Diamond Vision 348
dichroic mirror 299
diffraction effects 303
diffraction grating 303, 319
diffractive interference 340
diffuse density 67
diffuse transmittance 67
digital convergence 271
Digital Light Processing 336
digital mirror device 336
direct addressing 363
direct-view storage tube 198
discharge cell 376
discrete frequency CRT measurement 205
dispersion 46
display components 1

display geometry 207
display period 378
distortion of field 186
distribution coefficients 76
dithering 344
Dolby AC-3 24
dominant wavelength 48
dot pitch 151, 237
doubly diffuse transmittances 68
dynamic convergence 271
dynamic focus 231
dynode 278

E

effective resolution 51
E-field vector 330
Eidophor 317
Einzel lens 185
electroluminescence 11
electroluminescent display 34
electromagnetic deflection 220
electromagnetic interference shielding 257
electromagnetic radiation 45
electron beam 163
electron beam deflection 213
electron gun 178
electron lens 264
electron optics 195, 239
electron-beam addressing 317
Electronics Industries Association 6
electrophoretic screening 168
electrostatic deflection 214
electrostatic field 178
electrostatic focus 184
emission uniformity 382
energy distribution curve 46
equalizing pulses 113
equienergy stimulus 84
equivalent color reproduction 148
Ernst Alexanderson 5
exact color reproduction 148
excitation purity 49
extraordinary ray 304

F

FD Trinitron Wega 282
Fermat's principle of least time 289
Ferry-Porter law 59, 138
field emission display 379
field frequency 51
field scan 112
film quality 22
first principal point 296
first-order theory 297
flat CRT 33
flat panel devices 359
flat panel display 37
flat screen CRT 269
flat tension mask 256
flat-face distortion 222, 229
flicker 51
flicker effect 59
flood beam CRT 346
flood gun 196
fluorescence 168, 175
focal length 293
focus 207
focusing anode 185
footcandle 62
footlambert 63
footprint (CRT) 200
Fourier transform 204
fovea centralis 49
frame 111 - 112
frame scanner 324
frequency sweep signal 153
front projection display 306

G

gamma 149
gamma correctors 123
gas discharge 370
gas firing voltage 371
geometric nesting 246
geometric optics 289
glare 136
glass cockpit 38
glow discharge 372
GLV pixel 340

graphics projector 287
Grassmann's laws 74
Grating Light Valve 340
grayscale performance 368
guard band 244

H

half power width measurement 204
HDTV imaging 19
HDTV monitor 255
HDTV projector 288
HiBi electron gun 262
high bipotential electron gun 262
high resolution 253
high-definition television 10
homeotropic LCLV 328
horizontal address vertical deflection CRT 280
horizontal blanking 114
horizontal resolution 13, 143
Howard Vollum 9
HSV model 109
hue 48, 78
hum bars 112
human visual system 45, 69, 71
Huygen's principle 289

I

iconoscope 3
ILA light valve 334
illuminance 61
illuminant D6500 95
image distance 293
image detail 130
Image Light Amplifier 334
image sharpness 127, 130
imaging sensors 31
implosion protection 166
index of refraction 290, 304
infrared sensors 31
infinite focus 324
in-line electron gun 239
in-neck yoke 271
in-plane switching 366
integrated tube component 271
intercomponent confusion 145

interference filter 315
interference pattern 302
interlace 5, 112
interlace factor 127
interlaced scanning 131
interline flicker 131
International Commission on Illumination 52
inverse-square law 62
IRE units 116

J

Jumbotron 348

K

Karl Braun 2
Kell factor 127, 203
Kerr cell 304
Kerr effect 304
keystone scanning 310
kinescope 3
knife edge measurement 205

L

lambert 63
Lambert's cosine law 63
lambertian surface 307
Landolt ring 56
large screen projection displays 287
laser-addressed liquid crystal light valve 331
laser projection display 322
laser raster scanner 333
lateral chromatism 298
Law of Reflection 182, 290
Law of Refraction 290
Lawrence tube 240
LC panel temperature 326
LCD absorption mode 34
LCD polarization mode 34
LCD scattering mode 34
lens aberrations 297
letter-box cropping 27
light emitting diode 11
light valve display 317
light valve projection 288

lightness 78
limiting-aperture plane 246
line emitter 173
line image 298
line scanner 323
line width CRT measurement 200
linearity 207
linearity correction 229
liquid crystal display 34, 288, 325, 359
liquid crystal light valve 317
liquid crystal light valve (LCLV) projector 325
liquid crystals 360
liquid-cooled CRT 313
LoBi electron gun 262
longitudinal chromatism 298
low bipotential electron gun 262
lumen equivalent 172
lumens per square meter 61
luminance 49, 54
luminance condition 77
luminance range 51
luminance reproduction 51
luminance signal 117
luminosity curve 52
luminosity function 54, 66
luminous efficiency 33, 220
luminous emittance 63
luminous flux 54
luminous reflectance 49
luminous transmittance 49
lux 61

M

Macbeth color checker 105
magnetic focusing 188
magnetic lens 190
magnification 294
Maksutov corrector 296
mask transmission loss 239
mask wrinkle 282
match equation 73
matrix drive and deflection system 279
matrix-addressed liquid crystal panel 349
matrixing 346
Maxwell triangle 98

Maxwell's electromagnetic equations 289
meniscus lens 296
mercury-minus-red systen 30
mesopic region 54
metamerism 73
metamers 73
metercandle 62
microelectromechanical systems 337, 340
microtips 381
misregistration 265
modulation transfer function 177, 311
monochrome CRT 192
monochrome electron gun 193
monolayer screening 168
motion artifacts 158
MTF 177
multisensor displays 32
Munsell system 78

N

National Television Systems Committee 6
negative glow region 372
negative guard band 177, 245
negative resistance region 371
negative spherical aberration 188
nematic crystals 360
neutral density filter 301
Newton's rings 302
Newton's second law 178
Newtonian form 297
nit 213
nonspectral color 46
NTSC 114

O

object colors 49
object distance 293
oil film projection 11
onion skin phosphors 242
opponent-color theory 102
optic axis 304
optical distortion 312
optical projection system 310
optimal viewing distance 126
optimal viewing ratio 126

ordinary ray 304
orthogonal diffraction 319
oscilloscope 198
overscan 207

P

packing factors 246
PAL 114
pan and scan 27
paraboloid of revolution 295
passive matrix addressing 363
penning ionization 371
perceived chroma 78
perceptual color space 78
perception-threshold 69
persistence 175
persistence of vision 13
phantom image 23
phase 302
Philo Farnsworth 4
phosphor efficiency 169
phosphor screen 167, 239
phosphor settling 168
phosphorescence 168, 175
phosphors 167
photoelectric colorimeter 146
photometer 52
photometric measurement 52
photometric quantities 60
photon 64
photopic vision 49
picture definition 58
picture elements 111
picture monitor alignment 206
pincushion 187
pincushion distortion 234, 298
pixel 12, 111, 151
pixel density 151
pixel format 151
planar carbon emitter 381
plane-polarized 304
plasma display panel 11, 288, 345, 370
plasma-addressed liquid crystal 359
plasma-addressed liquid crystal display 367
plasma/gas discharge display 34

polarization axis 330
polarizer elements 325
polarizing beam-splitter 335
polyimide alignment layer 366
positive column region 372
positive guard band 177, 244
positive spherical aberration 187
post-deflection acceleration 214
post-deflection accelerator 220
preaccelerator 185
preferred color reproduction 148
primary image surface 298
priming particles 375
principal points 296
principle of superposition 301
projection CRT 313
projection system brightness 306
projective transformations 91
psychometric chroma 103
psychometric coordinates 103
psychometric hue angle 103
pulse-number-modulation 378
Purkinje region 54

Q

Q-spacing 245
quadripotential focus lens 263
quadrupole electron gun 256

R

radar 9, 199
radiant emittance 63
Radio Manufacturers Association 6
raster 112
raster line width 195
rear projection display 306
reference white 117
reference white point 102
reflective LCD 360
refraction 46
reproduce colors 95
reproduction distortions 265
resolution 56, 239, 247
resolving power 56
retinal illuminance 64

retrace 112
retrace time 226
rod vision 137

S

saturation 48, 78
Scanned Linear GLV 341
scanner 323
scanning frequencies 145
scene presence 288
Schlieren optics 318
Schmidt corrector 295
Schmidt reflective optical system 313
scotopic luminosity function 54
scotopic vision 49
screen brightness 307
screen burn 170
screen gain 307
SECAM 114
second principal point 296
self-converging deflection yoke 265
sensor fusion 32
settling time 228
shadow-mask CRT 237
sharpness 58
shrinking raster CRT measurement 200
shunted monochrome principal 7
sine law 291
sintered alumina 315
smear 130
SMPTE 206
Snell's law 182
Society of Motion Picture and Television Engineers 206
sound field 22
sound lateralization 23
sound localization 23
spatial light modulator 334, 341
speckle 322
spectral color reproduction 148
spectral energy distribution 172
spectral irradiance 72
spectral reflectance 301
spectral reflectance factor 72
spectral response distribution 65

spectroradiometric measurement 145
spectrum locus 85
specular beam 67
specular density 67
specular transmittance 67
spherical aberration 186, 262, 295, 297
spherical mirror 294
spherical properties 299
Spindt array 381
spiral accelerator 220
spiral distortion 190
spot distortion 265
spot knocking process 165
spot size 195
spurious resolution 202
square-well diffraction grating 341
Standard Observer 86
steradian 54
Stiles-Crawford effect 64
storage CRT 196
storage mesh 196
stray magnetic fields 252
stretched-taught shadow-mask 256
subjective CRT measurement 200
sub-pixel pitch 377
subtractive primaries 107
super twisted nematic crystal 360
surface of best focus 298
surround 52
surround sound 22
sustaining voltage 374
synchronous demodulation 118

T

Talaria 319
Talaria MLV-HDTV 320
Talbot-Plateau 60
taut shadow-mask 250
television 111
television lines 151
temporal dithering 344
tension mask 249
test field 52
tetrode electron gun 180
TFT-LCD 365

theater projection system 349
theater television 19
thermal quenching 170
thick lens 296, 299
thin film transistor 364
thin-film technology liquid crystal display 11
third-order theory 297
threshold frequency 65
threshold-of-vision 54
throw distance 312
total internal reflection prism 338
transmission LCD 360
transverse electromagnetic field 213
transverse electrostatic field 213
trichromacy of color vision 72
trichromatic theory 78
trigger electrode 375
trilinear coordinates 80
trinescope 5
Trinitron 239
Trinitron electron gun 264
tripotential lens 262
tristimulus values 73, 77
troland 59, 64
tube envelope 164
TV limiting resolution 202
TV lines/picture height 202
twisted nematic cell 360

U

underscan 207
UniBi electron gun 263
uniform chromaticity scales 99
unipotential gun 261
unipotential lens 183

V

vacuum evaporation 168
vacuum ultraviolet 376
value level (Munsel system) 80

vector equation 96
vectorscope display 118
vertical resolution 14
video projector 288
video transfer function 123
viewing angle 15
viewing envelope 366
viewing environment 28, 205
virtual image 23
visual acuity 56, 135
Vladimir Zworykin 3
von Kries model 149

W

wall voltage 374
Weber's fraction 55
Weber's law 55
Werner Flechsig 6
white balance 124, 149
white field brightness 175
white point 99
Wilhelm Konrad Roentgen 2
window limited process 177
writing gun 196

X

x-rays 253

Y

YAG phosphor screen 315
YAG projection system 315
yellow-minus-yellow system 30
YIQ color model 106
yttrium aluminum garnet 314

Z

zone pattern 152
zone plate 152
zone plate signals 156

About the Author

Jerry Whitaker is a technical writer based in Morgan Hill, California, where he operates the consulting firm *Technical Press*. Mr. Whitaker has been involved in various aspects of the communications industry for more than 25 years. He is a Fellow of the Society of Broadcast Engineers and an SBE-certified Professional Broadcast Engineer. He is also a Fellow of the Society of Motion Picture and Television Engineers, and a member of the Institute of Electrical and Electronics Engineers. Mr. Whitaker has written and lectured extensively on the topic of electronic systems installation and maintenance.

Mr. Whitaker is the former editorial director and associate publisher of *Broadcast Engineering* and *Video Systems* magazines. He is also a former radio station chief engineer and TV news producer.

Mr. Whitaker is the author of a number of books, including:

- Editor-in-Chief, *Standard Handbook of Video and Television Engineering*, 2nd ed., McGraw-Hill, 2000.
- Editor, *Television Engineers' Field Manual*, McGraw-Hill, 2000.
- *DTV: The Revolution in Electronic Imaging*, 2nd ed., McGraw-Hill, 1998.
- Editor-in-Chief, *NAB Engineering Handbook*, 9the ed., National Association of Broadcasters, 1998.
- Editor-in-Chief, *The Electronics Handbook*, CRC Press, 1996.
- Coauthor, *Communications Receivers: Principles and Design*, 2nd ed., McGraw-Hill, 1996.
- *Power Vacuum Tubes Handbook*, 2nd ed., CRC Press, 1999.
- *AC Power Systems*, 2nd ed., CRC Press, 1998.
- *Electronic Displays: Technology, Design, and Applications*, McGraw-Hill, 1994.
- Coauthor, *Interconnecting Electronic Systems*, CRC Press, 1992.
- Coeditor, *Information Age Dictionary*, Intertec/Bellcore, 1992.
- *Maintaining Electronic Systems*, CRC Press, 1991.
- *Radio Frequency Transmission Systems: Design and Operation*, McGraw-Hill, 1990.
- Coauthor, *Television and Audio Handbook for Technicians and Engineers*, McGraw-Hill, 1990.

Mr. Whitaker has twice received a Jesse H. Neal Award *Certificate of Merit* from the Association of Business Publishers for editorial excellence. He also has been recognized as *Educator of the Year* by the Society of Broadcast Engineers.

On the CD-ROM

The enclosed CD includes valuable background information and files relating to video display technologies. Included are chapters from the classic second edition of the *Television Engineering Handbook*, and high-resolution test waveforms and images.

Archive Chapters (Directory = \Archive)

The second edition of the *Television Engineering Handbook* examined conventional television systems in great detail. In recognition of the widespread deployment of digital television worldwide, much of the detailed data relating to the NTSC, PAL, and SECAM systems has not been repeated in the new current edition—retitled the *Standard Handbook of Video and Television Engineering*—in order to make room for new material. Because of the high quality of the classic material and the continuing need for it—albeit on a less-frequent basis—a number of important chapters have been scanned from the second edition and made available to readers in the form of the accompanying CD-ROM.

- "Monochrome and Color Image-Display Devices," by Donald L. Say, R. A. Hedler, L. L. Maninger, R. A. Momberger, and J. D. Robbins (CRT devices and systems) **\Display.pdf**

- "Standards and Recommended Practices," by Dalton H. Pritchard (tabular data, and equations relating to television signal transmission standards and equipment standards for conventional television systems) **\Standard.pdf**

- "Waveforms and Spectra of Composite Video Signals," by Joseph F. Fisher and Richard G. Clapp (detailed data on the generation and characteristics of composite video signals) **\Waveform.pdf**

High-Resolution Images (Directory = \Images)

High-resolution images of important test waveforms given in Chapter 5 are included on the CD-ROM for those readers wishing to use these files for test and performance verification purposes. The constraints of practical printing of a book of this type limit the resolution of scanned black-and-white images to approximately 266 dots/inch (DPI). This resolution (*line count*, in printing terms) produces images that are quite acceptable for all but the most critical applications. In the event that readers have the need for the very highest resolution test patterns and screen captures, high-resolution (1200 DPI) images are provided on the CD-ROM. The images are saved as TIFF files. The file numbers correspond to the figure numbers given on the appropriate printed page.

Important Note

Two file formats are used for the CD-ROM files: conventional uncompressed TIFF image files and Adobe Acrobat 4.0 Portable Document Format (PDF) files. The Acrobat reader is available from Adobe Systems (www.adobe.com). The enclosed CD-ROM is supplied "as is." No warranty or technical support is available for the CD.

On-Line Updates

Additional information and updates relating to video displays in general, and this book in particular, can be found at the *Standard Handbook of Video and Television Engineering* web site:

www.tvhandbook.com

The tvhandbook.com web site supports the professional video community with news, updates, and product information relating to the broadcast, post production, and business/industrial applications of digital video.

Check the site regularly for news, updated chapters, and special events related to video engineering. The technologies encompassed by *Video Display Engineering* are changing rapidly, with new developments and standards work announced each month. Changing market conditions and regulatory issues are adding to the rapid flow of news and information in this area.

Specific services found at **www.tvhandbook.com** include:

- **Video Technology News**. News reports and technical articles on the latest developments in digital television, both in the U.S. and around the world. Check in at least once a month to see what's happening in the fast-moving area of digital television.

- **Television Handbook Resource Center**. Check for the latest information on professional and broadcast video systems. The Resource Center provides updates on implementation and standardization efforts, plus links to related web sites.

- **tvhandbook.com Update Port**. Updated material for *Video Display Engineering* is posted on the site regularly. Material available includes updated sections and chapters in areas of rapidly advancing technologies.

- **tvhandbook.com Book Store**. Check to find the latest books on digital video and audio technologies. Direct links to authors and publishers are provided. You can also place secure orders from our on-line bookstore.

In addition to the resources outlined above, detailed information is available on other books in the McGraw-Hill Video/Audio Series.

www.tvhandbook.com

DISK WARRANTY

This software is protected by both United States copyright law and international copyright treaty provision. You must treat this software just like a book, except that you may copy it into a computer in order to be used and you may make archival copies of the software for the sole purpose of backing up our software and protecting your investment from loss.

By saying "just like a book," McGraw-Hill means, for example, that this software may be used by any number of people and may be freely moved from one computer location to another, so long as there is no possibility of its being used at one location or on one computer while it also is being used at another. Just as a book cannot be read by two different people in two different places at the same time, neither can the software be used by two different people in two different places at the same time (unless, of course, McGraw-Hill's copyright is being violated).

LIMITED WARRANTY

McGraw-Hill takes great care to provide you with top-quality software, thoroughly checked to prevent virus infections. McGraw-Hill warrants the physical diskette(s) contained herein to be free of defects in materials and workmanship for a period of sixty days from the purchase date. If McGraw-Hill receives written notification within the warranty period of defects in materials or workmanship, and such notification is determined by McGraw-Hill to be correct, McGraw-Hill will replace the defective diskette(s). Send requests to:

> McGraw-Hill
> Customer Services
> P.O. Box 545
> Blacklick, OH 43004-0545

The entire and exclusive liability and remedy for breach of this Limited Warranty shall be limited to replacement of defective diskette(s) and shall not include or extend to any claim for or right to cover any other damages, including but not limited to, loss of profit, data, or use of the software, or special, incidental, or consequential damages or other similar claims, even if McGraw-Hill has been specifically advised of the possibility of such damages. In no event will McGraw-Hill's liability for any damages to you or any other person ever exceed the lower of suggested list price or actual price paid for the license to use the software, regardless of any form of the claim.

McGRAW-HILL SPECIFICALLY DISCLAIMS ALL OTHER WARRANTIES, EXPRESS OR IMPLIED, INCLUDING, BUT NOT LIMITED TO, ANY IMPLIED WARRANTY OF MERCHANTABILITY OR FITNESS FOR A PARTICULAR PURPOSE.

Specifically, McGraw-Hill makes no representation or warranty that the software is fit for any particular purpose and any implied warranty of merchantability is limited to the sixty-day duration of the Limited Warranty covering the physical diskette(s) only (and not the software) and is otherwise expressly and specifically disclaimed.

This limited warranty gives you specific legal rights; you may have others which may vary from state to state. Some states do not allow the exclusion of incidental or consequential damages, or the limitation on how long an implied warranty lasts, so some of the above may not apply to you.